AF529350

Skills in
Mathematics for
JEE Main & Advanced

Play with Graphs

Skills in
Mathematics for
JEE Main &
Advanced

Play with Graphs

With Sessionwise Theory & Exercises...

Amit M. Agarwal

arihant
ARIHANT PRAKASHAN (Series), MEERUT

ARIHANT PRAKASHAN (Series), MEERUT

All Rights Reserved

© AUTHOR

No part of this publication may be re-produced, stored in a retrieval system or by any means, electronic mechanical, photocopying, recording, scanning, web or otherwise without the written permission of the publisher. Arihant has obtained all the information in this book from the sources believed to be reliable and true. However, Arihant or its editors or authors or illustrators don't take any responsibility for the absolute accuracy of any information published, and the damages or loss suffered thereupon.

All disputes subject to Meerut (UP) jurisdiction only.

卐 Administrative & Production Offices

Regd. Office

'Ramchhaya' 4577/15, Agarwal Road, Darya Ganj, New Delhi -110002
Tele: 011- 47630600, 43518550

卐 **Head Office**

Kalindi, TP Nagar, Meerut (UP) - 250002 Tel: 0121-7156203, 7156204

卐 Sales & Support Offices

Agra, Ahmedabad, Bengaluru, Bareilly, Chennai, Delhi, Guwahati, Hyderabad, Jaipur, Jhansi, Kolkata, Lucknow, Nagpur & Pune.

卐 ISBN : 978-93-26191-66-1

卐 PRICE : ₹235.00

PO No : TXT-XX-XXXXXXX-X-XX

Published by Arihant Publications (India) Ltd.

For further information about the books published by Arihant, log on to www.arihantbooks.com or e-mail at info@arihantbooks.com

Follow us on

PREFACE

It is a matter of great pleasure and pride for me to introduce to you this book "Play with Graphs". As a teacher, guiding the Engineering aspirants for over a decade now, I have always been in the lookout for right approach to understand various mathematical problems. I had always felt the need of a book that can develop and sharpen the ideas of the students within a very short span of time.

The book in your hands, aims to help you solve various mathematical problems by the use of graphs. Ways to draw different types of graphs are very easy and can be understood by even an average student. I feel glad to mention that the use of graphs in solving various mathematical problems has been well illustrated in this book.

I would like to take this opportunity to thank
M/s Arihant Prakashan for assigning this work to me.
It is their inspiration that has encouraged me to bring this book in this present form.

I would also like to thank Arihant DTP Unit for the nice laser typesetting.

Valuable suggestions from the students and teachers are always welcome, and these will find due places in the ensuing editions.

Amit M. Agarwal

CONTENTS

Chapter 1

INTRODUCTION OF GRAPHS

➡ **In this section, we shall revise some basic curves which are given as.**

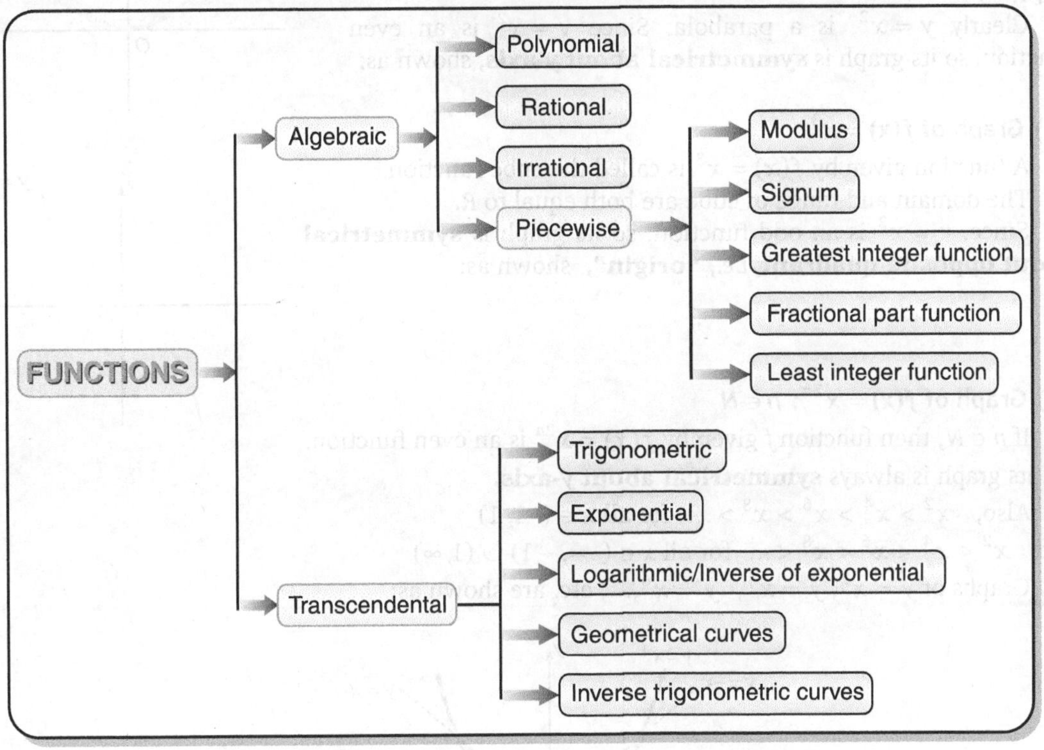

1.1 ALGEBRAIC FUNCTIONS

1. Polynomial Function

A function of the form:

$$f(x) = a_0 + a_1x + a_2x^2 + \ldots + a_nx^n ;$$

where $n \in N$ and $ga_0, a_1, a_2, \ldots, a_n \in R$.

Then, f is called a polynomial function. "$f(x)$ is also called polynomial in x".

Some of basic polynomial functions are

(i) Identity function/Graph of $f(x) = x$

A function f defined by $f(x) = x$ for all $x \in R$, is called the identity function.

Here, $y = x$ clearly represents a straight line passing through the origin and inclined at an angle of 45° with x-axis shown as:

The domain and range of identity functions are both equal to R.

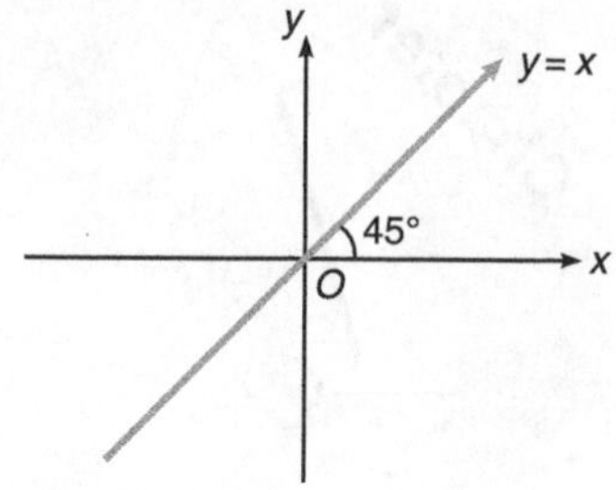

Fig. 1.1

(ii) Graph of $f(x) = x^2$

A function given by $f(x) = x^2$ is called the square function.

The domain of square function is R and its range is $R^+ \cup \{0\}$ or $[0, \infty)$

Clearly $y = x^2$, is a parabola. Since $y = x^2$ is an even function, so its graph is **symmetrical about y-axis**, shown as:

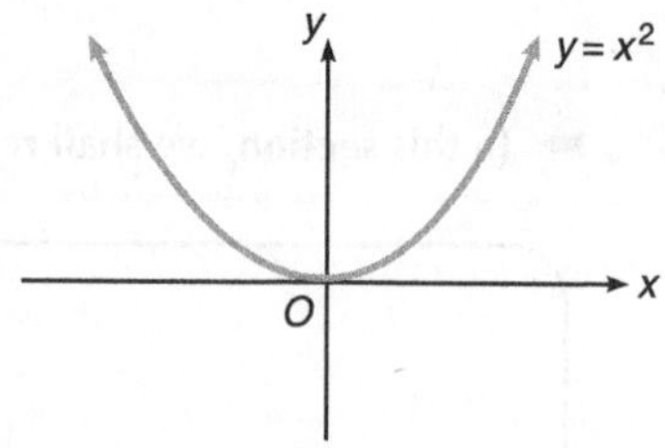

Fig. 1.2

(iii) Graph of $f(x) = x^3$

A function given by $f(x) = x^3$ is called the cube function.

The domain and range of cube are both equal to R.

Since, $y = x^3$ is an odd function, so its graph is **symmetrical about opposite quadrant**, *i.e.*, **"origin"**, shown as:

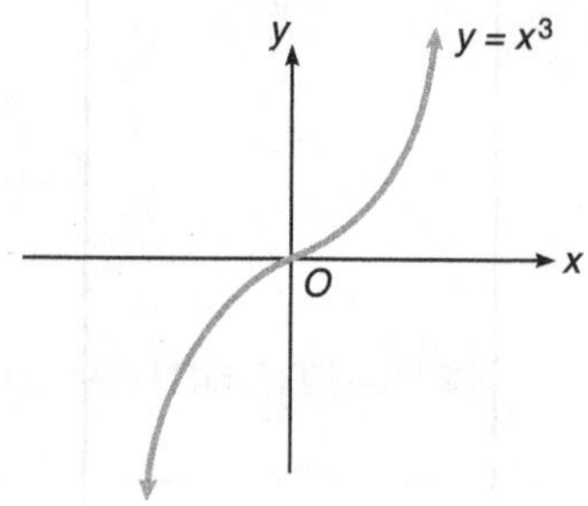

Fig. 1.3

(iv) Graph of $f(x) = x^{2n}$; $n \in N$

If $n \in N$, then function f given by $f(x) = x^{2n}$ is an even function. So, its graph is always **symmetrical about y-axis.**

Also, $x^2 > x^4 > x^6 > x^8 > \ldots$ for all $x \in (-1, 1)$

and $x^2 < x^4 < x^6 < x^8 < \ldots$ for all $x \in (-\infty, -1) \cup (1, \infty)$

Graphs of $y = x^2$, $y = x^4$, $y = x^6, \ldots$, etc. are shown as:

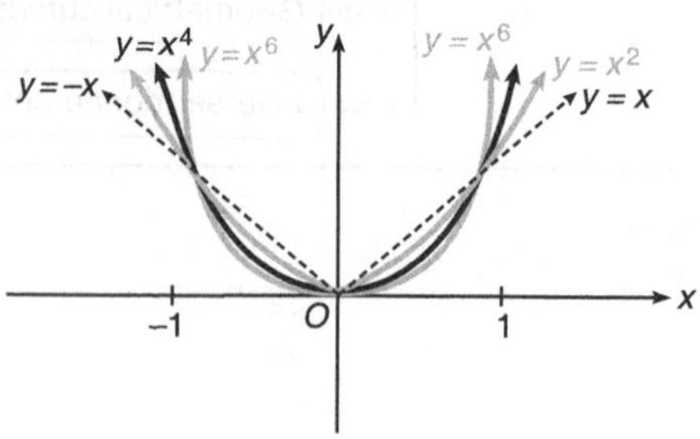

Fig. 1.4

(v) Graph of $f(x) = x^{2n-1}$; $n \in N$

If $n \in N$, then the function f given by $f(x) = x^{2n-1}$ is an odd function. So, its graph is **symmetrical about origin or opposite quadrants.**

Here, comparison of values of $x, x^3, x^5,\ldots$ for

$x \in (1, \infty)$ $\quad x < x^3 < x^5 < \ldots$

$x \in (0, 1)$ $\quad x > x^3 > x^5 > \ldots$

$x \in (-1, 0)$ $\quad x < x^3 < x^5 < \ldots$

$x \in (-\infty, -1)$ $\quad x > x^3 > x^5 > \ldots$

Graphs of $f(x) = x$, $f(x) = x^3$, $f(x) = x^5,\ldots$ are shown as in Fig. 1.5

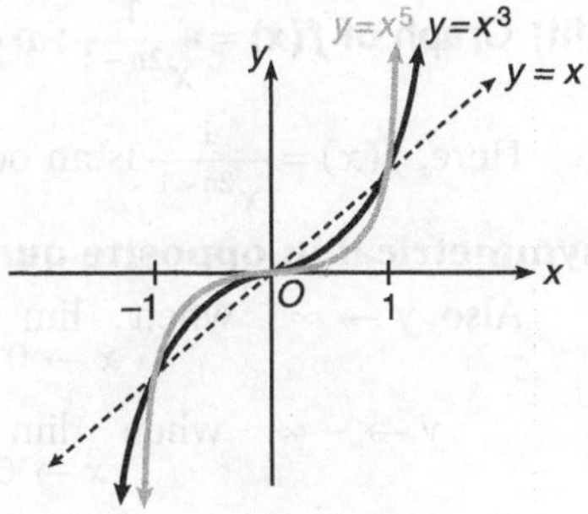

Fig. 1.5

2. Rational Expression

A function obtained by dividing a polynomial by another polynomial is called a rational function.

$$\Rightarrow \qquad f(x) = \frac{P(x)}{Q(x)}$$

$$\text{Domain} \in R - \{x \mid Q(x) = 0\}$$

i.e., domain $\in R$ except those points for which denominator = 0.

Graphs of some Simple Rational Functions

(i) Graph of $f(x) = \frac{1}{x}$

A function defined by $f(x) = \frac{1}{x}$ is called the reciprocal function or rectangular hyperbola, with coordinate axis as asymptotes. The domain and range of $f(x) = \frac{1}{x}$ is **R – {0}**.

Since, $f(x)$ is odd function, so its graph is **symmetrical about opposite quadrants.** Also, we observe

$$\lim_{x \to 0^+} f(x) = +\infty \quad \text{and} \quad \lim_{x \to 0^-} f(x) = -\infty .$$

and as $x \to \pm\infty \Rightarrow f(x) \to 0$

Thus, $f(x) = \frac{1}{x}$ could be shown as in Fig. 1.6.

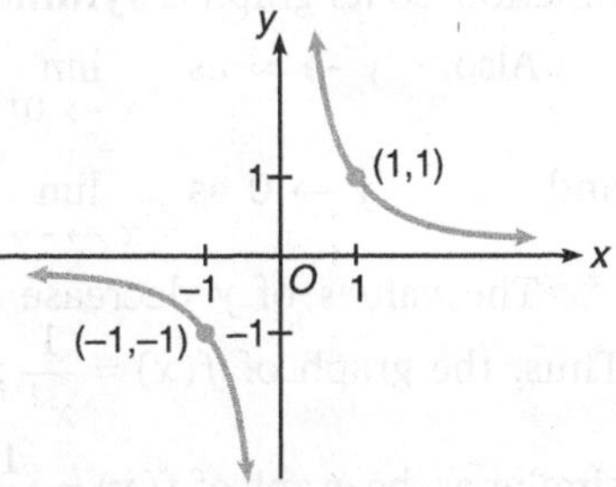

Fig. 1.6

(ii) Graph of $f(x) = \frac{1}{x^2}$

Here, $f(x) = \frac{1}{x^2}$ is an even function, so its graph is **symmetrical about y-axis.**

Domain of $f(x)$ is $R - \{0\}$ and range is $(0, \infty)$.

Also, as $y \to \infty$ as $\lim_{x \to 0^+} f(x)$ or $\lim_{x \to 0^-} f(x)$.

and $y \to 0$ as $\lim_{x \to \pm\infty} f(x)$.

Thus, $f(x) = \frac{1}{x^2}$ could be shown as in Fig. 1.7.

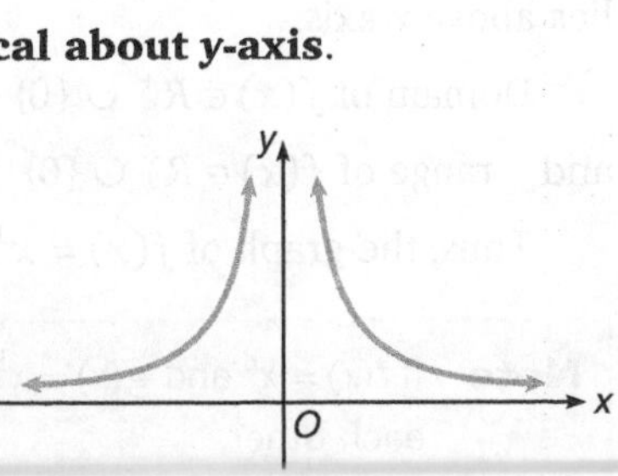

Fig. 1.7

(iii) Graph of $f(x) = \frac{1}{x^{2n-1}}$; $n \in N$

Here, $f(x) = \frac{1}{x^{2n-1}}$ is an odd function, so its graph is **symmetrical in opposite quadrants.**

Also, $y \to \infty$ when $\lim_{x \to 0^+} f(x)$ and

$y \to -\infty$ when $\lim_{x \to 0^-} f(x)$.

Thus, the graph for $f(x) = \frac{1}{x^3}$; $f(x) = \frac{1}{x^5}, \ldots$, etc. will be similar to the graph of $f(x) = \frac{1}{x}$ **which has asymptotes as coordinate axes,** shown as in Fig. 1.8

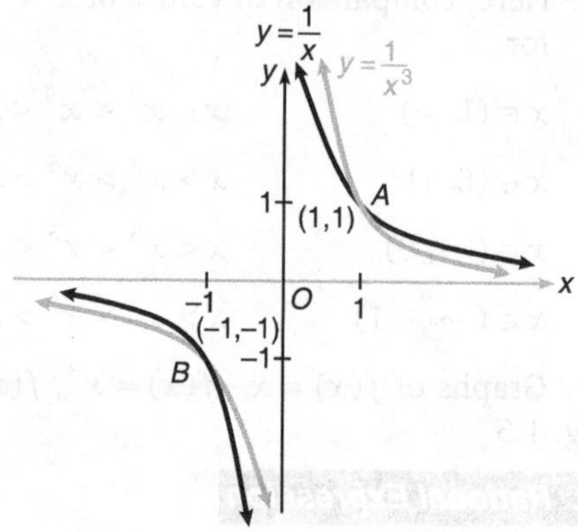

Fig. 1.8

(iv) Graph of $f(x) = \frac{1}{x^{2n}}$; $n \in N$

We observe that the function $f(x) = \frac{1}{x^{2n}}$ is an even function, so its graph is **symmetrical about y-axis.**

Also, $y \to \infty$ as $\lim_{x \to 0^+} f(x)$ or $\lim_{x \to 0^-} f(x)$

and $y \to 0$ as $\lim_{x \to -\infty} f(x)$ or $\lim_{x \to +\infty} f(x)$.

The values of y decrease as the values of x increase. Thus, the graph of $f(x) = \frac{1}{x^2}$; $f(x) = \frac{1}{x^4}, \ldots$, etc. will be similar as the graph of $f(x) = \frac{1}{x^2}$, which has **asymptotes as coordinate axis.** Shown as in Fig. 1.9.

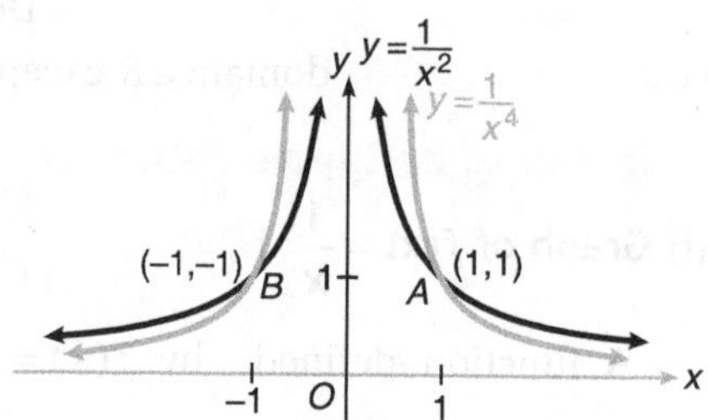

Fig. 1.9

3. Irrational Function

The algebraic function containing terms having non-integral rational powers of x are called irrational functions.

Graphs of Some Simple Irrational Functions

(i) Graph of $f(x) = x^{1/2}$

Here; $f(x) = \sqrt{x}$ is the portion of the parabola $y^2 = x$, which lies above x-axis.

Domain of $f(x) \in R^+ \cup \{0\}$ or $[0, \infty)$

and range of $f(x) \in R^+ \cup \{0\}$ or $[0, \infty)$.

Thus, the graph of $f(x) = x^{1/2}$ is shown as;

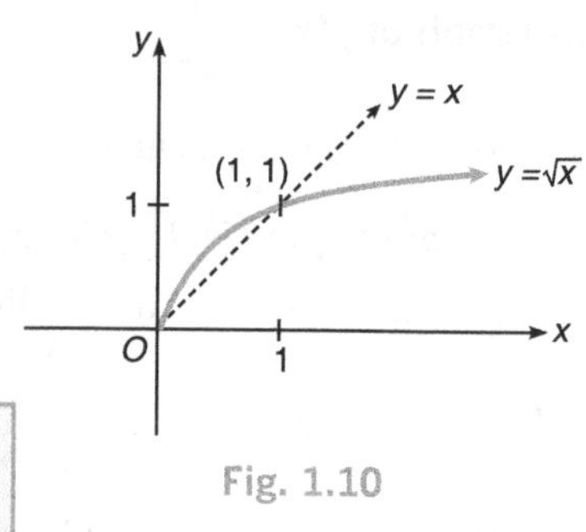

Fig. 1.10

> **Note** If $f(x) = x^n$ and $g(x) = x^{1/n}$, then $f(x)$ and $g(x)$ are inverse of each other.
> $\therefore$ $f(x) = x^n$ and $g(x) = x^{1/n}$ is the mirror image about $y = x$.

(ii) Graph of $f(x) = x^{1/3}$

As discussed above, if $g(x) = x^3$. Then $f(x) = x^{1/3}$ is image of $g(x)$ about $y = x$.

where domain $f(x) \in R$.

and range of $f(x) \in R$.

Thus, the graph of $f(x) = x^{1/3}$ is shown in Fig. 1.11;

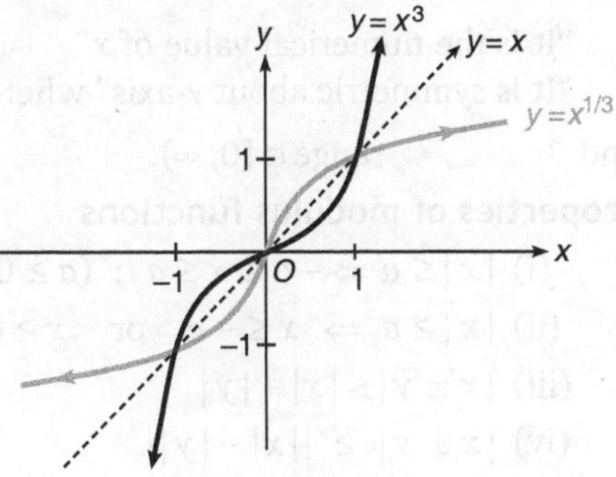

Fig. 1.11

(iii) Graph of $f(x) = x^{1/2n}$; $n \in N$

Here, $f(x) = x^{1/2n}$ is defined for all $x \in [0, \infty)$ and the values taken by $f(x)$ are positive.

So, domain and range of $f(x)$ are $[0, \infty)$.

Here, the graph of $f(x) = x^{1/2n}$ is the mirror image of the graph of $f(x) = x^{2n}$ about the line $y = x$, when $x \in [0, \infty)$.

Thus, $f(x) = x^{1/2}$, $f(x) = x^{1/4}$, ... are shown as;

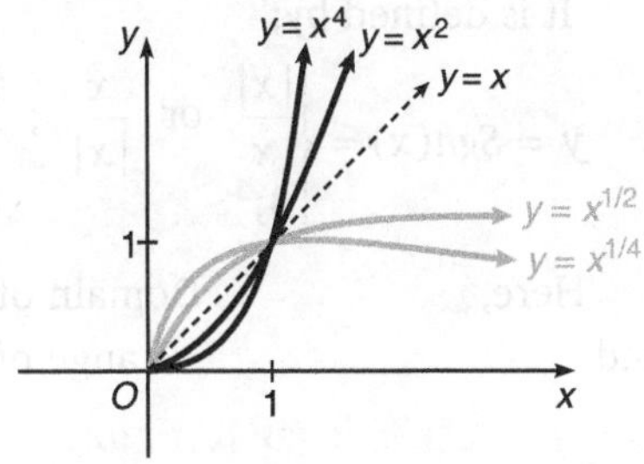

Fig. 1.12

(iv) Graph of $f(x) = x^{1/2n-1}$, when $n \in N$

Here, $f(x) = x^{1/2n-1}$ is defined for all $x \in R$. So, domain of $f(x) \in R$, and range of $f(x) \in R$. Also the graph of $f(x) = x^{1/2n-1}$ is the mirror image of the graph of $f(x) = x^{2n-1}$ about the line $y = x$ when $x \in R$.

Thus, $f(x) = x^{1/3}$, $f(x) = x^{1/5}$,..., are shown as;

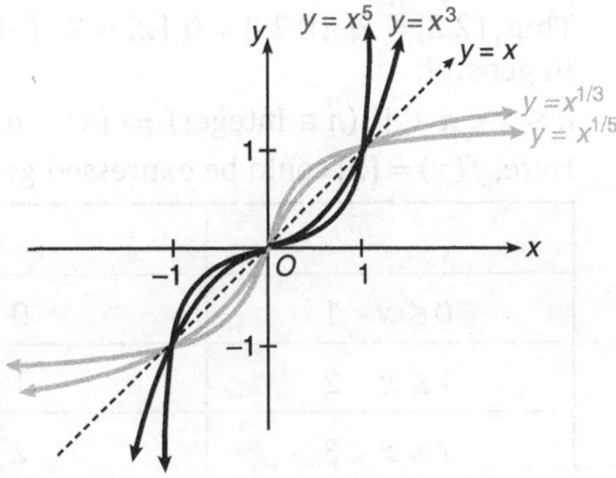

Fig. 1.13

> **Note** We have discussed some of the simple curves for Polynomial, Rational and Irrational functions. Graphs of the some more difficult rational functions will be discussed in chapter 3. Such as;
>
> $$y = \frac{x}{x+1}\quad y = \frac{1}{x^2-1},\quad y = \frac{x^2+x+1}{x^2-x+1},\ldots.$$

4. Piecewise Functions

As discussed piecewise functions are:

(a) Absolute value function (or modulus function),

(b) Signum function.

(c) Greatest integer function.

(d) Fractional part function.

(e) Least integer function.

(a) Absolute value function (or modulus function)

$$y = |x| = \begin{cases} x, & x \ge 0 \\ -x, & x < 0 \end{cases}$$

"It is the numerical value of x".

"It is symmetric about y-axis" where domain $\in R$

and range $\in [0, \infty)$.

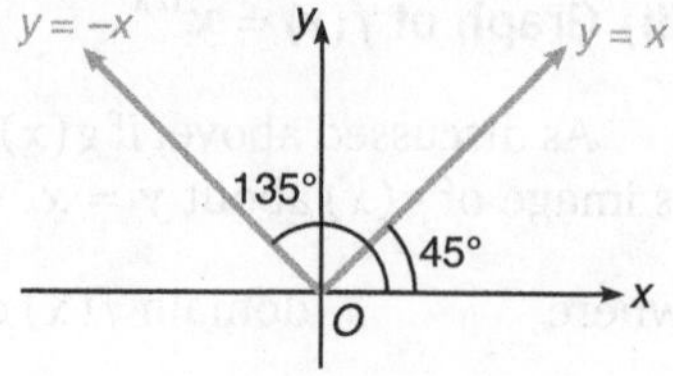

Fig. 1.14

Properties of modulus functions

(i) $|x| \le a \Rightarrow -a \le x \le a$; $(a \ge 0)$

(ii) $|x| \ge a \Rightarrow x \le -a$ or $x \ge a$; $(a \ge 0)$

(iii) $|x \pm y| \le |x| + |y|$

(iv) $|x \pm y| \ge ||x| - |y||$.

(b) Signum function; $y = Sgn(x)$

It is defined by;

$$y = Sgn(x) = \begin{cases} \dfrac{|x|}{x} \text{ or } \dfrac{x}{|x|}; & x \ne 0 \\ 0; & x = 0. \end{cases} = \begin{cases} +1, & \text{if } x > 0 \\ -1, & \text{if } x < 0 \\ 0, & \text{if } x = 0 \end{cases}$$

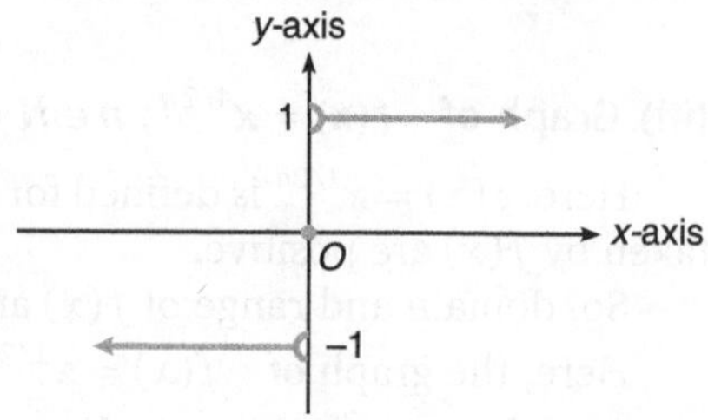

Fig. 1.15

Here, Domain of $f(x) \in R$.

and Range of $f(x) \in \{-1, 0, 1\}$.

(c) Greatest integer function

$[x]$ indicates the integral part of x which is nearest and smaller integer to x. It is also known as **floor of x**.

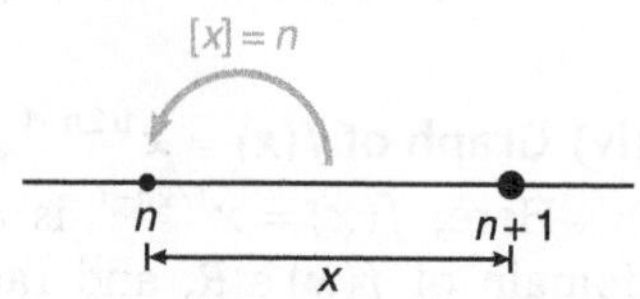

Fig. 1.16

Thus, $[2.3] = 2$, $[0.23] = 0$, $[2] = 2$, $[-8.0725] = -9$, ….

In general;

$n \le x < n + 1$ ($n \in$ Integer) $\Rightarrow [x] = n$.

Here, $f(x) = [x]$ could be expressed graphically as;

x	$[x]$
$0 \le x < 1$	0
$1 \le x < 2$	1
$2 \le x < 3$	2

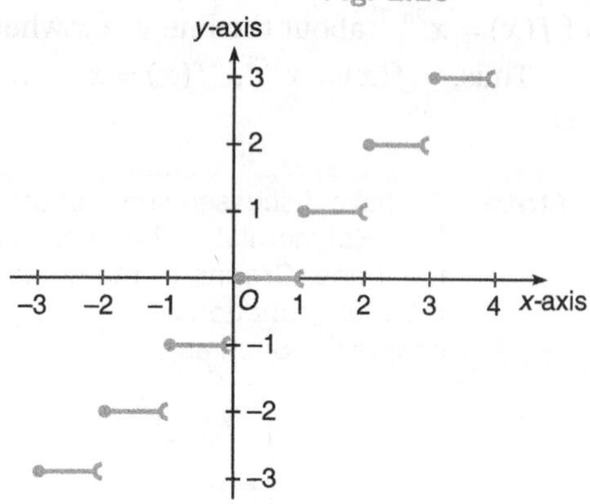

Fig. 1.17

Thus, $f(x) = [x]$ could be shown as;

Properties of greatest integer function

(i) $[x] = x$ holds, if x is integer.

(ii) $[x + I] = [x] + I$, if I is integer.

(iii) $[x + y] \ge [x] + [y]$.

(iv) If $[\phi(x)] \ge I$, then $\phi(x) \ge I$.

(v) If $[\phi(x)] \le I$, then $\phi(x) < I + 1$.

(vi) $[-x] = -[x]$, if $x \in$ integer.

(vii) $[-x] = -[x] - 1$, if $x \notin$ integer.

"It is also known as **stepwise function/floor of x.**"

(d) Fractional part of function

Here, {.} denotes the fractional part of x. Thus, in $y = \{x\}$.

$$x = [x] + \{x\} = I + f \; ; \text{ where } I = [x] \text{ and } f = \{x\}$$

$$\therefore \quad y = \{x\} = x - [x], \text{ where } 0 \le \{x\} < 1; \text{ shown as:}$$

x	$\{x\}$
$0 \le x < 1$	x
$1 \le x < 2$	$x - 1$
$2 \le x < 3$	$x - 2$
$-1 \le x < 0$	$x + 1$
$-2 \le x < -1$	$x + 2$

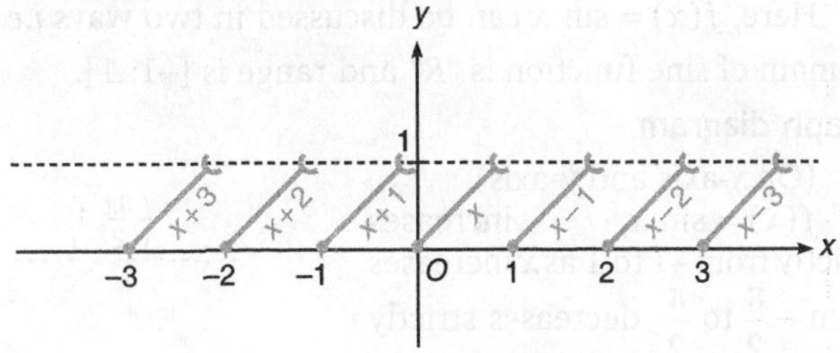

Fig. 1.18

Properties of fractional part of x

(i) $\{x\} = x$; if $0 \le x < 1$

(ii) $\{x\} = 0$; if $x \in$ integer.

(iii) $\{-x\} = 1 - \{x\}$; if $x \in$ integer.

(e) Least integer function

$$y = (x) = \lceil x \rceil,$$

(x) or $\lceil x \rceil$ indicates the integral part of x which is nearest and greatest integer to x.

It is known as ceiling of x.

Thus, $\lceil 2.3023 \rceil = 3$, $(0.23) = 1$, $(-8.0725) = -8$, $(-0.6) = 0$

In general, $n < x \le n + 1$ ($n \in$ integer))

i.e., $\lceil x \rceil$ or $(x) = n + 1$

shown as;

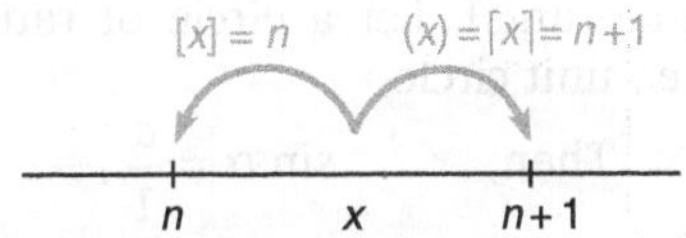

Fig. 1.19

Here, $f(x) = (x) = \lceil x \rceil$, can be expressed graphically as:

x	$\lceil x \rceil = (x)$
$-1 < x \le 0$	0
$0 < x \le 1$	1
$1 < x \le 2$	2
$-2 < x \le -1$	−1
$-3 < x \le -2$	−2

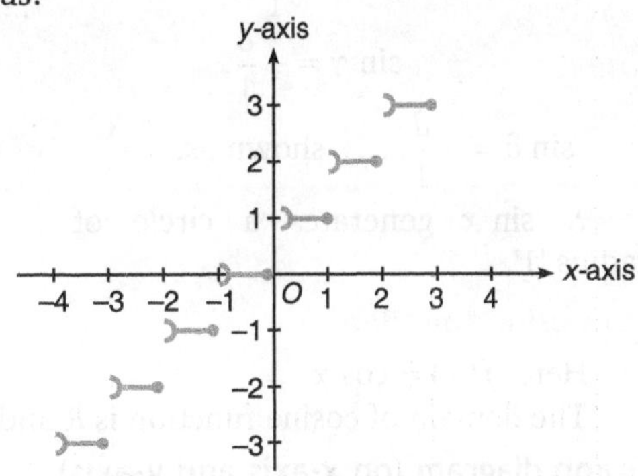

Fig. 1.20

Properties of least integer function

(i) $(x) = \lceil x \rceil = x$, if x is integer.

(ii) $(x + I) = \lceil x + I \rceil = (x) + I$; if $I \in$ integer.

(iii) Greatest integer converts $x = I + f$ to $[x] = I$ while $\lceil x \rceil$ converts to $(I + 1)$.

> **Note** We shall discuss the curves:
> $y = \{\sin x\}$, $y = \{x^3\}$, $y = \{\sin^{-1} (\sin x)\}$ $y = [\sin x]$, etc. in **chapter 2. (Curvature and Transformations)**.

1.2 TRANSCENDENTAL FUNCTIONS

1. Trigonometric Function

(a) Sine function

Here, $f(x) = \sin x$ can be discussed in two ways *i.e.,* Graph diagram and Circle diagram where Domain of sine function is "R" and range is $[-1, 1]$.

Graph diagram

(On x-axis and y-axis)

$f(x) = \sin x$, increases strictly from -1 to 1 as x increases from $-\frac{\pi}{2}$ to $\frac{\pi}{2}$, decreases strictly from 1 to -1 as x increases from $\frac{\pi}{2}$ to $\frac{3\pi}{2}$ and so on. We have graph as;

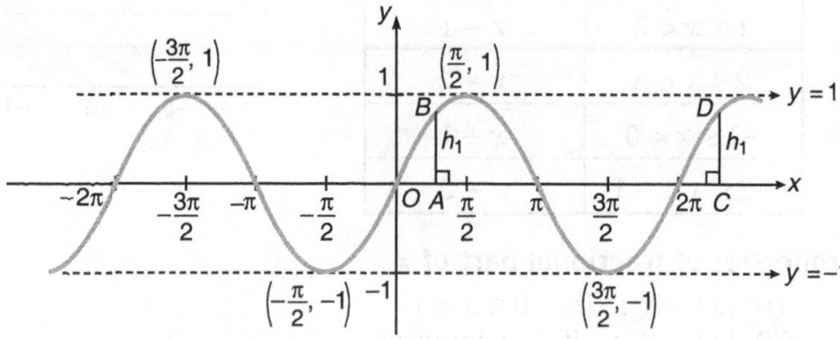

Fig. 1.21

Here, the height is same after every interval of 2π. (*i.e.,* In above figure, $AB = CD$ after every interval of 2π).

$\therefore$ $\sin x$ is called periodic function with period 2π.

Circle diagram

(On trigonometric plane or using quadrants). Let a circle of radius '1', *i.e.,* unit circle.

Then, $\sin\alpha = \frac{a}{1}$,

$\sin\beta = \frac{b}{1}$,

$\sin\gamma = -\frac{c}{1}$,

$\sin\delta = -\frac{d}{1}, \ldots$, shown as.

$\therefore$ $\sin x$ generates a circle of radius '1'.

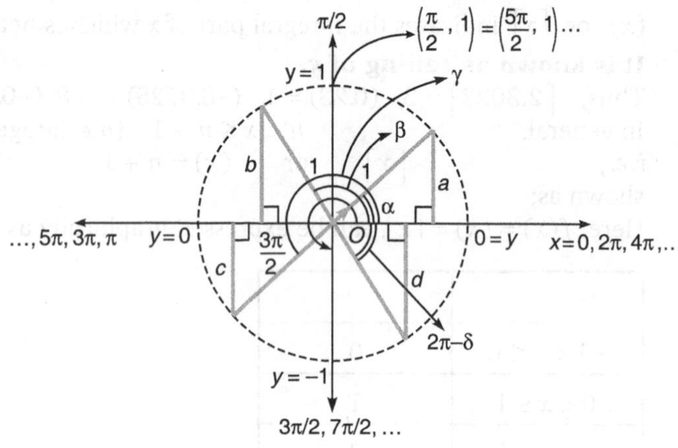

Fig. 1.22

(b) Cosine function

Here, $f(x) = \cos x$

The domain of cosine function is R and the range is $[-1, 1]$.

Graph diagram (on x-axis and y-axis)

As discussed, $\cos x$ decreases strictly from 1 to -1 as x increases from 0 to π, increases strictly from -1 to 1 as x increases from π to 2π and so on. Also, $\cos x$ is periodic with period 2π.

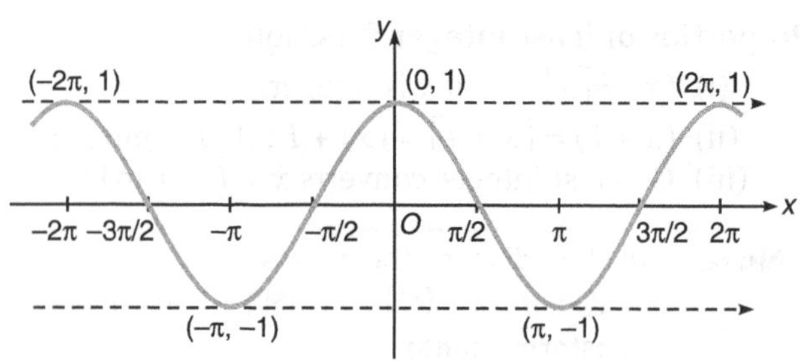

Fig. 1.23

Circle diagram

Let a circle of radius '1', *i.e.*, a unit circle.

Then, $\cos\alpha = \dfrac{a}{1}$, $\cos\beta = -\dfrac{b}{1}$,

$\cos\gamma = \dfrac{c}{1}$, $\cos\delta = -\dfrac{d}{1}$

$\therefore$ cos x generates a circle of **radius '1'**.

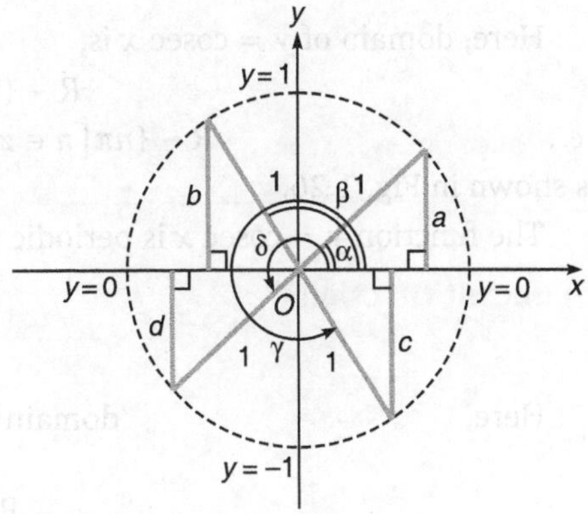

Fig. 1.24

(c) Tangent function

$f(x) = \tan x$

The domain of the function $y = \tan x$ is;

$R - \left\{\pm\dfrac{\pi}{2}, \pm\dfrac{3\pi}{2}, \pm\dfrac{5\pi}{2}, \ldots\right\}$

i.e., $R - \left\{(2n+1)\dfrac{\pi}{2}\right\}$

and Range $\in R$ or $(-\infty, \infty)$.

The function $y = \tan x$ increases strictly from $-\infty$ to $+\infty$ as x increases from

$-\dfrac{\pi}{2}$ to $\dfrac{\pi}{2}$, $\dfrac{\pi}{2}$ to $\dfrac{3\pi}{2}$, $\dfrac{3\pi}{2}$ to $\dfrac{5\pi}{2}$, ... and so on.

The graph is shown as :

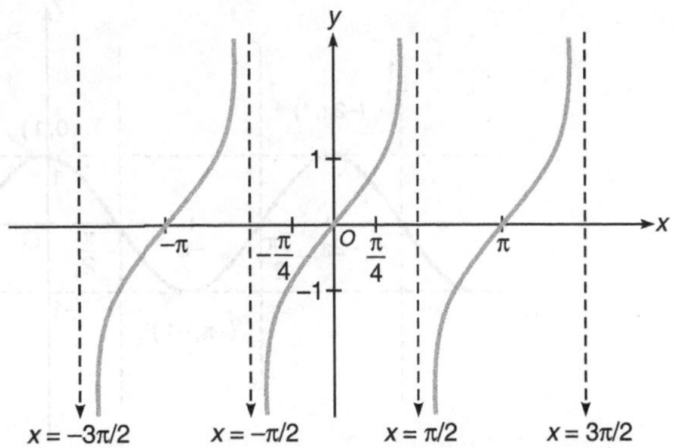

Fig. 1.25

Note Here, the curve tends to meet at $x = \pm\dfrac{\pi}{2}, \pm\dfrac{3\pi}{2} \pm\dfrac{5\pi}{2}, \ldots$ but never meets or tends to infinity.

$\therefore$ $x = \pm\dfrac{\pi}{2}, \pm\dfrac{3\pi}{2}, \pm\dfrac{5\pi}{2} \ldots$ are asymptotes to $y = \tan x$.

(d) Cosecant function

$f(x) = \text{cosec}\ x$

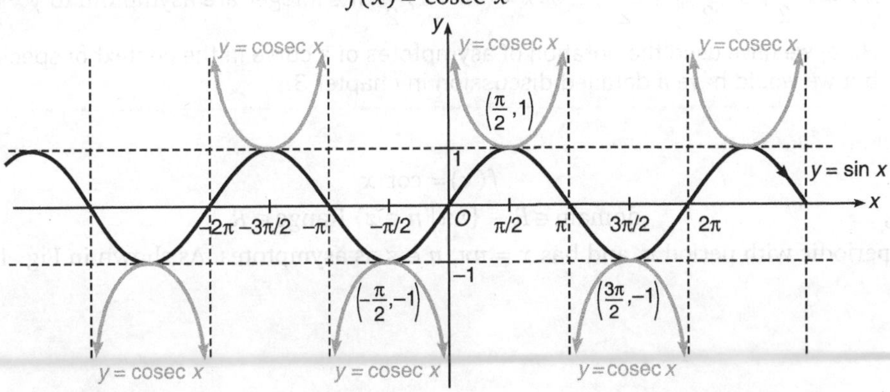

Fig. 1.26

Here, domain of y = cosec x is,

$$R - \{0, \pm\pi, \pm 2\pi, \pm 3\pi, \ldots\}$$

i.e., $R - \{n\pi \mid n \in z\}$ and range $\in R - (-1, 1)$.

as shown in Fig. 1.26.

The function y = cosec x is periodic with period 2π.

(e) Secant function

$$f(x) = \sec x$$

Here, $$\text{domain} \in R - \left\{(2n+1)\frac{\pi}{2} \,\middle|\, n \in z\right\}$$

$$\text{Range} \in R - (-1, 1)$$

Shown as:

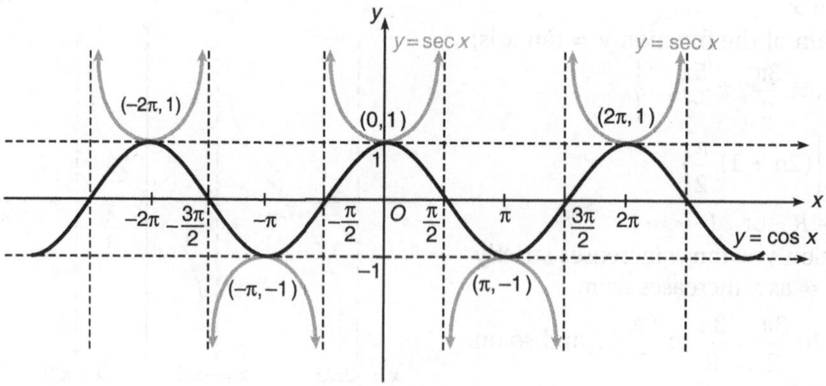

Fig. 1.27

The function $y = \sec x$ is periodic with period 2π.

> **Note** (i) The curve y = cosec x tends to meet at $x = 0, \pm\pi, \pm 2\pi, \ldots$ at infinity.
>
> $\therefore$ $x = 0, \pm\pi, \pm 2\pi, \ldots$
>
> or $x = n\pi$, $n \in$ integer are asymptote to y = cosec x.
>
> (ii) The curve $y = \sec x$ tends to meet at $x = \pm\frac{\pi}{2} \pm \frac{3\pi}{2}, \ldots$ at infinity.
>
> $\therefore x = \pm\frac{\pi}{2}, \pm\frac{3\pi}{2}, \pm\frac{5\pi}{2}, \ldots$ or $x = (2n+1)\frac{\pi}{2}$, $n \in$ integer are asymptote to y = cosec x.
>
> Here, we have used the notation of asymptotes of a curve in the context of special curves, but we would have a detailed discussion in chapter 3.

(f) Cotangent function

$$f(x) = \cot x$$

Here, domain $\in R - \{n\pi \mid n \in z\}$ Range $\in R$.

which is periodic with period π, and has $x = n\pi$, $n \in z$ as asymptotes. As shown in Fig. 1.28;

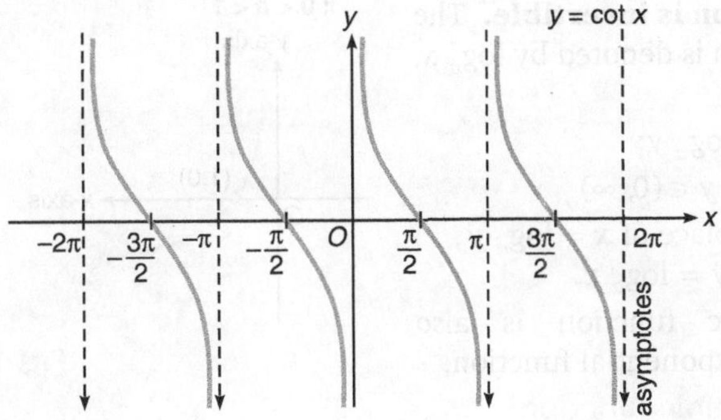

Fig. 1.28

2. Exponential Function

Here, $f(x) = a^x$, $a > 0$, $a \neq 1$, and $x \in R$, where domain $\in R$,

Range $\in (0, \infty)$.

Case I. $a > 1$

Here, $f(x) = y = a^x$ increase with the increase in x, *i.e.*, $f(x)$ is increasing function on R.

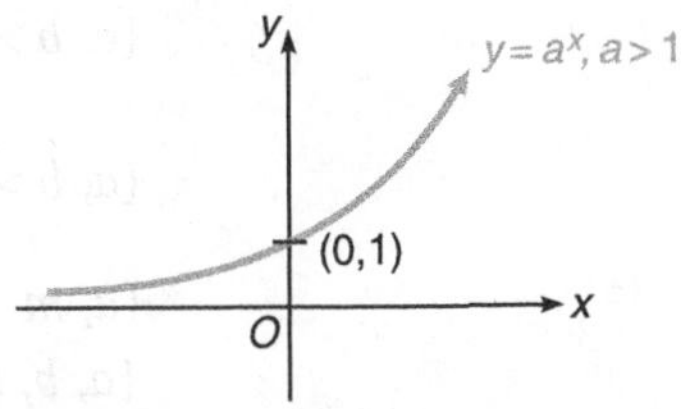

Fig. 1.29

For example;

$y = 2^x$, $y = 3^x$, $y = 4^x$,... have;

$\mathbf{2^x < 3^x < 4^x <}$ ... **for** $\mathbf{x > 1}$

and $\mathbf{2^x > 3^x > 4^x >}$ ... **for** $\mathbf{0 < x < 1}$.

and they can be shown as;

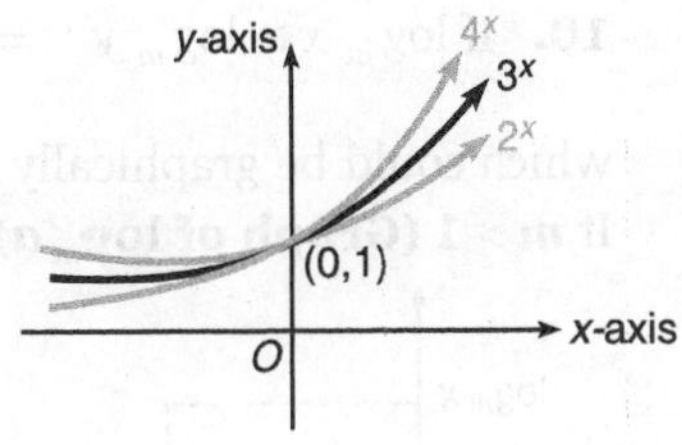

Fig. 1.30

Case II. $0 < a < 1$

Here, $f(x) = a^x$ decrease with the increase in x, *i.e.*, $f(x)$ is decreasing function on R.

"In general, exponential function increases or decreases as $(a > 1)$ or **$(0 < a < 1)$ respectively".**

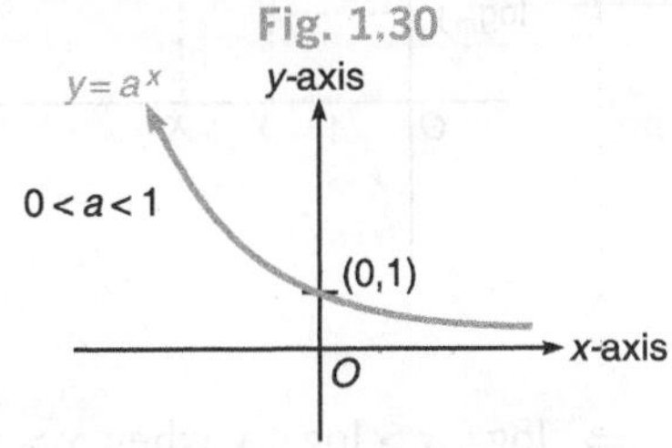

Fig. 1.31

3. Logarithmic Function

(Inverse of Exponential)

The function $f(x) = \log_a x$; $(x, a > 0)$ and $a \neq 1$ is a logarithmic function.

Thus, the domain of logarithmic function is all real positive numbers and their range is the set R of all real numbers.

We have seen that $y = a^x$ is strictly increasing **when $a > 1$** and strictly decreasing **when $0 < a < 1$.**

Thus, the function is invertible. The inverse of this function is denoted by $\log_a x$, we write

$$y = a^x \quad \Rightarrow \quad x = \log_a y;$$

where $x \in R$ and $y \in (0, \infty)$

writing $y = \log_a x$ in place of $x = \log_a y$, we have the graph of $y = \log_a x$.

Thus, logarithmic function is also known as inverse of exponential function.

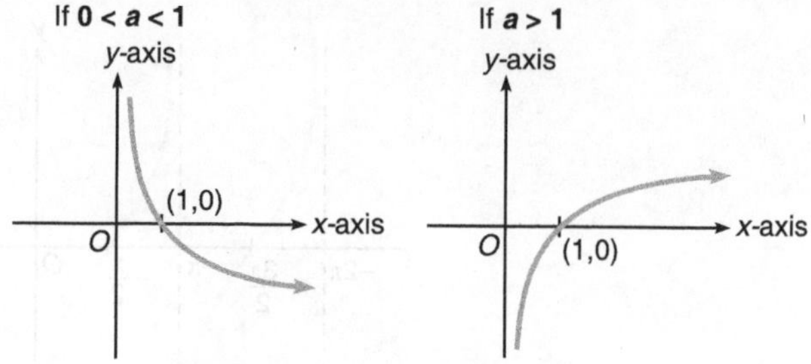

Fig. 1.32

Properties of logarithmic function

1. $\log_e (ab) = \log_e a + \log_e b$ $\quad \{a, b > 0\}$
2. $\log_e \left(\dfrac{a}{b}\right) = \log_e a - \log_e b$ $\quad \{a, b > 0\}$
3. $\log_e a^m = m \log_e a$ $\quad \{a > 0 \text{ and } m \in R\}$
4. $\log_a a = 1$ $\quad \{a > 0 \text{ and } a \neq 1\}$
5. $\log_{b^m} a = \dfrac{1}{m} \log_b a$ $\quad \{a, b > 0,\ b \neq 1 \text{ and } m \in R\}$
6. $\log_b a = \dfrac{1}{\log_a b}$ $\quad \{a, b > 0 \text{ and } a, b \neq 1\}$
7. $\log_b a = \dfrac{\log_m a}{\log_m b}$ $\quad \{a, b > 0 \neq \{1\} \text{ and } m > 0\}$
8. $a^{\log_a m} = m$ $\quad \{a, m > 0 \text{ and } a \neq 1\}$
9. $a^{\log_c b} = b^{\log_c a}$ $\quad \{a, b, c > 0 \text{ and } c \neq 1\}$
10. If $\log_m x > \log_m y \Rightarrow \begin{cases} x > y, & \text{if } m > 1 \\ x < y, & \text{if } 0 < m < 1 \end{cases}$ $\quad \{m, x, y, > 0 \text{ and } m \neq 1\}$

which could be graphically shown as;

If $m > 1$ **(Graph of $\log_m a$)**

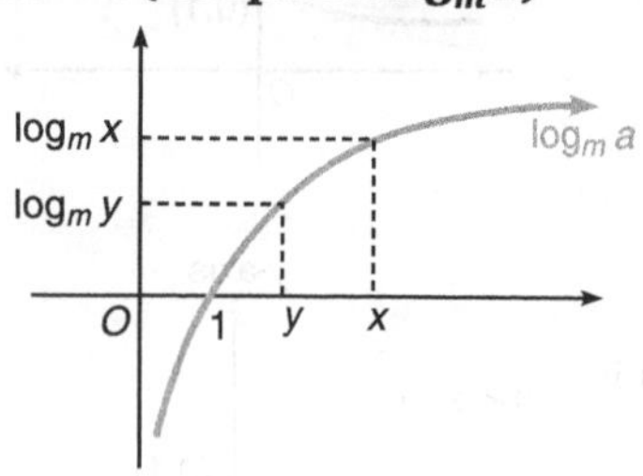

Fig. 1.33

$\Rightarrow \log_m x > \log_m y$ when $x > y$ and $m > 1$.

Again **if $0 < m < 1$. (Graph of $\log_m a$)**

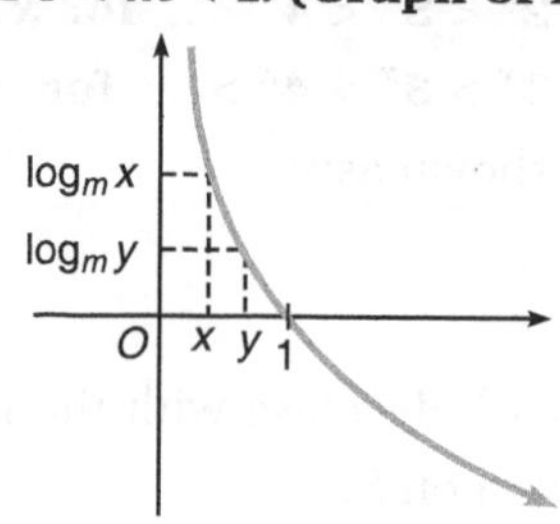

Fig. 1.34

$\Rightarrow \log_m x > \log_m y$; when $x < y$ and $0 < m < 1$.

11. $\log_m a = b \Rightarrow a = m^b$ $\quad \{a, m > 0\,;\ m \neq 1\,;\ b \in R\}$
12. $\log_m a > b \Rightarrow \begin{cases} a > m^b; & \text{if } m > 1 \\ a < m^b; & \text{if } 0 < m < 1 \end{cases}$
13. $\log_m a < b \Rightarrow \begin{cases} a < m^b; & \text{if } m > 1 \\ a > m^b; & \text{if } 0 < m < 1 \end{cases}$.

4. Geometrical Curves

(a) Straight line

$ax + by + c = 0$ (represents general equation of straight line). We know,

$$y = -\frac{c}{b} \qquad \text{when} \quad x = 0$$

and $$x = -\frac{c}{a} \qquad \text{when} \quad y = 0$$

joining above points we get required straight line.

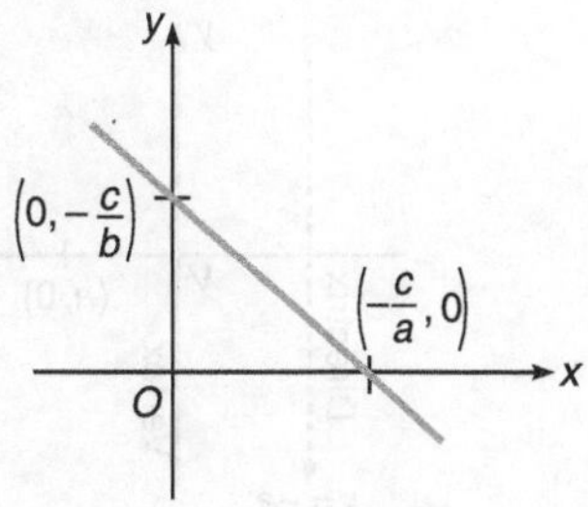

Fig. 1.35

(b) Circle

We know,

(i) $x^2 + y^2 = a^2$ is circle with centre (0, 0) and radius r.

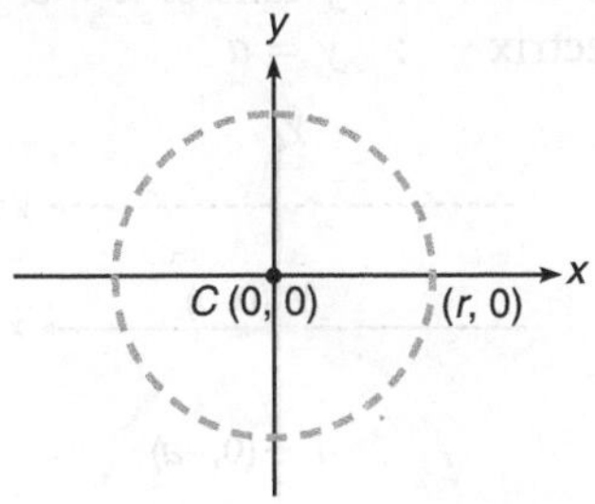

Fig. 1.36

(ii) $(x - a)^2 + (y - b)^2 = r^2$, circle with centre (a, b) and radius r.

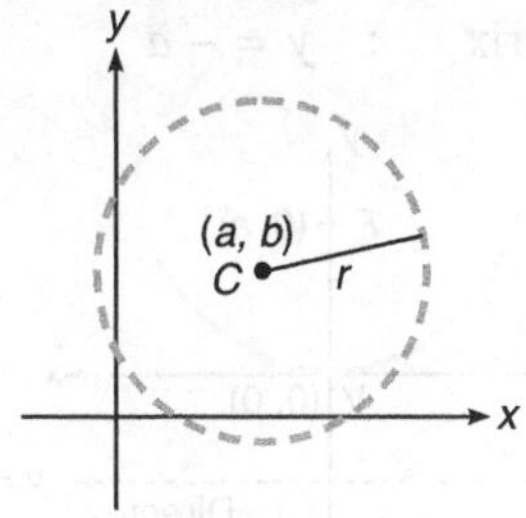

Fig. 1.37

(iii) $x^2 + y^2 + 2gx + 2fy + c = 0$;

centre $(-g, -f)$; radius $\sqrt{g^2 + f^2 - c}$.

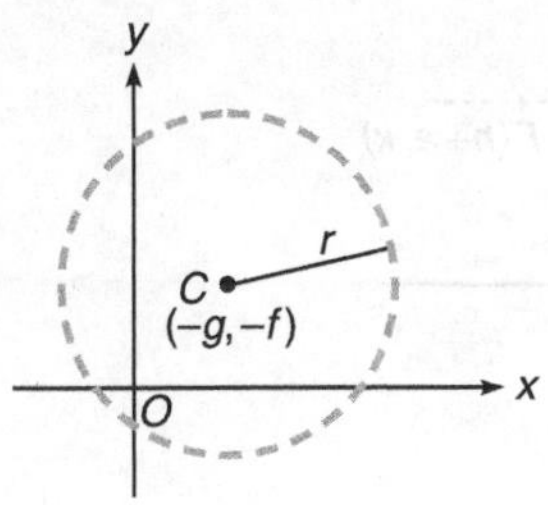

Fig. 1.38

(iv) $(x - x_1)(x - x_2) + (y - y_1)(y - y_2) = 0$;

End points of diameter are (x_1, y_1) and (x_2, y_2).

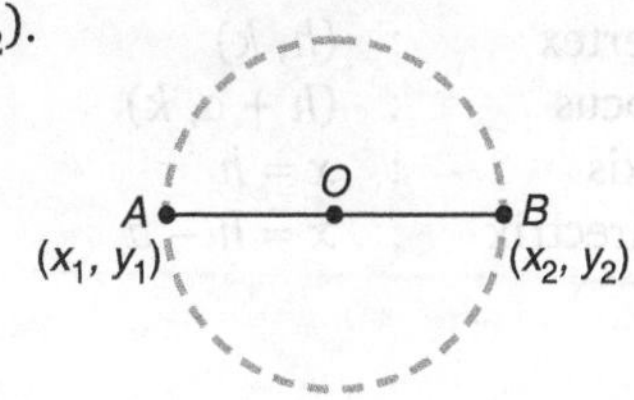

Fig. 1.39

(c) Parabola

(i) $y^2 = 4ax$

Vertex	:	(0, 0)
Focus	:	$(a, 0)$
Axis	:	x-axis or $y = 0$
Directrix	:	$x = -a$

(ii) $y^2 = -4ax$

Vertex	:	(0, 0)
Focus	:	$(-a, 0)$
Axis	:	x-axis or $y = 0$
Directrix	:	$x = a$

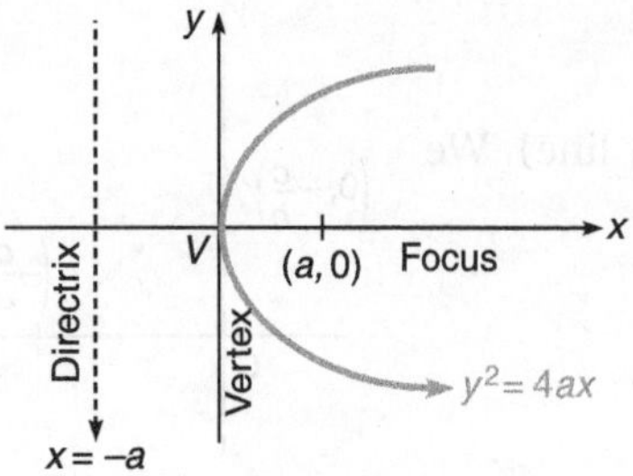

Fig. 1.40

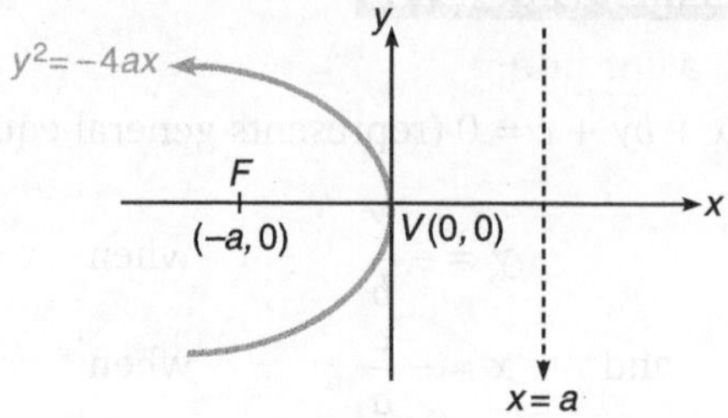

Fig. 1.41

(iii) $x^2 = 4ay$

Vextex	:	(0, 0)
Focus	:	$(0, a)$
Axis	:	y-axis or $x = 0$
Directrix	:	$y = -a$

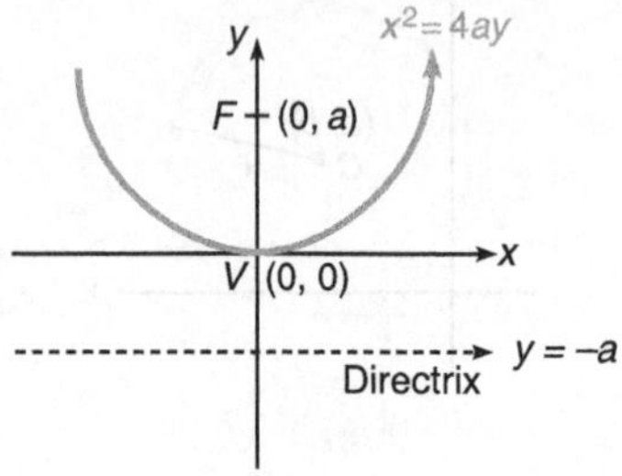

Fig. 1.42

(iv) $x^2 = -4ay$

Vertex	:	(0, 0)
Focus	:	$(0, -a)$
Axis	:	y-axis or $x = 0$
Directrix	:	$y = a$

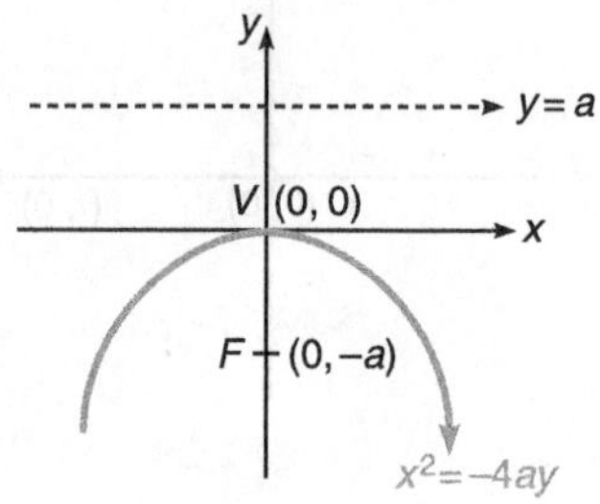

Fig. 1.43

(v) $(y - k)^2 = 4a(x - h)$

Vertex	:	(h, k)
Focus	:	$(h + a, k)$
Axis	:	$x = h$
Directrix	:	$x = h - a$

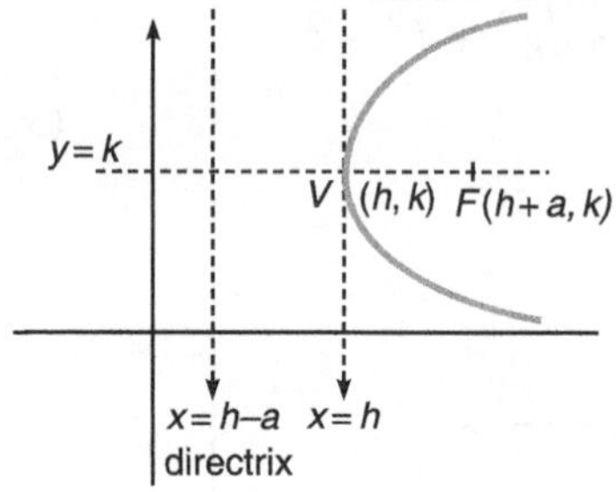

Fig. 1.44

(d) Ellipse

(i) $\dfrac{x^2}{a^2} + \dfrac{y^2}{b^2} = 1 \; (a^2 > b^2)$

Centre	:	(0, 0)
Focus	:	$(\pm ae, 0)$
Vertex	:	$(\pm a, 0)$
Eccentricity	:	$e = \sqrt{1 - \dfrac{b^2}{a^2}}$
Directrix	:	$x = \pm \dfrac{a}{e}$

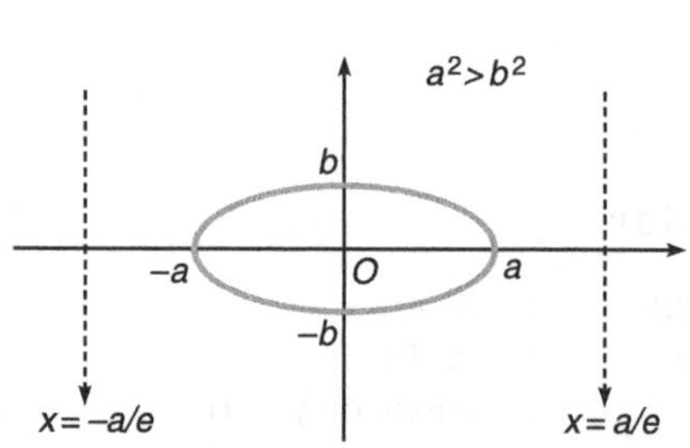

Fig. 1.45

(ii) $\dfrac{x^2}{a^2}+\dfrac{y^2}{b^2}=1 \quad (a^2<b^2)$

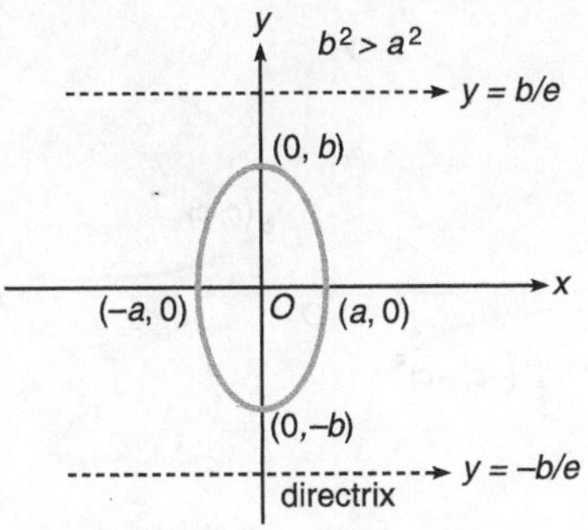

Fig. 1.46

(iii) $\dfrac{(x-h)^2}{a^2}+\dfrac{(y-k)^2}{b^2}=1 \quad (a^2>b^2)$

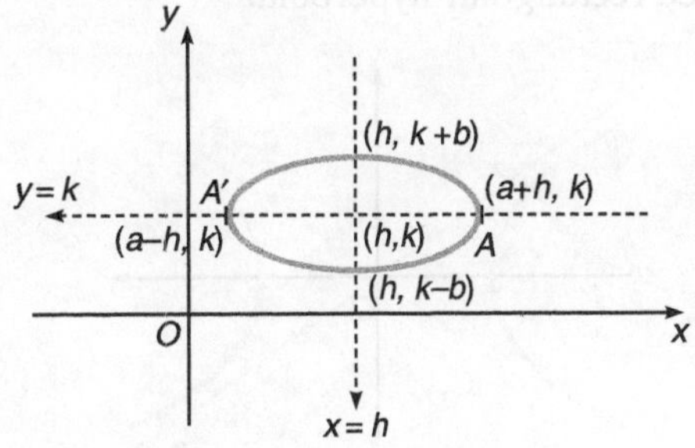

Fig. 1.47

(e) Hyperbola

(i) $\dfrac{x^2}{a^2}-\dfrac{y^2}{b^2}=1$

Centre : $(0, 0)$

Focus : $(\pm ae, 0)$

Vertices : $(\pm a, 0)$

Eccentricity : $e=\sqrt{1+\dfrac{b^2}{a^2}}$

Directrix : $x=\pm\dfrac{a}{e}$

In above figure asymptotes are $y=\pm\dfrac{b}{a}x.$

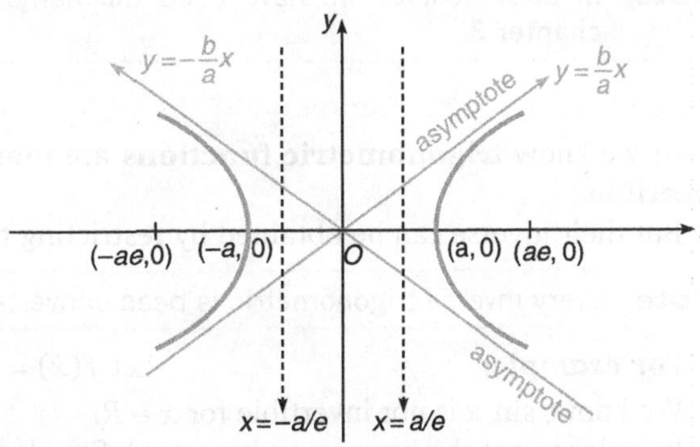

Fig. 1.48

(ii) $-\dfrac{x^2}{a^2}+\dfrac{y^2}{b^2}=1$

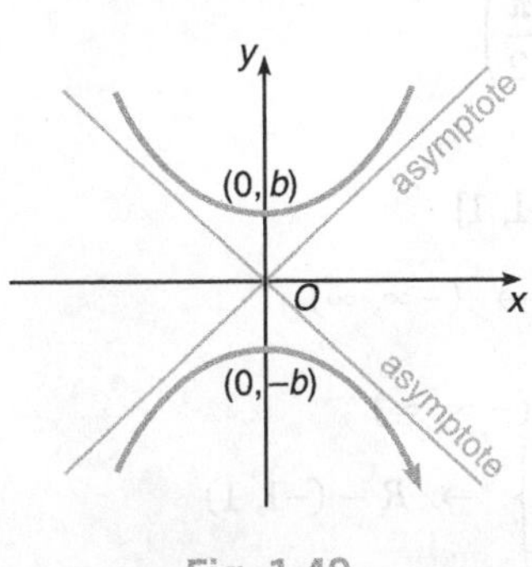

Fig. 1.49

(iii) $\dfrac{(x-h)^2}{a^2}-\dfrac{(y-k)^2}{b^2}=1$

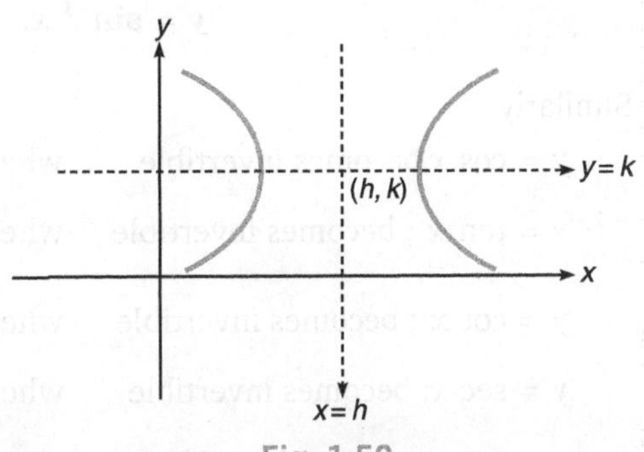

Fig. 1.50

(iv) $x^2 - y^2 = a^2$ (Rectangular hyperbola)

As asymptotes are perpendicular. Therefore, called rectangular hyperbola.

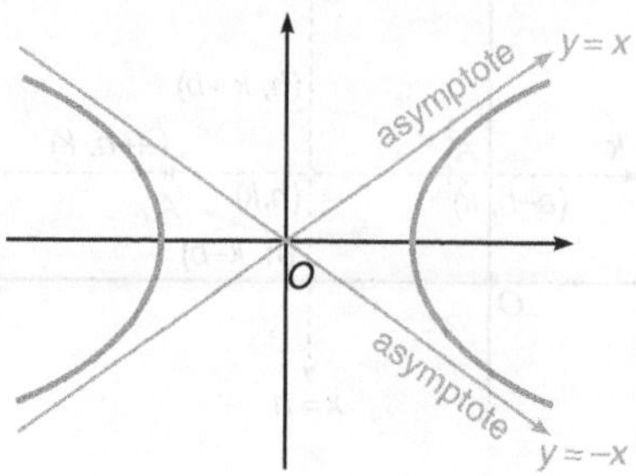

Fig. 1.51

(v) $xy = c^2$

Here, the asymptotes are x-axis and y-axis.

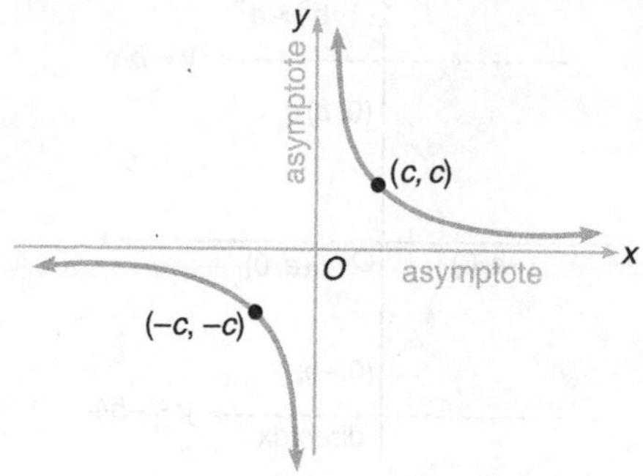

Fig. 1.52

Note In above curves we have used the name asymptotes for its complete definition see **chapter 3.**

Inverse Trigonometric Curves

As we know **trigonometric functions** are **many one** in their domain, **hence, they are not invertible.**

But their inverse can be obtained by restricting the domain so as to make invertible.

Note Every inverse trigonometric is been converted to a function by shortening the domain.

For example: Let $f(x) = \sin x$

We know, $\sin x$ is not invertible for $x \in R$.

In order to get the inverse we have to define domain as:

$$x \in \left[-\frac{\pi}{2}, \frac{\pi}{2}\right]$$

$\therefore$ If $f : \left[-\frac{\pi}{2}, \frac{\pi}{2}\right] \to [-1, 1]$ defined by $f(x) = \sin x$ is invertible and inverse can be represented by:

$$\boldsymbol{y = \sin^{-1} x}. \quad \left(-\frac{\pi}{2} \le \sin^{-1} x \le \frac{\pi}{2}\right)$$

Similarly,

$y = \cos x$ becomes invertible when $f : [0, \pi] \to [-1, 1]$

$y = \tan x$; becomes invertible when $f : \left(-\frac{\pi}{2}, \frac{\pi}{2}\right) \to (-\infty, \infty)$

$y = \cot x$; becomes invertible when $f : (0, \pi) \to (-\infty, \infty)$

$y = \sec x$; becomes invertible when $f : [0, \pi] - \left\{\frac{\pi}{2}\right\} \to R - (-1, 1)$

$y = \text{cosec } x$; becomes invertible when $f : \left[-\frac{\pi}{2}, \frac{\pi}{2}\right] - \{0\} \to R - (-1, 1)$

(i) Graph of $y = \sin^{-1} x$;

where,

$$x \in [-1, 1]$$

and $$y \in \left[-\frac{\pi}{2}, \frac{\pi}{2}\right]$$

As the graph of $f^{-1}(x)$ is mirror image of $f(x)$ about $y = x$.

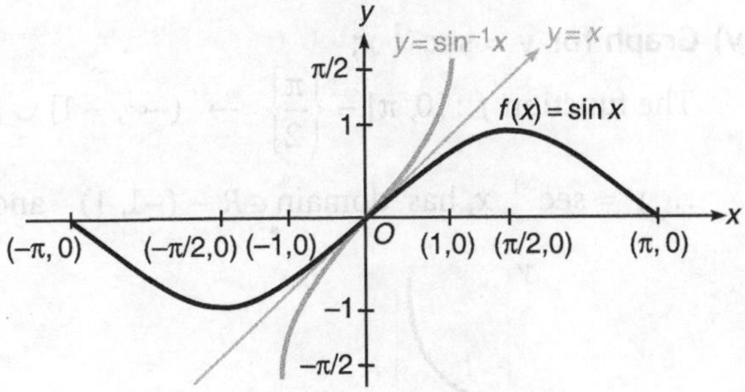

Fig. 1.53

(ii) Graph of $y = \cos^{-1} x$;

Here,

domain $\in [-1, 1]$

Range $\in [0, \pi]$

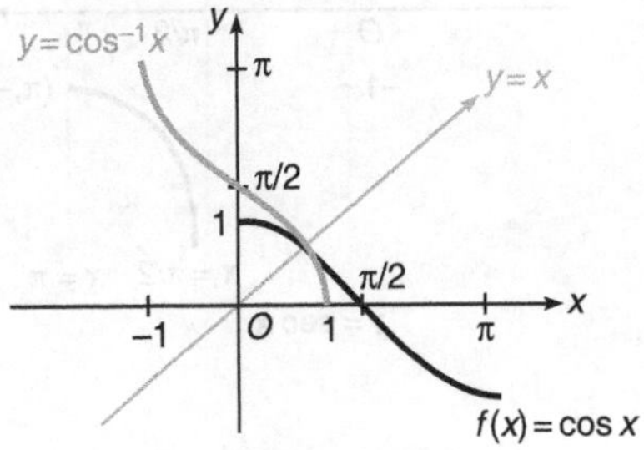

Fig. 1.54

(iii) Graph of $y = \tan^{-1} x$;

Here, domain $\in R$, Range $\in \left(-\frac{\pi}{2}, \frac{\pi}{2}\right)$.

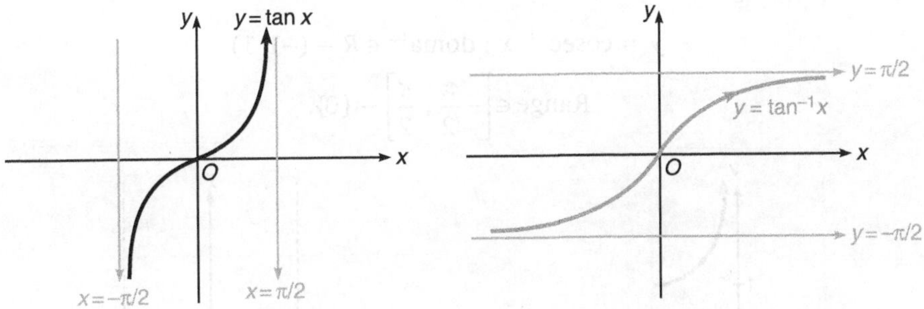

Fig. 1.55

As we have discussed earlier, "graph of inverse function is image of $f(x)$ about $y = x$" or "by interchanging the coordinate axes".

(iv) Graph of $y = \cot^{-1} x$;

We know that the function $f : (0, \pi) \to R$, given by $f(\theta) = \cot \theta$ is invertible.

∴ Thus, domain of $\cot^{-1} x \in R$ and Range $\in (0, \pi)$.

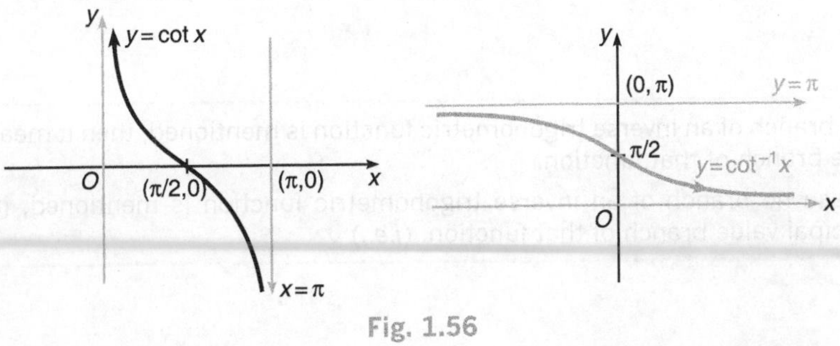

Fig. 1.56

(v) Graph for $y = \sec^{-1} x$;

The function $f : [0, \pi] - \left\{\frac{\pi}{2}\right\} \rightarrow (-\infty, -1] \cup [1, \infty)$ given by $f(\theta) = \sec\theta$ is invertible.

$\therefore$ $y = \sec^{-1} x$, has domain $\in R - (-1, 1)$ and range $\in [0, \pi] - \left\{\frac{\pi}{2}\right\}$: shown as

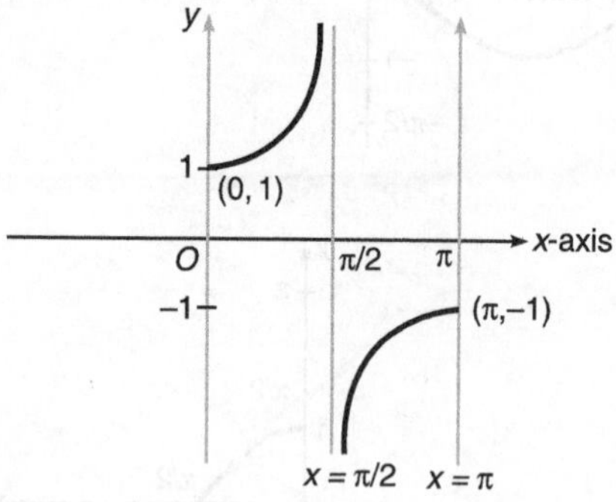

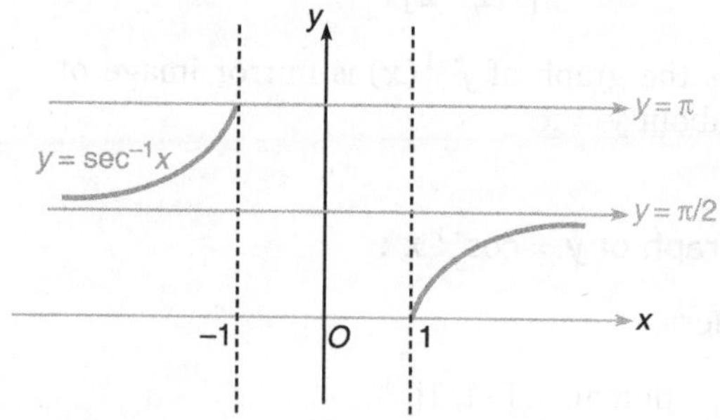

Fig. 1.57

(vi) Graph for $y = \text{cosec}^{-1}x$;

As we know, $f : \left[-\frac{\pi}{2}, \frac{\pi}{2}\right] - \{0\} \rightarrow R - (-1, 1)$ is invertible given by $f(\theta) = \cos\theta$.

$\therefore$ $y = \text{cosec}^{-1} x$; domain $\in R - (-1, 1)$

Range $\in \left[-\frac{\pi}{2}, \frac{\pi}{2}\right] - \{0\}$.

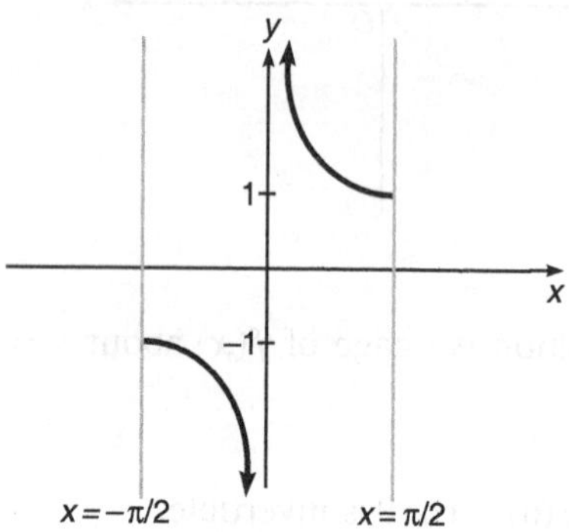

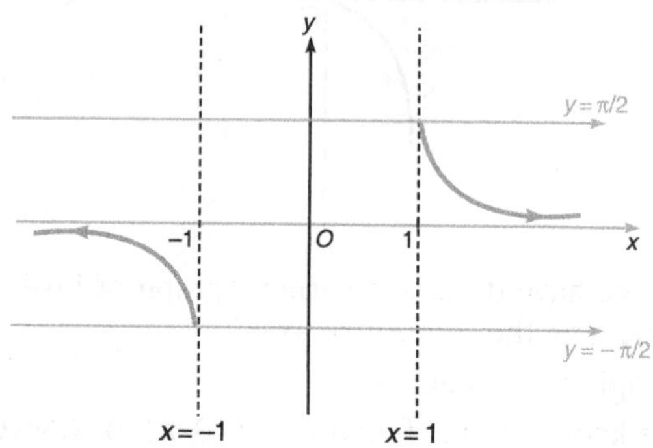

Fig. 1.58

> **Note** If no branch of an inverse trigonometric function is mentioned, then it means the principal value branch of that function.
> In case no branch of an inverse trigonometric function is mentioned, it will mean the principal value branch of that function. (*i.e.,*)

	Function	Domain	Range	Principal value branch
1.	$\sin^{-1} x$	$[-1, 1]$	$\left[-\frac{\pi}{2}, \frac{\pi}{2}\right]$	$-\frac{\pi}{2} \le y \le \frac{\pi}{2}$, where $y = \sin^{-1} x$
2.	$\cos^{-1} x$	$[-1, -1]$	$[0, \pi]$	$0 \le y \le \pi$, where $y = \cos^{-1} x$
3.	$\tan^{-1} x$	R	$\left(-\frac{\pi}{2}, \frac{\pi}{2}\right)$	$-\frac{\pi}{2} < y < \frac{\pi}{2}$, where $y = \tan^{-1} x$
4.	$\text{cosec}^{-1} x$	$(-\infty, -1] \cup [1, \infty)$	$\left[-\frac{\pi}{2}, \frac{\pi}{2}\right] - \{0\}$	$-\frac{\pi}{2} \le y \le \frac{\pi}{2}$; $y \ne 0$, where $y = \text{cosec}^{-1} x$
5.	$\sec^{-1} x$	$(-\infty, -1] \cup [1, \infty)$	$[0, \pi] - \left\{\frac{\pi}{2}\right\}$	$0 \le y \le \pi$; $y \ne \left\{\frac{\pi}{2}\right\}$, where $y = \sec^{-1} x$
6.	$\cot^{-1} x$	R	$(0, \pi)$	$0 < y < \pi$; where $y = \cot^{-1} x$.

1.3 TRIGONOMETRIC INEQUALITIES

To solve trigonometric inequalities including trigonometric functions, it is good to practice periodicity and monotonicity of functions.

Thus, first solve the inequality for the periodicity and then get the set of all solutions by adding numbers of the form $2n\pi$; $n \in z$, to each of the solutions obtained on that interval.

EXAMPLE 1 *Solve the inequality;* $\sin x > -\frac{1}{2}$.

SOLUTION As the function $\sin x$ has least positive period 2π. {That is why it is sufficient to solve inequality of the form $\sin x > a$, $\sin x \ge a$, $\sin x < a$, $\sin x \le a$ first on the interval of length 2π, and then get the solution set by adding numbers of the form $2\pi n$, $n \in z$, to each of the solutions obtained on that interval}. Thus, let us solve this inequality on the interval $\left[-\frac{\pi}{2}, \frac{3\pi}{2}\right]$, where graph of $y = \sin x$ and $y = -\frac{1}{2}$ are taken two curves on x-y plane.

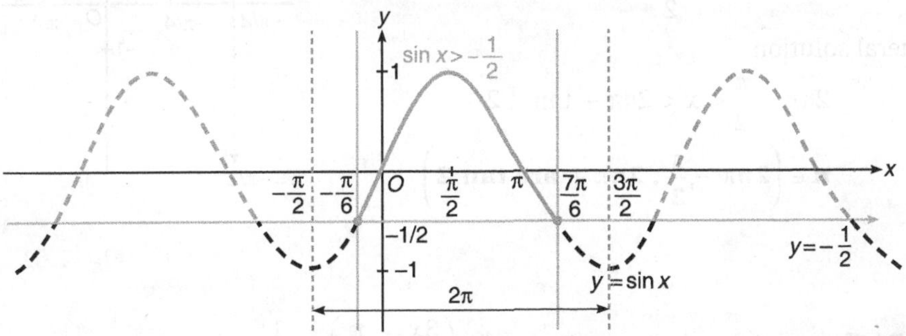

Fig. 1.59

$$\mathbf{y = \sin x} \quad \text{and} \quad \mathbf{y = -\frac{1}{2}}$$

From above figure, $\sin x > -\frac{1}{2}$ when $-\frac{\pi}{6} < x < \frac{7\pi}{6}$.

Thus, on generalising above solution;

$$\mathbf{2n\pi - \frac{\pi}{6} < x < 2n\pi + \frac{7\pi}{6};\ n \in z.}$$

which implies that those and only those values of x each of which satisfies these two inequalities for a certain $n \in z$ can serve as solutions to the original inequality.

EXAMPLE 2 *Solve the inequality:* $\cos x \le -\frac{1}{2}$.

SOLUTION As discussed in previous example, $\cos x$ is periodic with period 2π. So, to check the solution in $[0, 2\pi]$.

It is clear from figure, $\cos x \le -\frac{1}{2}$ when;

$\frac{2\pi}{3} \le x \le \frac{4\pi}{3}$.

On generalising above solution;

$$2n\pi + \frac{2\pi}{3} \le x \le 2n\pi + \frac{4\pi}{3};\ n \in z$$

$\therefore$ Solution of $\cos x \le -\frac{1}{2}$

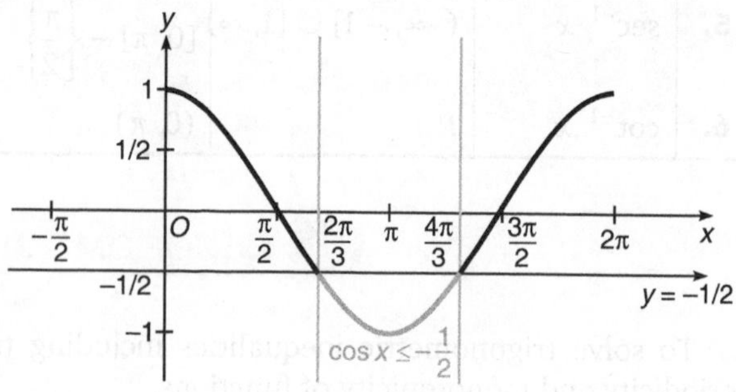

Fig. 1.60

$\Rightarrow$ $$x \in \left[2n\pi + \frac{2\pi}{3}, 2n\pi + \frac{4\pi}{3}\right];\ n \in z.$$

EXAMPLE 3 *Solve the inequality:* $\tan x < 2$.

SOLUTION We know $\tan x$ is periodic with period π. So, to check the solution on the interval $\left(-\frac{\pi}{2}, \frac{\pi}{2}\right)$.

It is clear from figure, $\tan x < 2$ when: $-\frac{\pi}{2} < x < \tan^{-1} 2$ or $-\frac{\pi}{2} < x < \text{arc}\tan 2$

$\Rightarrow$ General solution

$$2n\pi - \frac{\pi}{2} < x < 2n\pi + \tan^{-1} 2$$

$\Rightarrow$ $$n \in \left(2n\pi - \frac{\pi}{2}, 2n\pi + \text{arc}\tan 2\right)$$

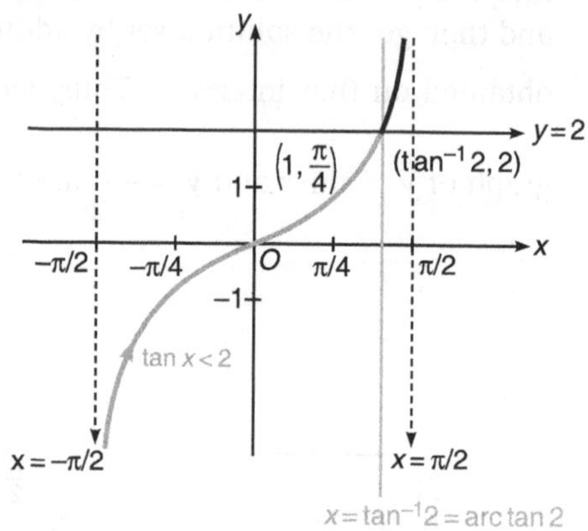

Fig. 1.61

EXAMPLE 4 *Solve the inequality:* $\sin\left(\frac{3x}{2} + \frac{\pi}{12}\right) < \frac{1}{\sqrt{2}}$.

SOLUTION Here, $\sin\left(\frac{3x}{2} + \frac{\pi}{12}\right) < \frac{1}{\sqrt{2}}$; put $\frac{3x}{2} + \frac{\pi}{12} = t$

$\therefore\ \sin t < \dfrac{1}{\sqrt{2}}$, now $\sin t$ is periodic with period 2π, thus to check on $\left[\dfrac{\pi}{2}, \dfrac{5\pi}{2}\right]$ or $\left[-\dfrac{\pi}{2}, \dfrac{3\pi}{2}\right]$

From figure,

$\sin t < \dfrac{1}{\sqrt{2}}$, when $\dfrac{3\pi}{4} < t < \dfrac{9\pi}{4}$.

$\therefore$ generalsolution

$2n\pi + \dfrac{3\pi}{4} < t < 2n\pi + \dfrac{9\pi}{4}$; $n \in z$

Substituting $t = \dfrac{3x}{2} + \dfrac{\pi}{12}$

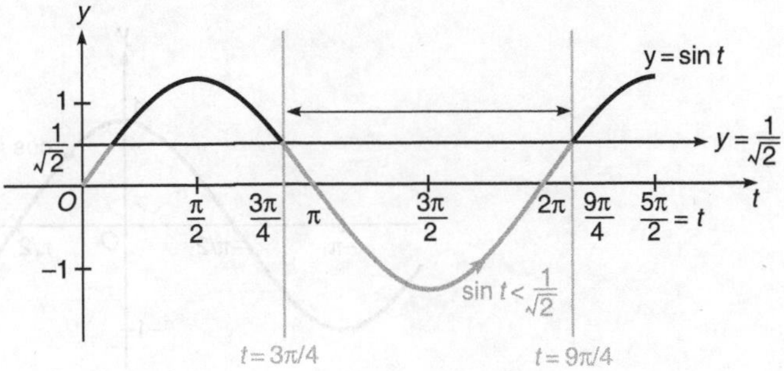

Fig. 1.62

$$2n\pi + \frac{3\pi}{4} < \frac{3x}{2} + \frac{\pi}{12} < 2n\pi + \frac{9\pi}{4}$$

$$\Rightarrow \quad \mathbf{\frac{4}{9}\pi + \frac{4}{3}\pi n < x < \frac{13}{9}\pi + \frac{4}{3}n\pi\ ;\ n \in z.}$$

EXAMPLE 5 *Solve the inequality :* $\cos 2x - \sin 2x \geq 0$.

SOLUTION Here, $\cos 2x - \sin 2x$ can be reduced to,

$$\sqrt{2}\left\{\frac{1}{\sqrt{2}}\cos 2x - \frac{1}{\sqrt{2}}\sin 2x\right\} \Rightarrow \sqrt{2}\left\{\cos\frac{\pi}{4}\cos 2x - \sin\frac{\pi}{4}\sin 2x\right\}$$

$$\Rightarrow \quad \sqrt{2}\cos\left(\frac{\pi}{4} + 2x\right)$$

$$\therefore \quad \cos 2x - \sin 2x \geq 0 \quad \text{or} \quad \cos\left(\frac{\pi}{4} + 2x\right) \geq 0\ ;\ \text{put } 2x + \frac{\pi}{4} = t$$

$\therefore\ \cos(t) \geq 0$, solving graphically,

Clearly; $\quad -\dfrac{\pi}{2} \leq t \leq \dfrac{\pi}{2}$

or $\quad 2n\pi - \dfrac{\pi}{2} < t < 2n\pi + \dfrac{\pi}{2}$

where $\quad t = 2n + \dfrac{\pi}{4}$

$\therefore\ 2n\pi - \dfrac{\pi}{2} \leq 2n + \dfrac{\pi}{4} \leq 2n\pi + \dfrac{\pi}{2}$

$\mathbf{n\pi - \dfrac{3\pi}{8} \leq n \leq n\pi + \dfrac{\pi}{8}\ ;\ n \in z.}$

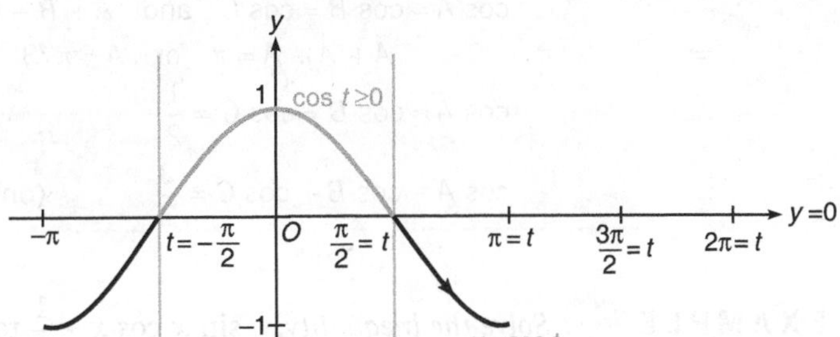

Fig. 1.63

EXAMPLE 6 *If* $A + B + C = \pi$, *then prove that;* $\cos A + \cos B + \cos C < \dfrac{3}{2}$; *where* A, B, C *are distinct.*

SOLUTION Here, we have the three trigonometric functions as $\cos A$, $\cos B$ and $\cos C$.

$\therefore$ let $f(x) = \cos x$; which can be plotted as;

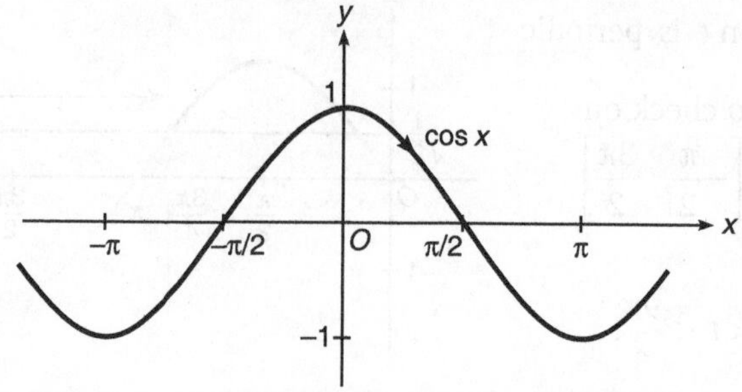

Fig. 1.64

Now, let us suppose any three points $x = A$, $x = B$, $x = C$ on $f(x) = \cos x$. So that $A + B + C = \pi$ or on the interval of length π.

where G, be centroid of Δ given by

$$\left(\frac{A+B+C}{3}, \frac{\cos A + \cos B + \cos C}{3}\right)$$

Thus, from figure points Q, G, P are collinear,

where; ordinate of $GQ <$ ordinate of PQ.

$$\frac{\cos A + \cos B + \cos C}{3} < \cos\left(\frac{A+B+C}{3}\right)$$

$$\Rightarrow \quad \cos A + \cos B + \cos C < 3\cos\left(\frac{\pi}{3}\right)$$

$$\Rightarrow \quad \mathbf{\cos A + \cos B + \cos C < \frac{3}{2}}$$

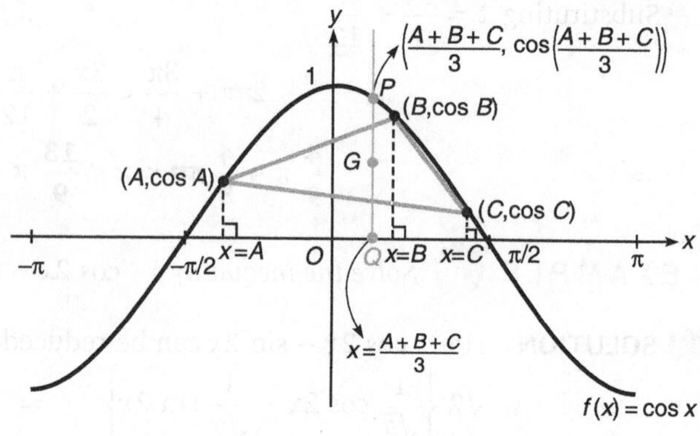

Fig. 1.65

Note Here, a particular case arises when $A = B = C$ (*i.e.*, when A, B, C are non-distinct)

$$\cos A = \cos B = \cos C \quad \text{and} \quad A + B + C = \pi$$

$$\Rightarrow \quad A + A + A = \pi \quad \text{or} \quad A = \pi/3.$$

$$\therefore \quad \cos A = \cos B = \cos C = \frac{1}{2}$$

$$\therefore \quad \cos A + \cos B + \cos C = \frac{3}{2} \qquad \text{(only when } A = B = C\text{)}$$

EXAMPLE 7 *Solve the inequality:* $\sin x \cos x + \frac{1}{2}\tan x \geq 1.$

SOLUTION Here; left hand side is defined for all x, except $x = n\pi + \frac{\pi}{2}$, where $n \in z$.

$$\therefore \quad 2\sin x\cos x + \tan x \geq 2$$

$$\Rightarrow \quad \frac{2\tan x}{1+\tan^2 x} + \tan x \geq 2 \qquad [\text{Let, } \tan x = y]$$

$$\Rightarrow \quad \frac{2y}{1+y^2} + y \geq 2$$

$$\Rightarrow \quad \frac{2y + y(1+y^2) - 2(1+y^2)}{(1+y^2)} \geq 0 \qquad \{\because 1 + y^2 \geq 0\}$$

$$\therefore \quad 2y + y(1+y^2) - 2(1+y^2) \geq 0$$

$$\Rightarrow \quad y^3 - 2y^2 + 3y - 2 \geq 0$$

$$\Rightarrow \quad y^2(y-1) - y(y-1) + 2(y-1) \geq 0$$

$$\text{or} \quad (y-1)(y^2 - y + 2) \geq 0$$

$$\Rightarrow \quad y - 1 \geq 0$$

$$\left\{\because y^2 - y + 2 = \left(y - \frac{1}{2}\right)^2 + \frac{7}{4} > 0, \text{ for all } y\right\}$$

$\therefore$ $\tan x \geq 1$, shown as:

from given figure;

$$\frac{\pi}{4} \leq x < \frac{\pi}{2}$$

$$\text{or} \quad n\pi + \frac{\pi}{4} \leq x < n\pi + \frac{\pi}{2} \quad ; \quad n \in z$$

$$\therefore \quad \boldsymbol{x \in \left[n\pi + \frac{\pi}{4}, n\pi + \frac{\pi}{2}\right]; \; n \in z}$$

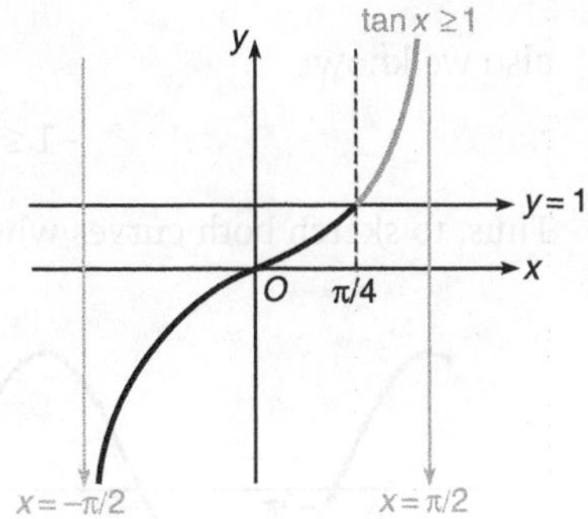

Fig. 1.66

EXAMPLE 8 *If $A + B + C = \pi$, then prove that*

$$\tan^2\frac{A}{2} + \tan^2\frac{B}{2} + \tan^2\frac{C}{2} > 1.$$

SOLUTION Here, $\tan^2\frac{A}{2}$, $\tan^2\frac{B}{2}$ and $\tan^2\frac{C}{2}$ are three same function. So consider $f(x) = \tan^2\frac{x}{2}$, whose period is 2π.

$\therefore$ plotting $\tan^2\frac{x}{2}$ for $x \in (-\pi, \pi)$.

In given curve let us consider any three points A, B, C such that

$$A + B + C = \pi.$$

Now, centroid of ΔRST;

$$G\left(\frac{A+B+C}{3}, \frac{\tan^2\frac{A}{2} + \tan^2\frac{B}{2} + \tan^2\frac{C}{2}}{3}\right)$$

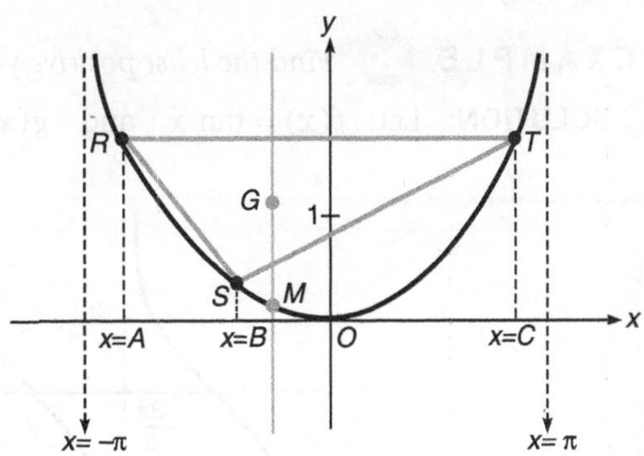

Fig. 1.67

also, $\quad M\left(\frac{A+B+C}{3}, \tan^2\left(\frac{A+B+C}{2(3)}\right)\right)$ where; $GN > MN$.

$$\Rightarrow \quad \frac{\tan^2\frac{A}{2} + \tan^2\frac{B}{2} + \tan^2\frac{C}{2}}{3} > \tan^2\left(\frac{A+B+C}{6}\right) \quad \Rightarrow \quad \boldsymbol{\tan^2\frac{A}{2} + \tan^2\frac{B}{2} + \tan^2\frac{C}{2} > 1.}$$

1.4 SOLVING EQUATIONS GRAPHICALLY

Here, we sketch both left hand and right hand side of equality and the numbers of intersections are required solutions.24

EXAMPLE 9 *Find the number of solutions of;* $\sin x = \frac{x}{10}$.

SOLUTION Here, let $f(x) = \sin x$ and $g(x) = \frac{x}{10}$

also we know; $-1 \le \sin x \le 1$

$$\therefore \quad -1 \le \frac{x}{10} \le 1 \Rightarrow -10 \le x \le 10$$

Thus, to sketch both curves when $x \in [-10, 10]$

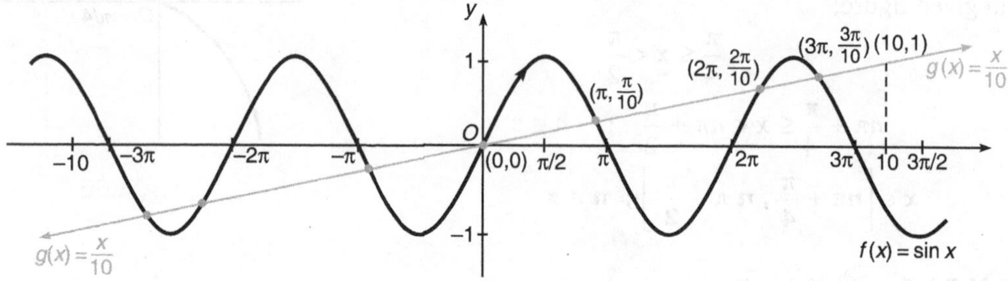

Fig. 1.68

From above figure $f(x) = \sin x$ and $g(x) = \frac{x}{10}$ intersect at 7 points. So, **numbers of solutions are 7.**

EXAMPLE 10 *Find the least positive value of x, satisfying* $\tan x = x + 1$ *lies in the interval.*

SOLUTION Let; $f(x) = \tan x$ and $g(x) = x + 1$; which could be shown as:

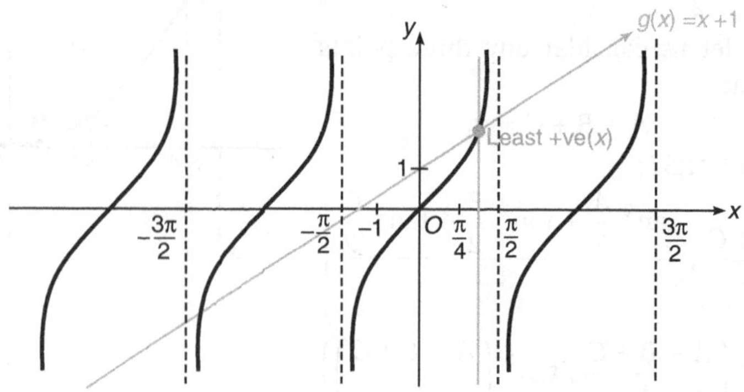

Fig. 1.69

From the above figure $\tan x = x + 1$ has infinitly many solutions but the least positive value of $\boldsymbol{x} \in \left(\frac{\pi}{4}, \frac{\pi}{2}\right)$.

EXAMPLE 11 *Find the number of solutions of the equation,*

$$\sin x = x^2 + x + 1.$$

SOLUTION Let; $f(x) = \sin x$ and $g(x) = x^2 + x + 1 = \left(x + \frac{1}{2}\right)^2 + \frac{3}{4}$

which could be shown as;

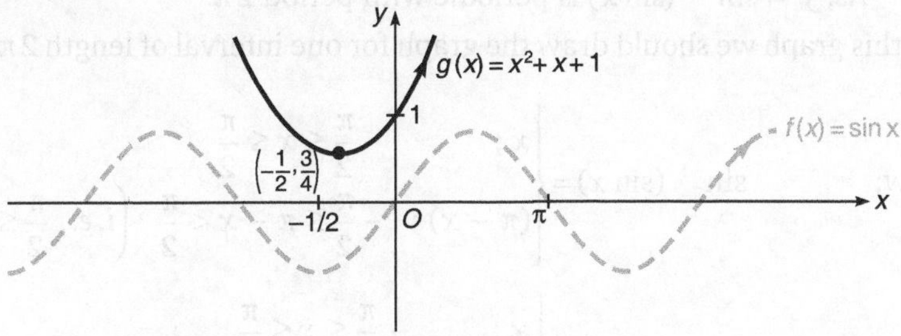

Fig. 1.70

which do not intersect at any point, therefore **no solution.**

EXAMPLE 12 *Find the number of solutions of:* $e^x = x^4$.

SOLUTION Let; $f(x) = e^x$ and $g(x) = x^4$, which could be shown as;

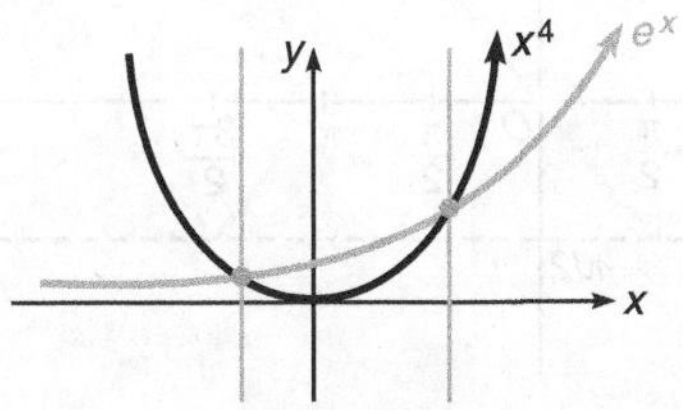

Fig. 1.71

From the figure, it is clear they intersect at two points, therefore **two solutions**.

EXAMPLE 13 *Find the number of solutions of;* $\log_{10} x = x$.

SOLUTION Let; $f(x) = \log_{10} x$ and $g(x) = x$; which could be shown as;

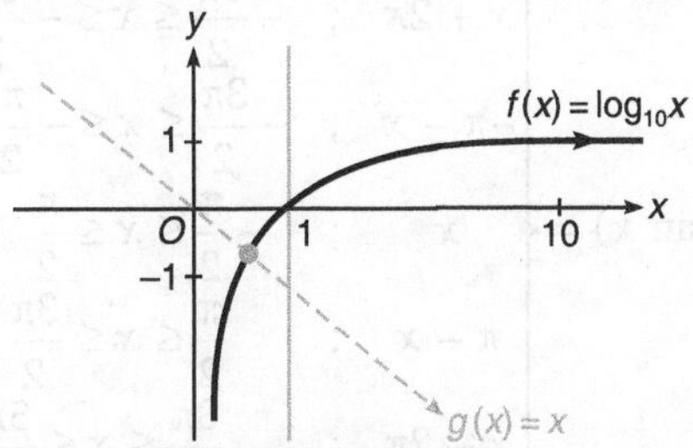

Fig. 1.72

From above figure, it is clear they intersect at one points, therefore **1 solution**.

SOME MORE SOLVED EXAMPLES

EXAMPLE 1 *Sketch the graph for* $y = \sin^{-1}(\sin x)$.

SOLUTION As, $y = \sin^{-1}(\sin x)$ is periodic with period 2π.

$\therefore$ to draw this graph we should draw the graph for one interval of length 2π and repeat for entire values of x.

As we know; $$\sin^{-1}(\sin x) = \begin{cases} x; & -\dfrac{\pi}{2} \le x \le \dfrac{\pi}{2} \\ (\pi - x); & -\dfrac{\pi}{2} \le \pi - x < \dfrac{\pi}{2} \quad \left(i.e., \dfrac{\pi}{2} \le x \le \dfrac{3\pi}{2}\right) \end{cases}$$

or $$\sin^{-1}(\sin x) = \begin{cases} x, & -\dfrac{\pi}{2} \le x \le \dfrac{\pi}{2} \\ \pi - x, & \dfrac{\pi}{2} \le x \le \dfrac{3\pi}{2}, \end{cases}$$

which is defined for the interval of length 2π, plotted as;

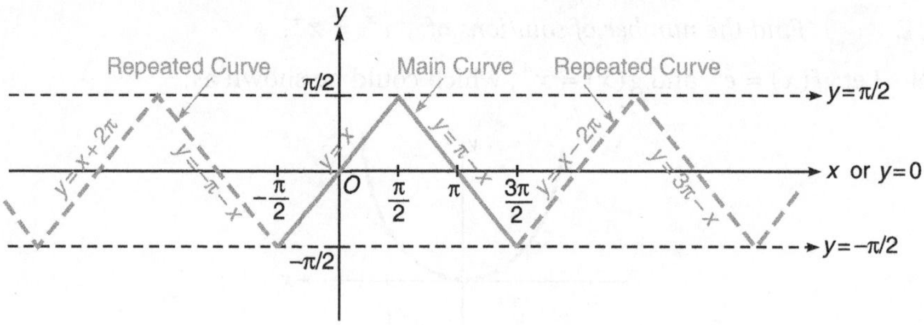

Fig. 1.73

Thus, the graph for $y = \sin^{-1}(\sin x)$, is a straight line up and a straight line down with slopes 1 and –1 respectively lying between $\left[-\dfrac{\pi}{2}, \dfrac{\pi}{2}\right]$.

Note : Students are adviced to learn the definition of $\sin^{-1}(\sin x)$ as,

$$y = \sin^{-1}(\sin x) = \begin{cases} x + 2\pi \;; & -\dfrac{5\pi}{2} \le x \le -\dfrac{3\pi}{2} \\ -\pi - x \;; & -\dfrac{3\pi}{2} \le x \le -\dfrac{\pi}{2} \\ x \;; & -\dfrac{\pi}{2} \le x \le \dfrac{\pi}{2} \\ \pi - x \;; & \dfrac{\pi}{2} \le x \le \dfrac{3\pi}{2} \\ x - 2\pi \;; & \dfrac{3\pi}{2} \le x \le \dfrac{5\pi}{2} \ldots \text{ and so on} \end{cases}$$

EXAMPLE 2 *Sketch the graph for $y = \cos^{-1}(\cos x)$.*

SOLUTION As, $y = \cos^{-1}(\cos x)$ is periodic with period 2π.

$\therefore$ to draw this graph we should draw the graph for one interval of length 2π and repeat for entire values of x of length 2π.

As we know;

$$\cos^{-1}(\cos x) = \begin{cases} x; & 0 \le x \le \pi \\ 2\pi - x; & 0 \le 2\pi - x \le \pi, \end{cases}$$

or

$$\cos^{-1}(\cos x) = \begin{cases} x; & 0 \le x \le \pi \\ 2\pi - x; & \pi \le x \le 2\pi. \end{cases}$$

Thus, it has been defined for $0 < x < 2\pi$ that has length 2π. So, its graph could be plotted as;

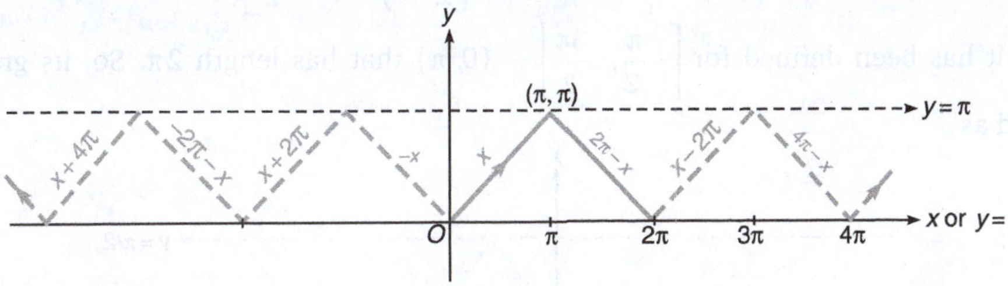

Fig. 1.74

Thus, the curve $y = \cos^{-1}(\cos x)$.

EXAMPLE 3 *Sketch the graph for $y = \tan^{-1}(\tan x)$.*

SOLUTION As $y = \tan^{-1}(\tan x)$ is periodic with period π.

$\therefore$ to draw this graph we should draw the graph for one interval of length π and repeat for entire values of x.

As we know; $\tan^{-1}(\tan x) = \left\{x\,;\ -\dfrac{\pi}{2} < x < \dfrac{\pi}{2}\right\}$

Thus, it has been defined for $-\dfrac{\pi}{2} < x < \dfrac{\pi}{2}$ that has length π. So, its graph could be plotted as;

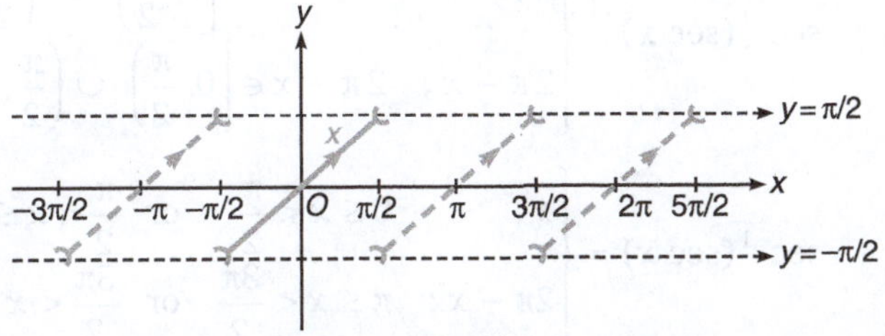

Fig. 1.75

Thus, the curve for $y = \tan^{-1}(\tan x)$, where y is not defined for $x \in (2n+1)\dfrac{\pi}{2}$.

EXAMPLE 4 *Sketch the graph for $y = \operatorname{cosec}^{-1}(\operatorname{cosec} x)$.*

SOLUTION As $y = \operatorname{cosec}^{-1}(\operatorname{cosec} x)$ is periodic with period 2π.

$\therefore$ to draw this graph we should draw the graph for one interval of length 2π and repeat for entire values of x.

As we know;

$$\operatorname{cosec}^{-1}(\operatorname{cosec} x) = \begin{cases} x\ ; & -\dfrac{\pi}{2} \le x < 0 \quad \text{or} \quad 0 < x \le \dfrac{\pi}{2} \\ \pi - x; & -\dfrac{\pi}{2} \le \pi - x < 0 \quad \text{or} \quad 0 < \pi - x \le \dfrac{\pi}{2} \end{cases}$$

or
$$\operatorname{cosec}^{-1}(\operatorname{cosec} x) = \begin{cases} x; & x \in \left[-\dfrac{\pi}{2}, 0\right) \cup \left(0, \dfrac{\pi}{2}\right] \\ \pi - x; & x \in \left[\dfrac{\pi}{2}, \pi\right) \cup \left(\pi, \dfrac{3\pi}{2}\right] \end{cases}$$

Thus, it has been defined for $\left[-\dfrac{\pi}{2}, \dfrac{3\pi}{2}\right] - \{0, \pi\}$ that has length 2π. So, its graph could be plotted as

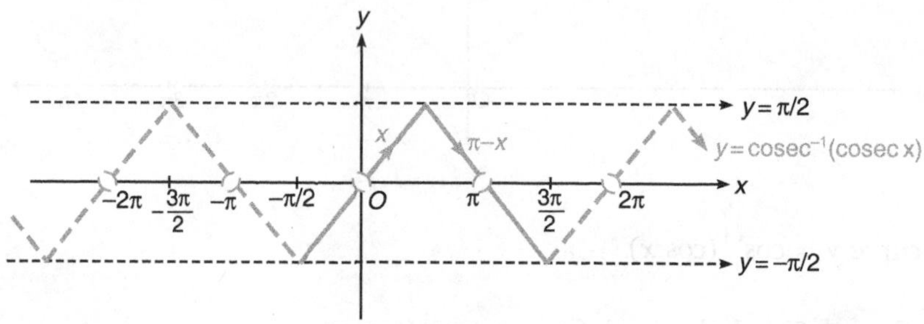

Fig. 1.76

EXAMPLE 5 *Sketch the graph for $y = \sec^{-1}(\sec x)$.*

SOLUTION As $y = \sec^{-1}(\sec x)$ is periodic with period 2π.

$\therefore$ to draw this graph we should draw the graph for one interval of length 2π and repeat for entire values of x.

As we know;

$$\sec^{-1}(\sec x) = \begin{cases} x\ ; & x \in \left[0, \dfrac{\pi}{2}\right) \cup \left(\dfrac{\pi}{2}, \pi\right] \\ 2\pi - x; & 2\pi - x \in \left[0, \dfrac{\pi}{2}\right) \cup \left(\dfrac{\pi}{2}, \pi\right] \end{cases}$$

or
$$\sec^{-1}(\sec x) = \begin{cases} x\ ; & 0 \le x < \dfrac{\pi}{2} \quad \text{or} \quad \dfrac{\pi}{2} < x \le \pi \\ 2\pi - x\ ; & \pi \le x < \dfrac{3\pi}{2} \quad \text{or} \quad \dfrac{3\pi}{2} < x \le 2\pi \end{cases}.$$

Thus, it has been defined for $[0, 2\pi] - \left\{\dfrac{\pi}{2}, \dfrac{3\pi}{2}\right\}$ that has length 2π. So, its graph could be plotted as;

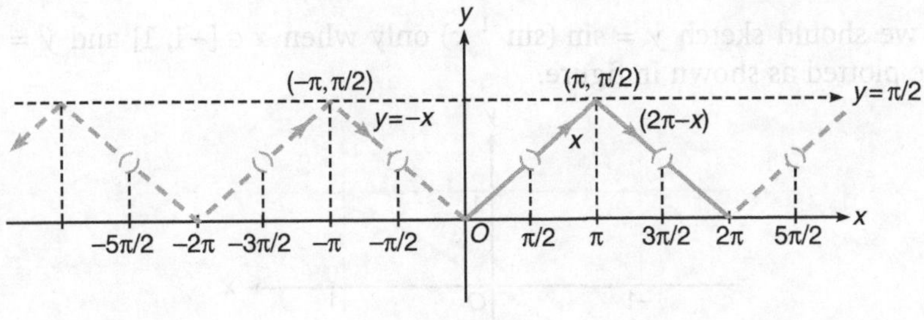

Fig. 1.77

Thus, the curve for $y = \sec^{-1}(\sec x)$.

EXAMPLE 6 *Sketch the graph for* $y = \cot^{-1}(\cot x)$.

SOLUTION As, $y = \cot^{-1}(\cot x)$ is periodic with period π.

$\therefore$ to draw this graph we should draw the graph for one interval of length π and repeat for entire values of x.

As we know

$$\cot^{-1}(\cot x) = \{x;\ 0 < x < \pi\}$$

which is defined for length π, *i.e.,* $x \in (0, \pi)$ and $x \notin \{n\pi, n \in z\}$.

So, its graph could be plotted as;

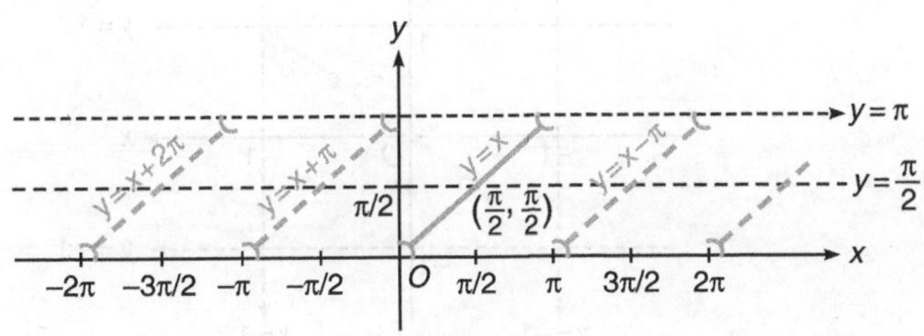

Fig. 1.78

Thus, the curve for $\mathbf{y = \cot^{-1}(\cot x)}$.

EXAMPLE 7 *Sketch the graph for:*

(i) $\sin(\sin^{-1} x)$ (ii) $\cos(\cos^{-1} x)$ (iii) $\tan(\tan^{-1} x)$

(iv) $\text{cosec}(\text{cosec}^{-1} x)$ (v) $\sec(\sec^{-1} x)$ (vi) $\cot(\cot^{-1} x)$

SOLUTION As we know, all the above mentioned six curves are **non-periodic, but have restricted domain and range.**

So, we shall first define each curve for its domain and range and then sketch these curves.

(i) Sketch for $y = \sin(\sin^{-1} x)$

We know; domain, $x \in [-1, 1]$ (*i.e.,* $-1 \le x \le 1$)

and range $y = x \Rightarrow y \in [-1, 1]$

Hence, we should sketch $y = \sin(\sin^{-1} x)$ only when $x \in [-1, 1]$ and $y = x$. So, its graph could be plotted as shown in figure.

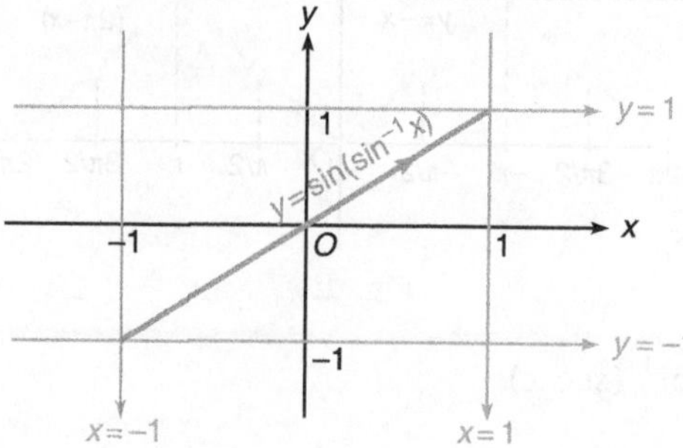

Fig. 1.79

Thus, the graph for $y = \sin(\sin^{-1} x)$.

(ii) Sketch for the curve $y = \cos(\cos^{-1} x)$.

We know, domain, $x \in [-1, 1]$ (*i.e.*, $-1 \le x \le 1$)

and range $y = x \Rightarrow y \in [-1, 1]$

Hence, we should sketch $y = \cos(\cos^{-1} x) = x$ only when $x \in [-1, 1]$. So, its graph could be plotted as shown in Fig. 1.80.

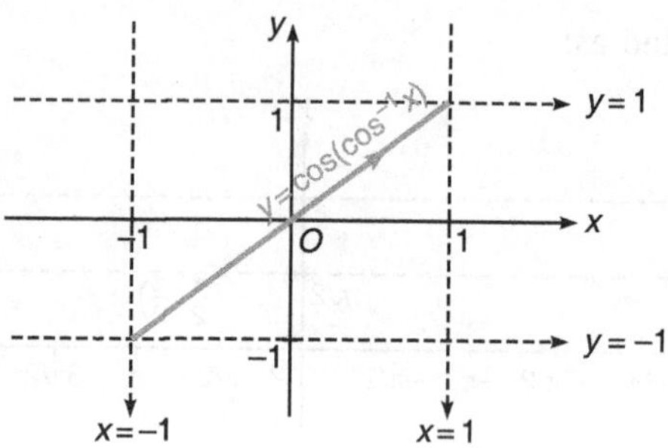

Fig. 1.80

Thus, the graph for $y = \cos(\cos^{-1} x)$.

(iii) Sketch for the curve $y = \tan(\tan^{-1} x)$

We know,

domain, $x \in R$ (*i.e.*, $-\infty < x < \infty$) and

Range $y = x \Rightarrow y \in R$.

Hence, we should sketch

$$y = \tan(\tan^{-1} x) = x, \forall\ x \in R.$$

So, its graph could be plotted as shown;

Thus, the graph for $y = \tan(\tan^{-1} x)$.

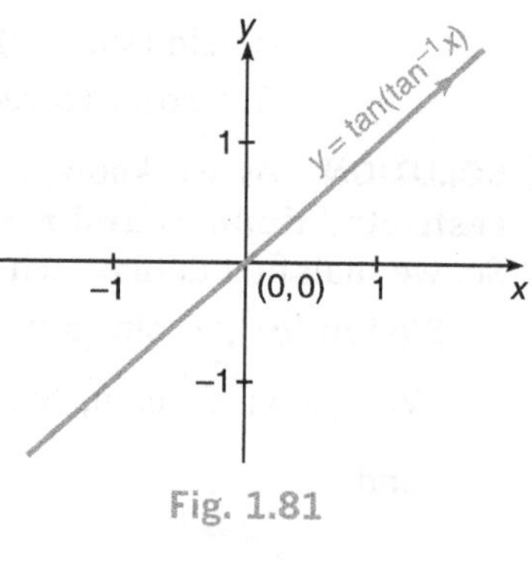

Fig. 1.81

(iv) Sketch for $y = \text{cosec}\,(\text{cosec}^{-1}x)$

We know;

domain $\in R - (-1, 1)$

$$(i.e., -\infty < x \le -1 \text{ or } 1 \le x < \infty)$$

and range $y = x \Rightarrow y \in R - (-1, 1)$.

Hence, we should sketch

$y = \text{cosec}\,(\text{cosec}^{-1}x) = x$ only when $x \in (-\infty, -1] \cup [1, \infty)$.

So, its graph could be plotted as shown in Fig. 1.82;

Thus, the graph for

$$y = \text{cosec}\,(\text{cosec}^{-1}x).$$

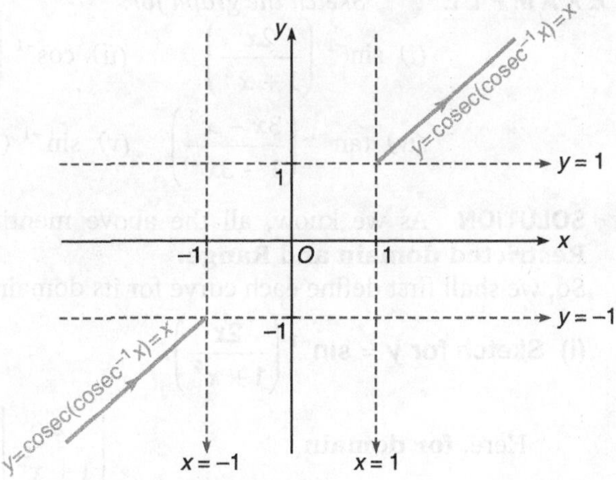

Fig 1.82

(v) Sketch for $y = \sec\,(\sec^{-1}x)$

We know, domain $\in R - (-1, 1)$

$(i.e., -\infty < x \le -1$ or $1 \le x < \infty)$

and range $y = x \Rightarrow y \in R - (-1, 1)$.

Hence, we should sketch

$$y = \sec\,(\sec^{-1}x) = x,$$

only when $x \in (-\infty, -1] \cup [1, \infty)$

So, its graph could be plotted as shown in Fig. 1.83.

Thus, the graph for

$$y = \sec\,(\sec^{-1}x) = x$$

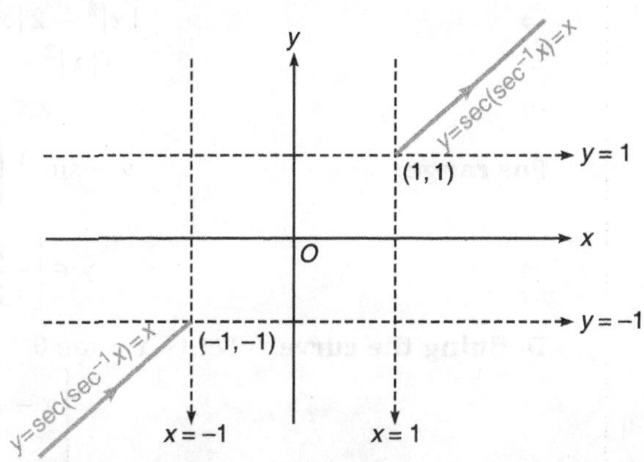

Fig. 1.83

(vi) Sketch for $y = \cot\,(\cot^{-1}x)$

We know; Domain $\in R$ $(i.e., -\infty < x < \infty)$

and Range $y = x \Rightarrow y = R$.

Hence, we should sketch

$y = \cot\,(\cot^{-1}x) = x, \forall\, x \in R$.

Shown as in Fig. 1.84.

Thus, the graph for $\boldsymbol{y = \cot\,(\cot^{-1}x)}$.

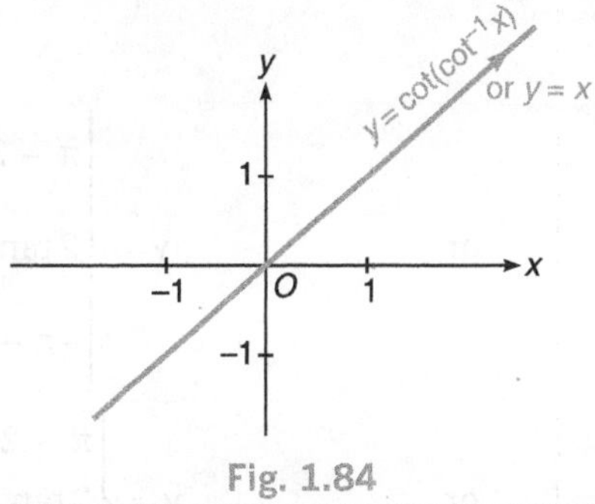

Fig. 1.84

Note From previous discussions, we learn that if:

(i) The function is periodic then find period and trace the curve.

(ii) If non-periodic, then define for their domain and find range to trace the curve.

Now, before going ahead you must revise previous curves of inverse trigonometry as;

$y = \sin^{-1}x$, $y = \cos^{-1}x$, $y = \tan^{-1}x$, $y = \cot^{-1}x$, $y = \text{cosec}^{-1}x$, $y = \sec^{-1}x$

with their domain and range.

EXAMPLE 8 *Sketch the graph for:*

(i) $\sin^{-1}\left(\frac{2x}{1+x^2}\right)$ (ii) $\cos^{-1}\left(\frac{1-x^2}{1+x^2}\right)$ (iii) $\tan^{-1}\left(\frac{2x}{1-x^2}\right)$

(iv) $\tan^{-1}\left(\frac{3x-x^3}{1-3x^2}\right)$ (v) $\sin^{-1}(3x-4x^3)$ (vi) $\cos^{-1}(4x^3-3x)$.

SOLUTION As we know, all the above mentioned six curves are **non-periodic,** but have **Restricted domain and Range.**

So, we shall first define each curve for its domain and range and then sketch these curves.

(i) Sketch for $y = \sin^{-1}\left(\frac{2x}{1+x^2}\right)$.

Here, **for domain** $\left|\frac{2x}{1+x^2}\right| \le 1$

$\Rightarrow \quad 2|x| \le 1 + x^2 \quad \{\because 1 + x^2 > 0 \text{ for all } x\}$

$\Rightarrow \quad |x|^2 - 2|x| + 1 \ge 0 \quad \{\because x^2 = |x|^2\}$

$\Rightarrow \quad (|x|^2 - 1)^2 \ge 0$

$\Rightarrow \quad x \in R.$

For range: $y = \sin^{-1}\left(\frac{2x}{1+x^2}\right)$

$\Rightarrow \quad y \in \left[-\frac{\pi}{2}, \frac{\pi}{2}\right] \quad \left\{\text{as}; \ y = \sin^{-1}\theta \Rightarrow y \in \left[-\frac{\pi}{2}, \frac{\pi}{2}\right]\right\}$

Defining the curve: Let, $x = \tan\theta$

$$\Rightarrow \quad y = \sin^{-1}(\sin 2\theta) = \begin{cases} \pi - 2\theta & ; \ 2\theta > \frac{\pi}{2} \\ 2\theta & ; \ -\frac{\pi}{2} \le 2\theta \le \frac{\pi}{2} \\ -\pi - 2\theta & ; \ 2\theta < -\frac{\pi}{2} \end{cases} \quad \{\text{See Ex. 1}\}$$

$$\text{or} \quad y = \begin{cases} \pi - 2\tan^{-1}x & ; \ \tan^{-1}x > \frac{\pi}{4} \\ 2\tan^{-1}x & ; \ -\frac{\pi}{4} \le \tan^{-1}x \le \frac{\pi}{4} \\ -\pi - 2\tan^{-1}x & ; \ \tan^{-1}x < -\frac{\pi}{4} \end{cases} \quad \{\because \tan\theta = x \Rightarrow \theta = \tan^{-1}x\}$$

$$\text{or} \quad y = \begin{cases} \pi - 2\tan^{-1}x & ; \ x > 1 \\ 2\tan^{-1}x & ; \ -1 \le x \le 1 \\ -\pi - 2\tan^{-1}x & ; \ x < -1 \end{cases} \quad \ldots\text{(i)}$$

Thus, $y = \sin^{-1}\left(\frac{2x}{1+x^2}\right)$ is defined for $x \in R$, where $y \in \left[-\frac{\pi}{2}, \frac{\pi}{2}\right]$, so the graph for Eq. (i) could be shown as in Fig. 1.85.

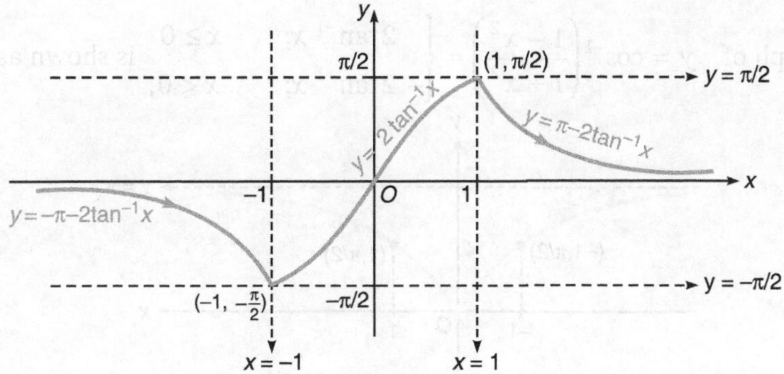

Fig. 1.85

Thus, the graph for $y = \sin^{-1}\left(\frac{2x}{1+x^2}\right)$.

> **Note** As in later section (*i.e.*, chapter 2) we shall discuss that functions having sharp edges and gaps are not differentiable at that point.
>
> So, in previous curve $y = \sin^{-1}\left(\frac{2x}{1+x^2}\right)$, we know it has sharp edge at $x = -1$ and $x = 1$.
>
> So, not differentiable.

(ii) Sketch for $y = \cos^{-1}\left(\frac{1-x^2}{1+x^2}\right)$

Here, **for domain** $\left|\frac{1-x^2}{1+x^2}\right| \le 1$

$\Rightarrow \quad |1-x^2| \le 1 + x^2 \qquad \{\because\ 1 + x^2 > 0,\ \forall\ x \in R\}$

which is true for all x; as $1 + x^2 > 1 - x^2$

$\therefore \quad \boldsymbol{x \in R}$

For range: $y = \cos^{-1}\left(\frac{1-x^2}{1+x^2}\right) \Rightarrow \boldsymbol{y \in (0, \pi)}$

Define the curve : Let, $x = \tan\theta$

$$\therefore \quad y = \cos^{-1}\left(\frac{1-\tan^2\theta}{1+\tan^2\theta}\right) = \cos^{-1}(\cos 2\theta)$$

$$= \begin{cases} 2\theta; & 2\theta \ge 0 \\ -2\theta; & 2\theta < 0 \end{cases} \qquad \{\text{See Example 2}\}$$

$$\Rightarrow \quad y = \begin{cases} 2\tan^{-1} x; & \tan^{-1} x \ge 0 \\ -2\tan^{-1} x; & \tan^{-1} x < 0 \end{cases} \qquad \{\because\ \tan\theta = x \Rightarrow \theta = \tan^{-1} x\}$$

So, the graph of $y = \cos^{-1}\left(\dfrac{1-x^2}{1+x^2}\right) = \begin{cases} 2\tan^{-1} x; & x \ge 0 \\ -2\tan^{-1} x; & x < 0, \end{cases}$ is shown as:

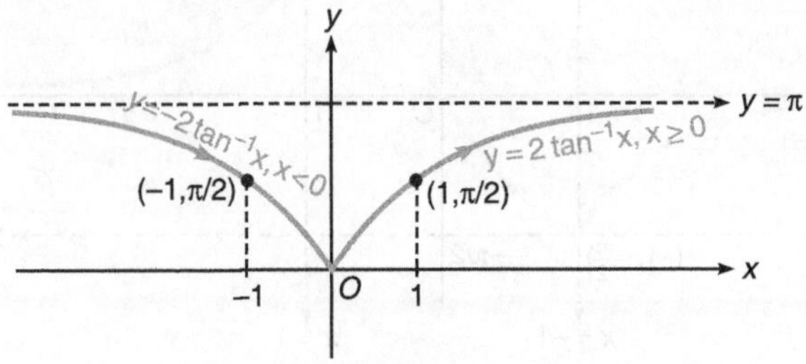

Fig. 1.86

Thus, the graph for $y = \cos^{-1}\left(\dfrac{1-x^2}{1+x^2}\right) = \begin{cases} 2\tan^{-1} x, & x \ge 0 \\ -2\tan^{-1} x, & x < 0 \end{cases}$.

From above figure it is clear $y = \cos^{-1}\left(\dfrac{1-x^2}{1+x^2}\right)$ is not differentiable at $x = 0$.

(iii) Sketch for $y = \tan^{-1}\left(\dfrac{2x}{1-x^2}\right)$

Here, **for domain** $\dfrac{2x}{1-x^2} \in R$ except; $1 - x^2 = 0$

i.e., $x \ne \pm 1$ or $\mathbf{x \in R - \{1, -1\}}$

For range $y = \tan^{-1}\left(\dfrac{2x}{1-x^2}\right)$

$\Rightarrow$ $y \in \left(-\dfrac{\pi}{2}, \dfrac{\pi}{2}\right)$ $\left\{\text{as } y = \tan^{-1}\theta \Rightarrow y \in \left(-\dfrac{\pi}{2}, \dfrac{\pi}{2}\right)\right\}$

Defining the curve

Let $x = \tan\theta$

$$\Rightarrow \quad y = \tan^{-1}\left(\frac{2\tan\theta}{1-\tan^2\theta}\right) = \tan^{-1}(\tan 2\theta) = \begin{cases} \pi + 2\theta; & 2\theta < -\dfrac{\pi}{2} \\ 2\theta; & -\dfrac{\pi}{2} < 2\theta < \dfrac{\pi}{2} \\ -\pi + 2\theta; & 2\theta > \dfrac{\pi}{2} \end{cases} \quad \{\text{See Example 3}\}$$

$$= \begin{cases} \pi + 2\tan^{-1} x; & \tan^{-1} x < -\dfrac{\pi}{4} \\ 2\tan^{-1} x; & -\dfrac{\pi}{4} < \tan^{-1} x < \dfrac{\pi}{4} \\ -\pi + 2\tan^{-1} x; & \tan^{-1} x > \dfrac{\pi}{4} \end{cases} \quad \{\text{as } \tan\theta = x \Rightarrow \theta = \tan^{-1} x\}$$

$$= \begin{cases} \pi + 2\tan^{-1} x; & x < -1 \\ 2\tan^{-1} x; & -1 < x < 1 \\ -\pi + 2\tan^{-1} x; & x > 1 \end{cases}$$

So, the graph of;

$$y = \tan^{-1}\left(\frac{2x}{1-x^2}\right) = \begin{cases} \pi + 2\tan^{-1} x; & x < -1 \\ 2\tan^{-1} x; & -1 < x < 1 \text{ is shown as;} \\ -\pi + 2\tan^{-1} x; & x > 1 \end{cases}$$

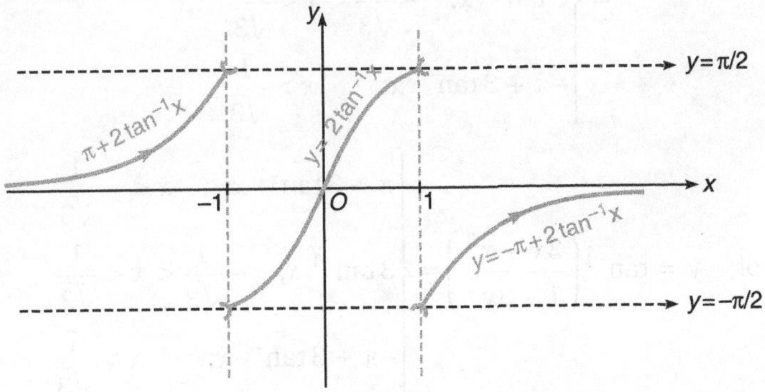

Fig. 1.87

Thus, the graph for $y = \tan^{-1}\left(\frac{2x}{1-x^2}\right) = \begin{cases} \pi + 2\tan^{-1} x, & x < -1 \\ 2\tan^{-1} x; & -1 < x < 1 \\ -\pi + 2\tan^{-1} x, & x > 1 \end{cases}$

which is neither continuous nor differentiable at $x = \{-1, 1\}$.

(v) Sketch for the curve $y = \tan^{-1}\left(\frac{3x - x^3}{1 - 3x^2}\right)$.

Here, **for domain** $\quad y = \tan^{-1}\left(\frac{3x - x^3}{1 - 3x^2}\right)$

$\Rightarrow \quad x \in R$ except $\quad 1 - 3x^2 = 0 \quad \Rightarrow \quad x \neq \pm \frac{1}{\sqrt{3}}$

$\therefore \quad \mathbf{x \in R - \left\{\pm \frac{1}{\sqrt{3}}\right\}}$

For range $\quad y = \tan^{-1}\left(\frac{3x - x^3}{1 - 3x^2}\right)$

$\Rightarrow \quad \mathbf{y \in \left(-\frac{\pi}{2}, \frac{\pi}{2}\right)} \quad \left\{\text{as } y = \tan^{-1}\theta \Rightarrow y \in \left(-\frac{\pi}{2}, \frac{\pi}{2}\right)\right\}$

Defining the curve: Let; $\quad x = \tan\theta$

$$\Rightarrow \quad y = \tan^{-1}(\tan 3\theta) = \begin{cases} \pi + 3\theta; & 3\theta < -\dfrac{\pi}{2} \\ 3\theta; & -\dfrac{\pi}{2} < 3\theta < \dfrac{\pi}{2} \\ -\pi + 3\theta; & 3\theta > \dfrac{\pi}{2} \end{cases} = \begin{cases} \pi + 3\tan^{-1} x; & \tan^{-1} x < -\dfrac{\pi}{6} \\ 3\tan^{-1} x; & -\dfrac{\pi}{6} < \tan^{-1} x < \dfrac{\pi}{6} \\ -\pi + 3\tan^{-1} x; & \tan^{-1} x > \dfrac{\pi}{6} \end{cases}$$

$$= \begin{cases} \pi + 3\tan^{-1} x; & x < -\dfrac{1}{\sqrt{3}} \\ 3\tan^{-1} x; & -\dfrac{1}{\sqrt{3}} < x < \dfrac{1}{\sqrt{3}} \\ -\pi + 3\tan^{-1} x; & x > \dfrac{1}{\sqrt{3}} \end{cases}$$

So, the graph of; $y = \tan^{-1}\left(\dfrac{3x - x^3}{1 - 3x^2}\right) = \begin{cases} \pi + 3\tan^{-1} x; & x < -\dfrac{1}{\sqrt{3}} \\ 3\tan^{-1} x; & -\dfrac{1}{\sqrt{3}} < x < \dfrac{1}{\sqrt{3}} \\ -\pi + 3\tan^{-1} x; & x > \dfrac{1}{\sqrt{3}} \end{cases}$

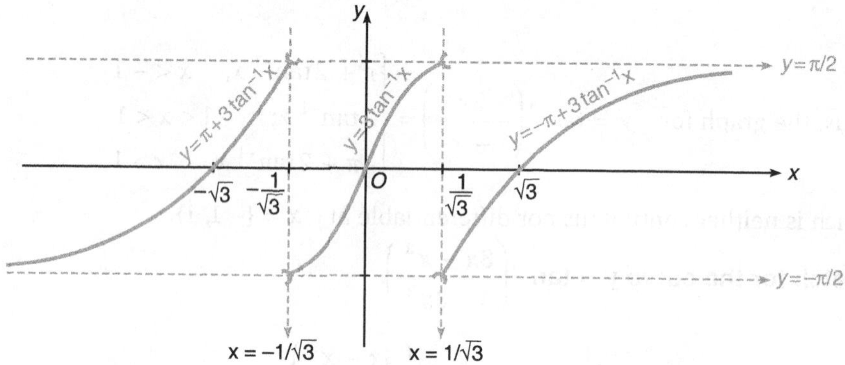

Fig. 1.88

Thus, the curve for $y = \tan^{-1}\left(\dfrac{3x - x^3}{1 - 3x^2}\right) = \begin{cases} \pi + 3\tan^{-1} x, & x < -\dfrac{1}{\sqrt{3}} \\ 3\tan^{-1} x, & -\dfrac{1}{\sqrt{3}} < x < \dfrac{1}{\sqrt{3}} \\ -\pi + 3\tan^{-1} x; & x > \dfrac{1}{\sqrt{3}} \end{cases}$

which is neither continuous nor differentiable for $\boldsymbol{x} = \left\{\pm \dfrac{\mathbf{1}}{\sqrt{\mathbf{3}}}\right\}$.

(v) Sketch the curve $y = \sin^{-1}(3x - 4x^3)$

Defining the curve: Let $x = \sin\theta$,

$$\Rightarrow \quad y = \sin^{-1}(\sin 3\theta) = \begin{cases} \pi - 3\theta; & \dfrac{\pi}{2} \le 3\theta \le \dfrac{3\pi}{2} \\ 3\theta; & -\dfrac{\pi}{2} \le 3\theta \le \dfrac{\pi}{2} \\ -\pi - 3\theta; & -\dfrac{3\pi}{2} \le 3\theta \le -\dfrac{\pi}{2} \end{cases} = \begin{cases} \pi - 3\sin^{-1} x; & \dfrac{\pi}{6} \le \sin^{-1} x \le \dfrac{\pi}{2} \\ 3\sin^{-1} x; & -\dfrac{\pi}{6} \le \sin^{-1} x \le \dfrac{\pi}{6} \\ -\pi - 3\sin^{-1} x; & -\dfrac{\pi}{2} \le \sin^{-1} x \le -\dfrac{\pi}{6} \end{cases}$$

$$\therefore \quad y = \sin^{-1}(3x - 4x^3) = \begin{cases} \pi - 3\sin^{-1} x; & \dfrac{1}{2} \le x \le 1 \\ 3\sin^{-1} x; & -\dfrac{1}{2} \le x \le \dfrac{1}{2} \\ -\pi - 3\sin^{-1} x; & -1 \le x \le -\dfrac{1}{2} \end{cases}$$

For domain $y = \sin^{-1}(3x - 4x^3) \quad \Rightarrow \quad$ $\boldsymbol{x \in [-1, 1]}$

For range $y = \sin^{-1}(3x - 4x^3) \quad \Rightarrow \quad y \in \left[-\dfrac{\pi}{2}, \dfrac{\pi}{2}\right]$

So, the graph of; $y = \sin^{-1}(3x - 4x^3) = \begin{cases} \pi - 3\sin^{-1} x; & \dfrac{1}{2} \le x \le 1 \\ 3\sin^{-1} x; & -\dfrac{1}{2} \le x \le \dfrac{1}{2} \\ -\pi - 3\sin^{-1} x; & -1 \le x \le -\dfrac{1}{2} \end{cases}$

is shown as:

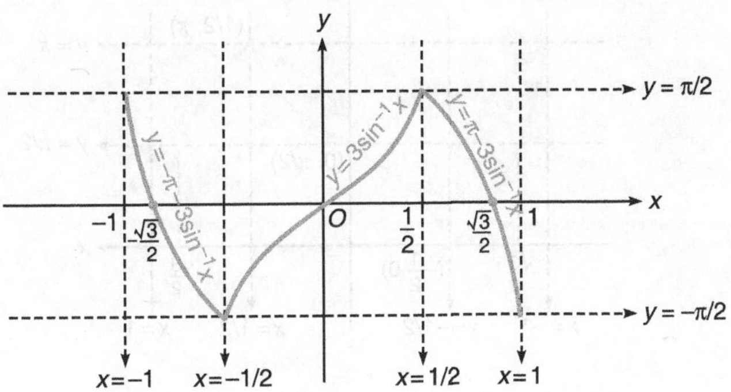

Fig. 1.89

Thus, the curve for $\quad y = \sin^{-1}(3x - 4x^3)$

which is not differentiable at $\quad x = \left\{\pm \dfrac{1}{2}\right\}$.

(vi) Sketch the curve $y = \cos^{-1}(4x^3 - 3x)$

Here, **domain** $\in [-1, 1]$ **range** $\in [0, \pi]$

Now, **defining the curve**

Let $\quad x = \cos\theta$

$$\Rightarrow \; y = \cos^{-1}(\cos 3\theta) = \begin{cases} 2\pi - 3\theta; & \pi \le 3\theta \le 2\pi \\ 3\theta; & 0 \le 3\theta \le \pi \\ -2\pi + 3\theta; & -\pi \le 3\theta \le 0 \end{cases} = \begin{cases} 2\pi - 3\cos^{-1} x; & \dfrac{\pi}{3} \le \cos^{-1} x \le \dfrac{2\pi}{3} \\ 3\cos^{-1} x; & 0 \le \cos^{-1} x \le \dfrac{\pi}{3} \\ -2\pi + 3\cos^{-1} x; & -\dfrac{\pi}{3} \le \cos^{-1} x \le 0 \end{cases}$$

$$= \begin{cases} 2\pi - 3\cos^{-1} x; & -\dfrac{1}{2} \le x \le \dfrac{1}{2} \\ 3\cos^{-1} x; & \dfrac{1}{2} \le x \le 1 \\ -2\pi + 3\cos^{-1} x; & -1 \le x \le -\dfrac{1}{2} \end{cases}$$

{∵ If $0 \le \theta \le \dfrac{\pi}{3} \Rightarrow \cos\dfrac{\pi}{3} \le \cos\theta \le \cos 0$ or $\dfrac{1}{2} \le \cos\theta \le 1$. Here, the interval changed since, $\cos x$ is decreasing in $[0, \pi]$}

So, the graph of; $y = \cos^{-1}(4x^3 - 3x) = \begin{cases} 2\pi - 3\cos^{-1} x; & -\dfrac{1}{2} \le x \le \dfrac{1}{2} \\ 3\cos^{-1} x; & \dfrac{1}{2} \le x \le 1 \\ -2\pi + 3\cos^{-1} x; & -1 \le x \le -\dfrac{1}{2}, \end{cases}$

is shown as;

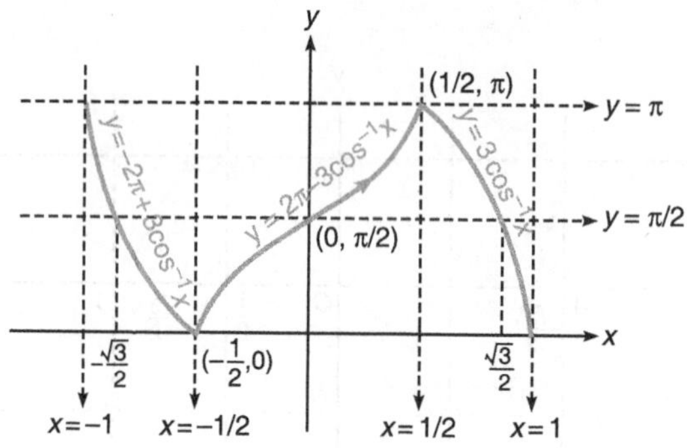

Fig. 1.90

Thus, the curve for $y = \cos^{-1}(4x^3 - 3x)$,

which is not differentiable at $x = \left\{\pm \dfrac{1}{2}\right\}$.

EXAMPLE 9 *Sketch the graph for:*

(i) $\sin x . \text{cosec } x$ *(ii)* $\cos x \cdot \sec x$ *(iii)* $\tan x \cdot \cot x$

SOLUTION As we know for the above curves each is equal to 1, but for different domain as;

(i) $y = \sin x \cdot \text{cosec } x = 1; \; \forall \; x \in R - \{n\pi; \; n \in z\}$

(ii) $y = \cos x \cdot \sec x = 1;\qquad x \in R - \left\{(2n+1)\dfrac{\pi}{2};\ n \in z\right\}$

(iii) $y = \tan x \cdot \cot x = 1;\qquad x \in R - \left\{n\pi,\ n\pi + \dfrac{\pi}{2};\ n \in z\right\}.$

Thus, they could be plotted as:

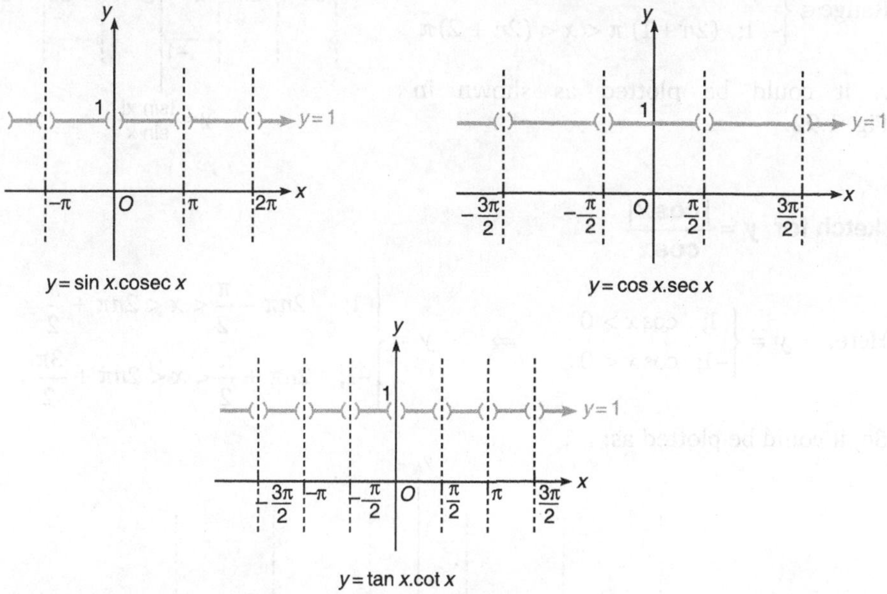

Fig. 1.91

> **Note** From above example it becomes clear that $y = \sin x \cdot \operatorname{cosec} x = 1$, $y = \cos x \cdot \sec x = 1$, $y = \tan x \cdot \cot x = 1$ but they are not equal, as their domains are different.
>
> $\therefore$ **Equal functions :** Those functions which have same domain and range are equal functions.

EXAMPLE 10 *Sketch the graph for:*

(i) $\dfrac{|\sin x|}{\sin x}$ *(ii)* $\dfrac{|\cos x|}{\cos x}$ *(iii)* $\sin^{-1}\left(\dfrac{1+x^2}{2x}\right)$

(iv) $\log_{1/4}\left(x - \dfrac{1}{4}\right) + \dfrac{1}{2}\log_4(16x^2 - 8x + 1)$

(v) $1 + 3(\log|\sin x| + \log|\operatorname{cosec} x|)$ *(vi)* $1 + 3(\log \sin x + \log \operatorname{cosec} x)$

SOLUTION As we know, to plot above curves we must check periodicity domain and range;

(i) $y = \dfrac{|\sin x|}{\sin x}$

Here, $$y = \begin{cases} 1; & \sin x > 0 \\ -1; & \sin x < 0 \end{cases}$$

$$y = \begin{cases} 1; & 2n\pi < x < (2n+1)\pi; \quad n \in z \\ -1; & (2n+1)\pi < x < (2n+2)\pi; \quad n \in z \end{cases}$$

So, from above,

domain $\in R - \{n\pi;\ n \in z\}$

$$\text{Range} \in \begin{cases} 1; & 2n\pi < x < (2n+1)\pi \\ -1; & (2n+1)\pi < x < (2n+2)\pi \end{cases}$$

$\therefore$ it could be plotted as shown in Fig. 1.92.

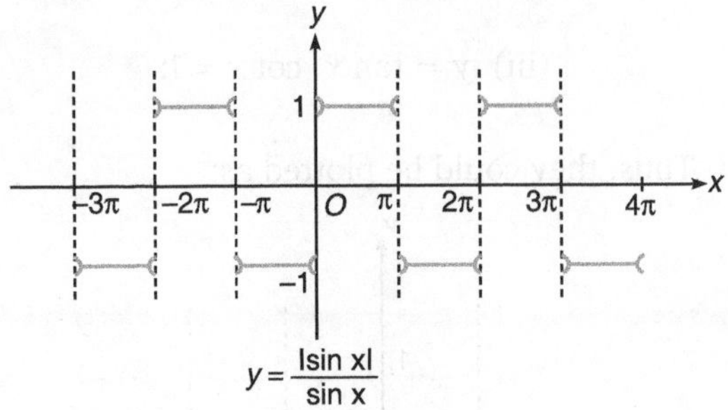

Fig. 1.92

(ii) Sketch for $y = \dfrac{|\cos x|}{\cos x}$

Here, $y = \begin{cases} 1; & \cos x > 0 \\ -1; & \cos x < 0 \end{cases} \quad \Rightarrow \quad y = \begin{cases} 1; & 2n\pi - \dfrac{\pi}{2} < x < 2n\pi + \dfrac{\pi}{2} \\ -1; & 2n\pi + \dfrac{\pi}{2} < x < 2n\pi + \dfrac{3\pi}{2} \end{cases}$

So, it could be plotted as:

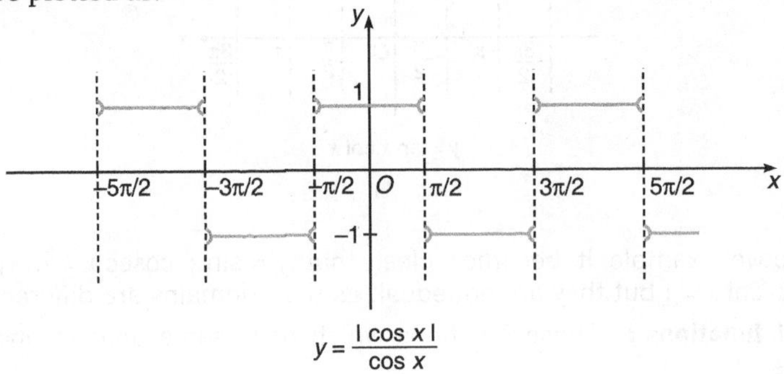

Fig. 1.93

(iii) Sketch for $y = \sin^{-1}\left(\dfrac{1+x^2}{2x}\right)$

Here $\quad y = \sin^{-1}\left(\dfrac{1+x^2}{2x}\right)$ is defined;

when; $\quad \left|\dfrac{1+x^2}{2x}\right| \le 1 \quad$ {as; $\sin^{-1} x$ is defined when $|x| \le 1$}

$\Rightarrow \quad 1 + x^2 \le 2|x| \quad$ {as; $1 + x^2 > 0$}

$\Rightarrow \quad x^2 - 2|x| + 1 \le 0 \quad$ {as; $x^2 = |x|^2$}

$\Rightarrow \quad |x|^2 - 2|x| + 1 \le 0$

$\Rightarrow \quad (|x| - 1)^2 \le 0$

$\Rightarrow \quad (|x| - 1)^2 = 0 \quad$ {as; $(|x| - 1)^2 < 0$ is not possible}

$\Rightarrow \quad x = \pm 1$

$\therefore$ **Domain** $\in \{\pm 1\}$

$\therefore$ For range $y = \sin^{-1}\left(\dfrac{1+x^2}{2x}\right)$, where $x = +1, -1$

$\therefore \quad y = \sin^{-1}(1)$ and $y = \sin^{-1}(-1)$

$\Rightarrow \quad y = \pm\dfrac{\pi}{2}$

$\therefore$ **Range** $\in \left\{\pm \dfrac{\pi}{2}\right\}$

Hence, the graph for $y = \sin^{-1}\left(\dfrac{1+x^2}{2x}\right)$ is **only two points.** Shown as:

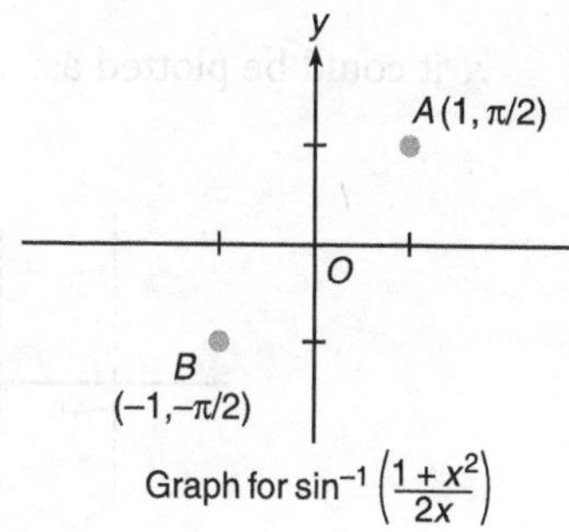

Fig. 1.94

Thus, the sketch for $y = \sin^{-1}\left(\dfrac{1+x^2}{2x}\right)$ is only two points ***A*** and ***B***.

(iv) Sketch for $y = \log_{1/4}\left(x - \frac{1}{4}\right) + \frac{1}{2}\log_4(16x^2 - 8x + 1)$

Here, $y = \log_{1/4}\left(x - \dfrac{1}{4}\right) + \dfrac{1}{2}\log_4(4x-1)^2$

$\Rightarrow \quad y = \log_{1/4}\left(x - \dfrac{1}{4}\right) + \dfrac{1}{2}\log_4 16 + \dfrac{1}{2}\log_4\left(x - \dfrac{1}{4}\right)^2$

or $\quad y = -\log_4\left(x - \dfrac{1}{4}\right) + \dfrac{1}{2}\log_4 4^2 + \dfrac{2}{2}\log_4\left(x - \dfrac{1}{4}\right) \quad \left\{\text{as; } \log_{b^n} a^m = \dfrac{m}{n}\log_b a\right\}$

$\Rightarrow \quad y = -\log_4\left(x - \dfrac{1}{4}\right) + \log_4\left(x - \dfrac{1}{4}\right) + \dfrac{2}{2}\log_4 4$

$\Rightarrow y = 1$, whenever; $\left(x - \dfrac{1}{4}\right) > 0$ {as; $\log_a x$ exists only when $a, x > 0$ and $a \neq 1$}

Thus, $y = \log_{1/4}\left(x - \dfrac{1}{4}\right) + \dfrac{1}{2}\log_4(4x-1)^2$

$\Rightarrow$ **Domain** $\in \left(\dfrac{1}{4}, \infty\right)$

Range $\in \{1\}$

Thus, the graph is shown as:

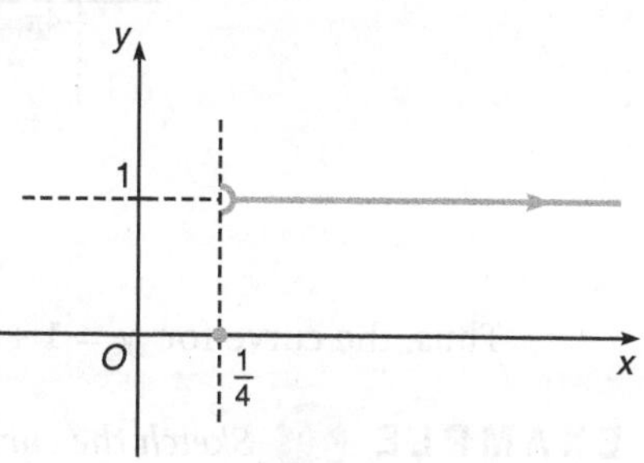

Fig. 1.95

Thus, the graph for

$$y = \log_{1/4}\left(x - \frac{1}{4}\right) + \frac{1}{2}\log_4(4x-1)^2.$$

(v) Sketch for $y = 1 + 3(\log|\sin x| + \log|\operatorname{cosec} x|)$

Here $\quad y = 1 + 3[\log(|\sin x| \cdot |\operatorname{cosec} x|)]$

whenever $|\sin x| \neq 0$ and $|\text{cosec } x| \neq 0$

i.e., $y = 1 + 3(\log 1);$ whenever $x \notin n\pi;\ n \in z.$

$\Rightarrow$ $y = 1$ $\{\text{as; } \log 1 = 0\}$

$\therefore$ **Domain** $\in R - \{n\pi;\ n \in z\}$

Range $\in \{1\}$

$\therefore$ it could be plotted as:

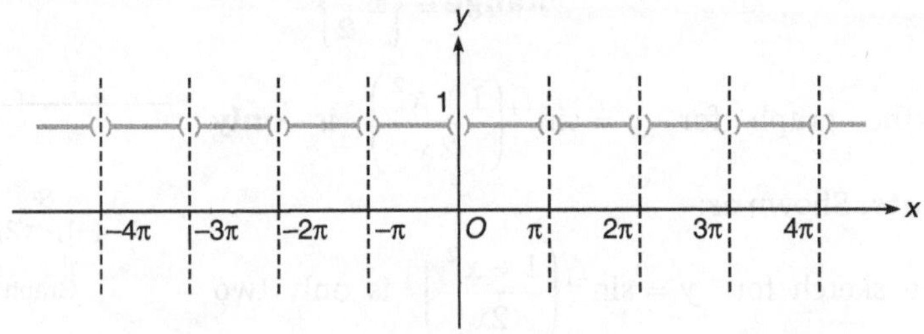

Fig. 1.96

Thus, the curve for $y = 1 + 3(\log|\sin x| + \log|\text{cosec } x|)$.

(vi) Sketch for $y = 1 + 3(\log \sin x + \log \text{cosec } x)$

Here $y = 1 + 3(\log \sin x \cdot \text{cosec } x)$ whenever $\sin x > 0$ and $\text{cosec } x > 0$

$\Rightarrow$ $y = 1 + 3 \log 1;$ $x \in (2n\pi, (2n+1)\pi)$

or $y = 1;$ whenever $x \in (2n\pi, (2n+1)\pi)$

$\therefore$ $y = 1 + 3(\log \sin x + \log \text{cosec } x) = \{1;\ 2n\pi < x < (2n+1)\pi$ is shown as;

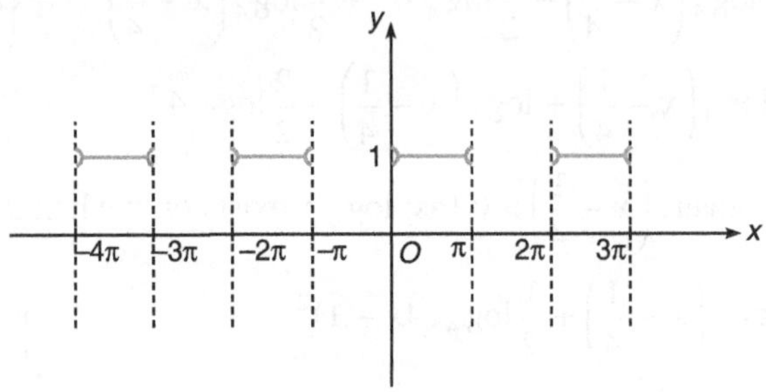

Fig. 1.97

Thus, the curve for $y = 1 + 3(\log \sin x + \log \text{cosec } x)$

EXAMPLE 11 *Sketch the curve for* $\cos y = \cos x$.

SOLUTION Here, $\cos y = \cos x \Rightarrow y = 2n\pi \pm x;\ n \in z$

$\therefore$ $\cos y = \cos x$, represents two straight lines;

$$y = \begin{cases} x + 2n\pi; & n \in z \\ -x + 2n\pi; & n \in z \end{cases}$$

i.e., two infinite set of perpendicular straight lines which could be shown as:

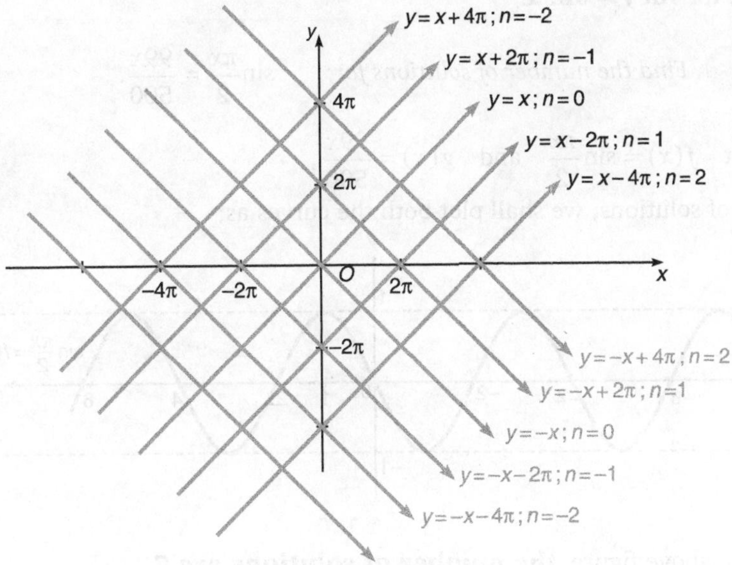

Fig. 1.98

Thus, graph for $\cos y = \cos x$; represents two infinite set of perpendicular straight lines which have infinite number of points of intersections; (So, if asked number of solutions then they are infinite).

EXAMPLE 12 *Sketch the curve for* $\sin y = \sin x$.

SOLUTION Here $\sin y = \sin x \quad \Rightarrow \quad y = n\pi + (-1)^n x;\ n \in z$

$\therefore$ $\sin y = \sin x$; represent two straight lines;

$$y = \begin{cases} n\pi + x; & n \text{ even integer} \\ n\pi - x; & n \text{ odd integer} \end{cases}$$

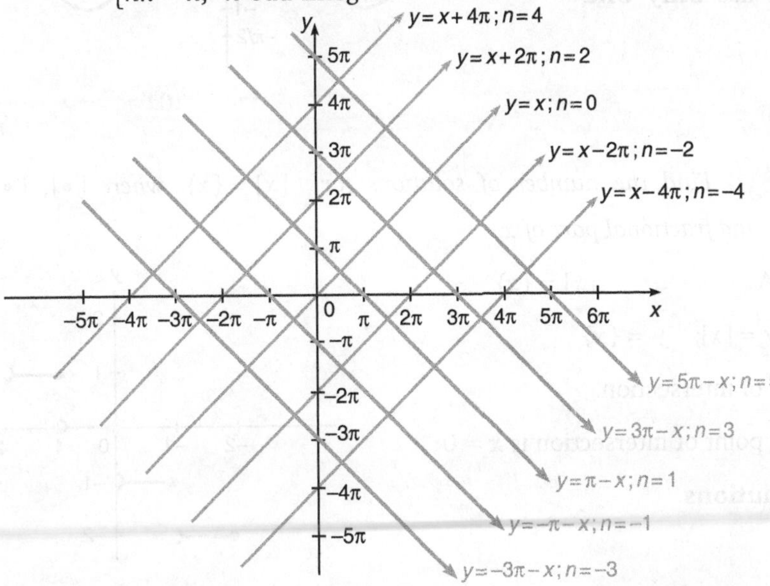

Fig. 1.99

i.e., two infinite set of perpendicular straight lines as shown in Fig. 1.99:
Thus, the graph for **sin y = sin x**.

EXAMPLE 13 *Find the number of solutions for;* $\sin\frac{\pi x}{2} = \frac{99x}{500}$.

SOLUTION Let $f(x) = \sin\frac{\pi x}{2}$ and $g(x) = \frac{99x}{500}$,

to find number of solutions; we shall plot both the curves as;

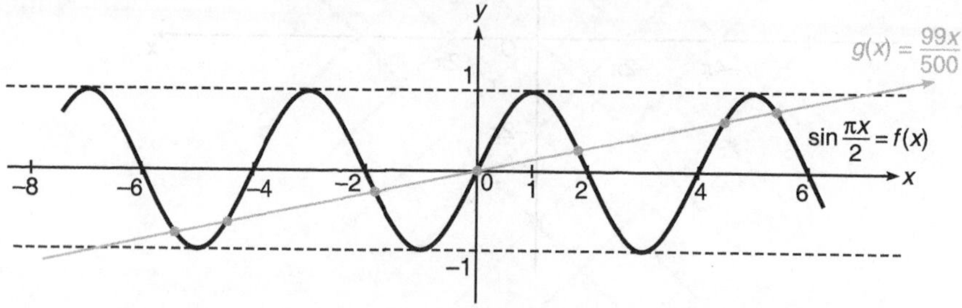

Fig. 1.100

Clearly, from the above figure, **the number of solutions are 7.**

EXAMPLE 14 *Find the number of solutions for;* $\cos x = x$

SOLUTION As, $\cos x = x$

$\therefore$ to plot the curve for $y = \cos x$; $y = x$ and find the point of intersection as to obtain number of solutions.

Here, the two curves intersect at a point A.

So, $\cos x = x$ has **only one solution.**

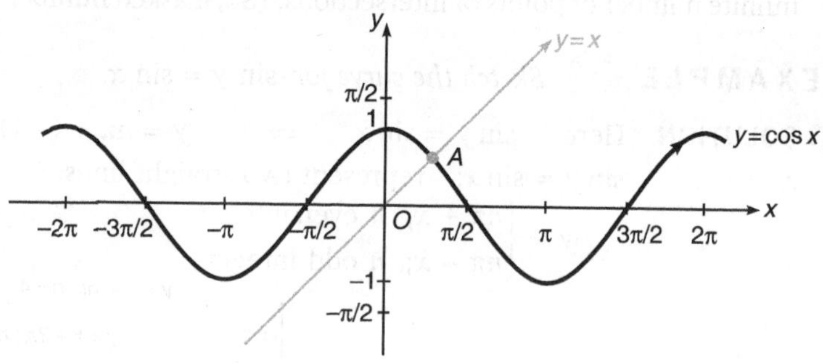

Fig. 1.101

EXAMPLE 15 *Find the number of solutions for;* $[x] = \{x\}$. *where* $[\bullet]$, $\{\bullet\}$ *represents greatest integer and fractional part of x.*

SOLUTION As, $[x] = \{x\}$

$\therefore$ to plot $y = [x]$; $y = \{x\}$

and find point of intersection.

Here, the only point of intersection is $x = 0$,

$\therefore$ only one **solutions**.

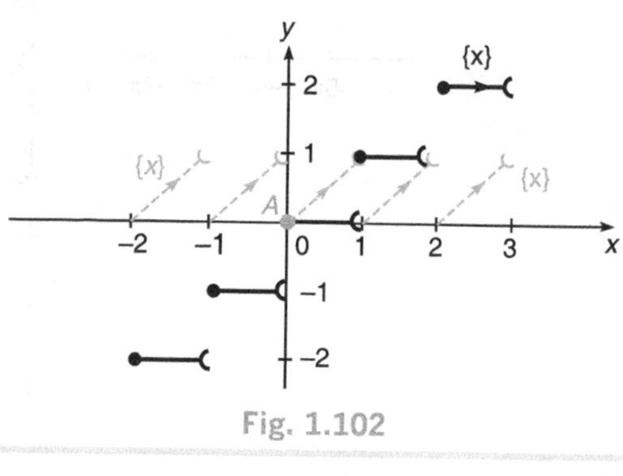

Fig. 1.102

EXAMPLE 16 *Find the number of solutions of*

$$4\{x\} = x + [x]$$

where $\{\cdot\}$, $[\cdot]$ represents fractional part and greatest integer function.

SOLUTION As we know, to find number of solutions of two curves we should find the point of intersection of two curves.

$\therefore \quad 4\{x\} = x + [x]$

$\Rightarrow \quad 4(x - [x]) = x + [x]$

$\{\because x = [x] + \{x\}\}$

$\Rightarrow \quad 4x - x = 4[x] + [x]$

$\Rightarrow \quad 3x = 5[x]$

$\Rightarrow \quad [x] = \dfrac{3}{5}x \quad \ldots(\text{i})$

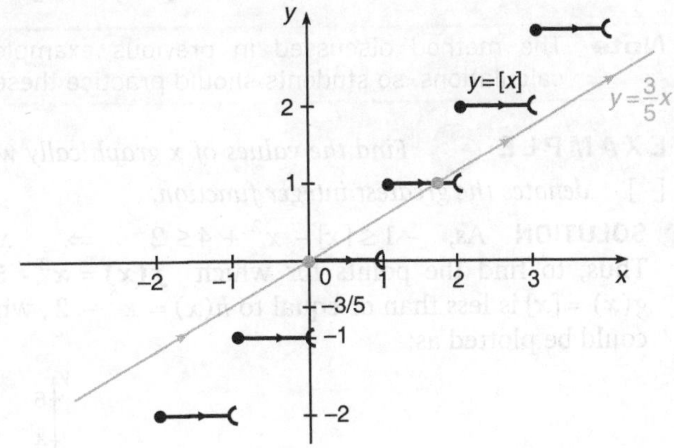

Fig. 1.103

$\therefore$ To plot the graph of both

$$y = [x] \quad \text{and} \quad y = \frac{3}{5}x.$$

Clearly, the two graphs intersects when

$$[x] = 0 \quad \textbf{and} \quad [x] = 1 \qquad \ldots(\text{ii})$$

$\therefore \quad x = \dfrac{5}{3}[x]$ [from Eqs. (i) and (ii)]

$$x = \frac{5}{3}\cdot 0 \quad \text{and} \quad x = \frac{5}{3}(1)$$

$\therefore x = 0$ and $x = \dfrac{5}{3}$ are the only two **solutions**.

EXAMPLE 17 *Find the value of x graphically satisfying; $[x] - 1 + x^2 \geq 0$; where $[\cdot]$ denotes the greatest integer function.*

SOLUTION As, $[x] - 1 + x^2 \geq 0 \quad \Rightarrow \quad x^2 - 1 \geq -[x]$

Thus, to find the points for which $f(x) = x^2 - 1$ is greater than or equals to

$$g(x) = -[x].$$

where two functions $f(x)$ and $g(x)$ could be plotted as shown in Fig. 1.104;

From the adjoining figure; the solution set lies when

$$x \leq A \quad \text{or} \quad x \geq B.$$

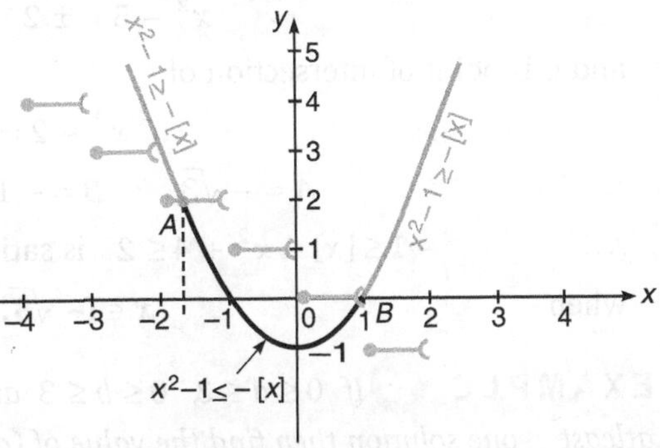

Fig. 1.104

Thus, to find A and B.

It is clear that $f(x)$ and $g(x)$ intersects when;

$-[x] = 2$.

$\therefore \quad x^2 - 1 = 2$

$x = \pm\sqrt{3} \quad \Rightarrow \quad x = -\sqrt{3}$

(neglecting $x = +\sqrt{3}$ as A lies for $x < 0$)

Thus, **for A :** $x = -\sqrt{3}$ and **for B :** $x = 1$.

$\therefore$ Solution set for which $x^2 - 1 \geq -[x]$ holds.

$\Rightarrow$ $\boldsymbol{x \in (-\infty, -\sqrt{3}] \cup [1, \infty)}$.

> **Note** The method discussed in previous example is very important as it reduces your calculations, so students should practice these forms.

EXAMPLE 18 *Find the values of x graphically which satisfy;* $-1 \leq [x] - x^2 + 4 \leq 2$; *where* $[\cdot]$ *denotes the greatest integer function.*

SOLUTION As, $-1 \leq [x] - x^2 + 4 \leq 2 \quad \Rightarrow \quad x^2 - 5 \leq [x] \leq x^2 - 2$

Thus, to find the points for which $\boldsymbol{f(x) = x^2 - 5}$ is less than or equal to $\boldsymbol{g(x) = [x]}$ and $g(x) = [x]$ is less than or equal to $h(x) = x^2 - 2$, where the three functions $f(x)$, $g(x)$ and $h(x)$ could be plotted as;

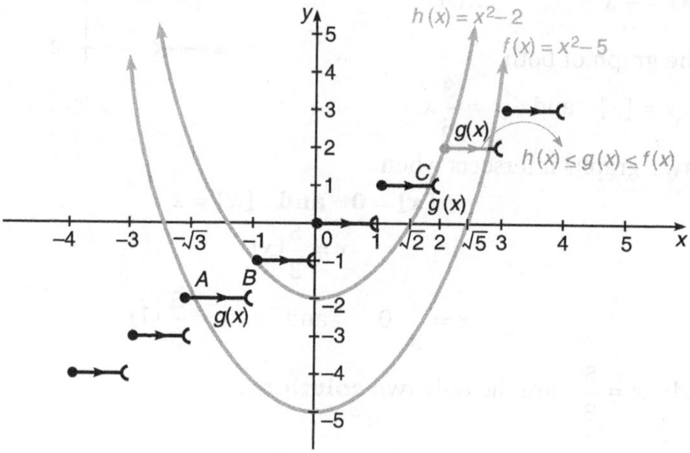

Fig. 1.105

Thus, from the above graph;

$$x^2 - 5 \leq [x] \leq x^2 - 2 \qquad \textbf{when} \quad \boldsymbol{x \in [A, B] \cup [C, D]}$$

where A and D is the point of intersection if;

$$x^2 - 5 = \pm 2 \quad \Rightarrow \quad x = -\sqrt{3}, \sqrt{7}$$

and C is point of intersection of

$$x^2 - 2 = 1 \quad \Rightarrow \quad x = \sqrt{3}.$$

$\therefore$ $\quad A = -\sqrt{3}, \quad B = -1, \quad C = \sqrt{3}$ and $D = \sqrt{7}$.

$\therefore$ $\quad -1 \leq [x] - x^2 + 4 \leq 2$ is satisfied;

when $\quad \boldsymbol{x \in [-\sqrt{3}, -1] \cup [\sqrt{3}, \sqrt{7}]}$.

EXAMPLE 19 *If* $0 \leq a \leq 3$, $0 \leq b \leq 3$ *and the equation;* $x^2 + 4 + 3\cos(ax + b) = 2x$ *has atleast one solution then find the value of* $(a + b)$.

SOLUTION Here, $x^2 + 4 + 3\cos(ax + b) = 2x$ or $x^2 - 2x + 4 = -3\cos(ax + b)$

$\Rightarrow$ $\quad (x - 1)^2 + 3 = -3\cos(ax + b)$

for above equation to have atleast one solution; plot $f(x) = (x-1)^2 + 3$ and $g(x) = -3\cos(ax+b)$ in such a way that they touch each other.

From figure the two curves could atmost touch at one point only when $-3\cos(a+b) = 3$

$\Rightarrow$ $\cos(a+b) = -1$

$\Rightarrow$ $a + b = \pi, 3\pi, 5\pi, \ldots$

But $3\pi > 6$

$\Rightarrow$ $\boldsymbol{a+b} = \pi$ as $0 \le a, b \le 3$.

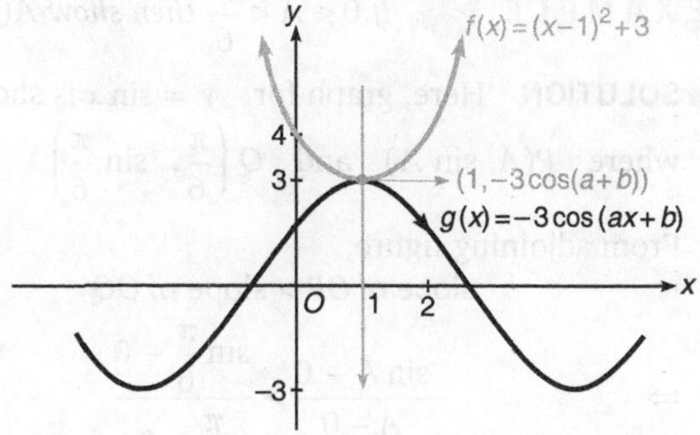

Fig. 1.106

EXAMPLE 20 *If $A + B + C = \pi$ and A, B, C are angles of Δ; then show*

$$\sin A + \sin B + \sin C < \frac{3\sqrt{3}}{2}.$$

SOLUTION Here; we have three trigonometric ratios $\sin A$, $\sin B$, $\sin C$.

$\therefore$ Let $y = \sin x$, on which there are three points $x = A$, $x = B$ and $x = C$ shown as;

As from the figure; **In ΔPQR,**

Centroid of Δ formed by $P(A, \sin A)$ $Q(B, \sin B)$ $R(C, \sin C)$ is;

$$G = \left(\frac{A+B+C}{3}, \frac{\sin A + \sin B + \sin C}{3}\right)$$

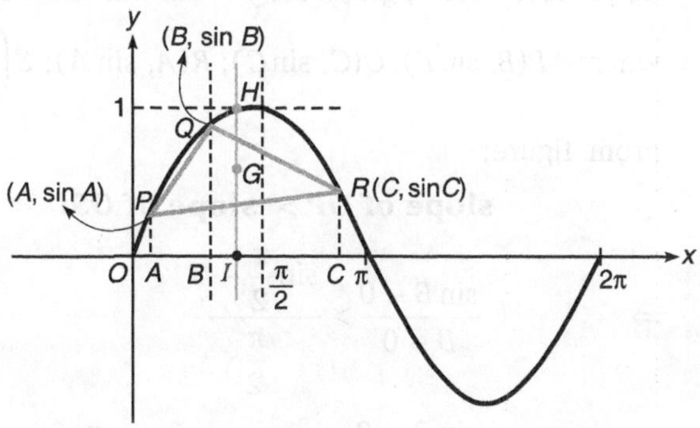

Fig. 1.107

where; G, H and I are collinear

$$I\left(\frac{A+B+C}{3}, 0\right); \; G\left(\frac{A+B+C}{3}, \frac{\sin A + \sin B + \sin C}{3}\right)$$

and $H\left(\frac{A+B+C}{3}, \sin\left(\frac{A+B+C}{3}\right)\right)$

From figure; $HI > GI$

i.e., ordinate of $H >$ ordinate of G

$$\Rightarrow \quad \sin\left(\frac{A+B+C}{3}\right) > \frac{\sin A + \sin B + \sin C}{3}$$

$$\Rightarrow \quad \frac{3\sqrt{3}}{2} > \sin A + \sin B + \sin C.$$

EXAMPLE 21 *If* $0 < A < \frac{\pi}{6}$, *then show* $A(\operatorname{cosec} A) < \frac{\pi}{3}$.

SOLUTION Here, graph for $y = \sin x$ is shown as;

where $P(A, \sin A)$ and $Q\left(\frac{\pi}{6}, \sin\frac{\pi}{6}\right)$

From adjoining figure;

$$\text{slope of } OP > \text{slope of } OQ$$

$$\Rightarrow \quad \frac{\sin A - 0}{A - 0} > \frac{\sin\frac{\pi}{6} - 0}{\frac{\pi}{6} - 0}$$

$$\Rightarrow \quad \frac{\sin A}{A} > \frac{3}{\pi} \quad \text{or} \quad \frac{A}{\sin A} < \frac{\pi}{3}$$

or $$\mathbf{A(\operatorname{cosec} A) < \frac{\pi}{3}.}$$

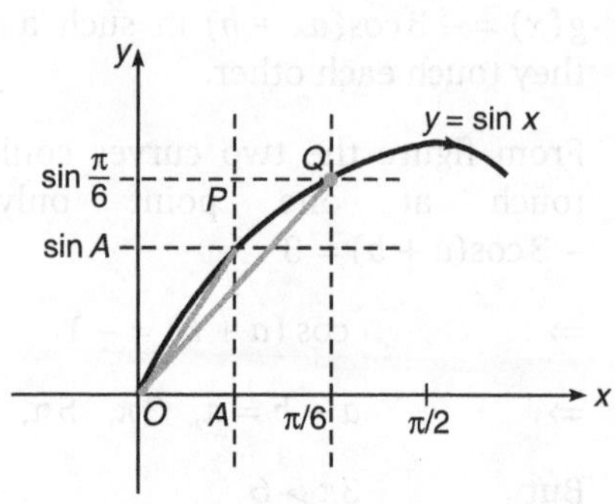

Fig. 1.108

EXAMPLE 22 *If* $0 < A, B, C < \frac{\pi}{2}$, *then show that:* $A \operatorname{cosec} A + B \operatorname{cosec} B + C \operatorname{cosec} C < \frac{3\pi}{2}$.

SOLUTION Here, graph for $y = \sin x$ is shown as;

where $P(B, \sin B); Q(C, \sin C); R(A, \sin A); S\left(\frac{\pi}{2}, \sin\frac{\pi}{2}\right)$.

From figure;

$$\textbf{slope of } \boldsymbol{OP} > \textbf{ slope of } \boldsymbol{OS}\textbf{:}$$

$$\Rightarrow \quad \frac{\sin B - 0}{B - 0} > \frac{\sin\frac{\pi}{2} - 0}{\frac{\pi}{2}}$$

$$\Rightarrow \quad \frac{\sin B}{B} > \frac{2}{\pi} \quad \text{or} \quad \frac{B}{\sin B} < \frac{\pi}{2}$$

or $$B \operatorname{cosec} B < \frac{\pi}{2} \qquad \ldots\text{(i)}$$

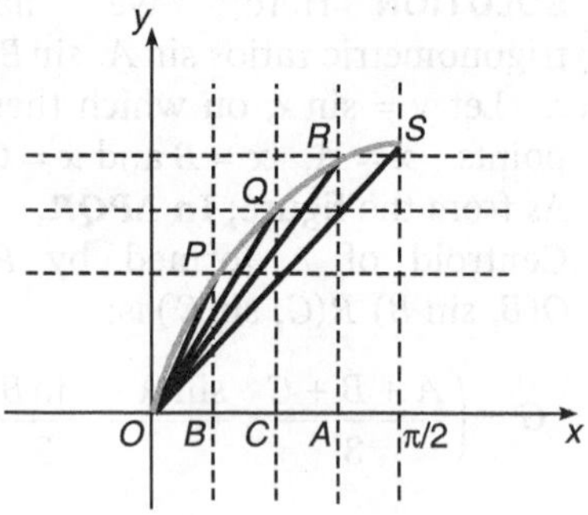

Fig. 1.109

Similarly, slope of $OQ <$ slope of OS and slope of $OR <$ slope of OS.

$$\Rightarrow \quad C \operatorname{cosec} C < \pi/2 \qquad \ldots\text{(ii)}$$

$$A \operatorname{cosec} A < \pi/2 \qquad \ldots\text{(iii)}$$

Adding Eqs. (i), (ii) and (iii), we get

$$A \operatorname{cosec} A + B \operatorname{cosec} B + C \operatorname{cosec} C < \frac{3\pi}{2}.$$

Note Students must practice above method in different questions of trigonometric inequality as it saves time.

EXERCISE

1. Construct the graph for;

   (i) $f(x)=\begin{cases} x-1; & x<0 \\ \dfrac{1}{4}; & x=0 \\ x^2; & x>0 \end{cases}$

   (ii) $f(x)=\begin{cases} 2x+3; & -3\le x<-2 \\ x+1; & -2\le x<0 \\ x+2; & 0\le x\le 1 \end{cases}$

2. Construct the graph of the function:

   (i) $f(x)=|x-1|+|x+1|$

   (ii) $f(x)=\begin{cases} 3^x; & -1\le x\le 1 \\ 4-x; & 1\le x<4 \end{cases}$

   (iii) $f(x)=[x]+|x-1|; \quad -1\le x\le 3$
   (where $[\cdot]$ denotes greatest integer function)

   (iv) $f(x)=\begin{cases} x^4; & x^2<1 \\ x; & x^2\ge 1 \end{cases}$

3. Is $f(x)=x^2+x+1$ invertible? If not in which region it is invertible.

4. If f is defined by $y=f(x)$; where $x=2t-|t|$, $y=t^2+t|t|$, $t\in R$. Then construct the graph for $f(x)$.

5. Construct the graph for $f(x)=[[x]-x]$; where $[\cdot]$ denotes greatest integer function.

6. If $0<\alpha<1$; then show; $\dfrac{\tan^{-1}\alpha}{\alpha}>\dfrac{\pi}{4}$.

7. Find the number of solutions for; $\cos^{-1}(\cos x)=[x]$ where $[\cdot]$ denotes the greatest integer function.

8. Find the number of solutions for; $[[x]-x]=\sin x$; where $[\cdot]$ denotes the greatest integer function.

9. Find the values of x graphically which satisfy $\left|\dfrac{x^2}{x-1}\right|\le 1$.

10. Find the value of x for which $x^3-[x]=3$, where $[\cdot]$ denotes the greatest integer function.

ANSWERS

3. $x\ge -\dfrac{1}{2}$ **7.** 5 solutions. **8.** infinite **9.** $x\in\left[\dfrac{-1-\sqrt{5}}{2}, \dfrac{-1+\sqrt{5}}{2}\right]$ **10.** $x=2^{2/3}$.

Myths About Tangent Lines

Remark 1

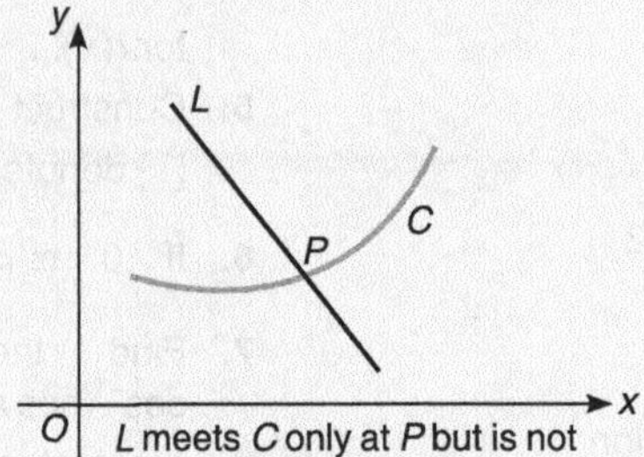

L meets *C* only at *P* but is not tangent to *C*.

Fig. 1

Remark 2

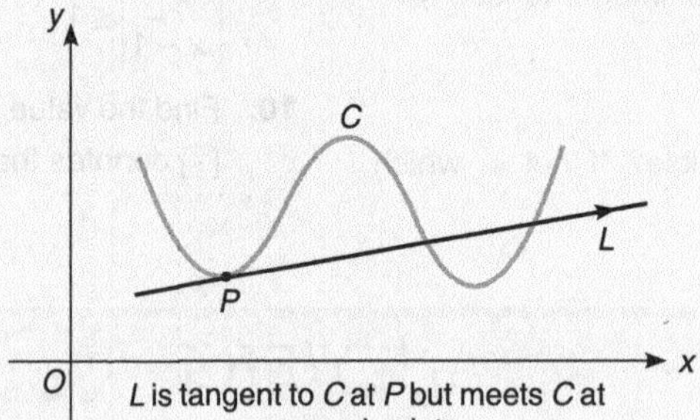

L is tangent to *C* at *P* but meets *C* at several points

Fig. 2

Remark 3

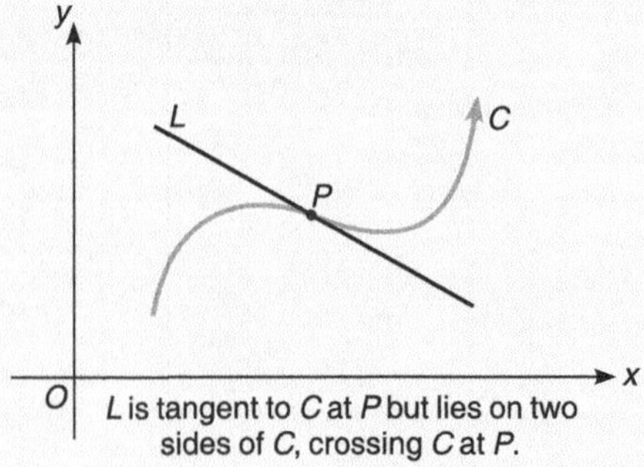

L is tangent to *C* at *P* but lies on two sides of *C*, crossing *C* at *P*.

Fig. 3

Chapter 2 CURVATURE AND TRANSFORMATIONS

In this chapter we shall study:

- The bending of curves at different points.
- Transformations of curves.

2.1 CURVATURE

"The study of bending of curves at different points is known as curvature." or **"Rate at which the curve curves"**.

Consider a curve and a point P on it and let Q be a point near P. Let A be a point on the curve.

$$\text{arc } AP = s, \qquad \text{arc } AQ = s + \delta s$$

$$\therefore \qquad \text{arc } PQ = \delta s$$

Let ψ and $\psi + \delta\psi$ be angles which the tangents at P and Q makes with x-axis.

$\therefore$ $\delta\psi$ is called **the total curvature of the arc PQ.**

$\dfrac{\delta\psi}{\delta s}$ is called **the average curvature of the arc PQ.**

$\lim\limits_{\delta s \to 0} \dfrac{\delta\psi}{\delta s} = \dfrac{d\psi}{ds}$ = **curvature of the curve at P.**

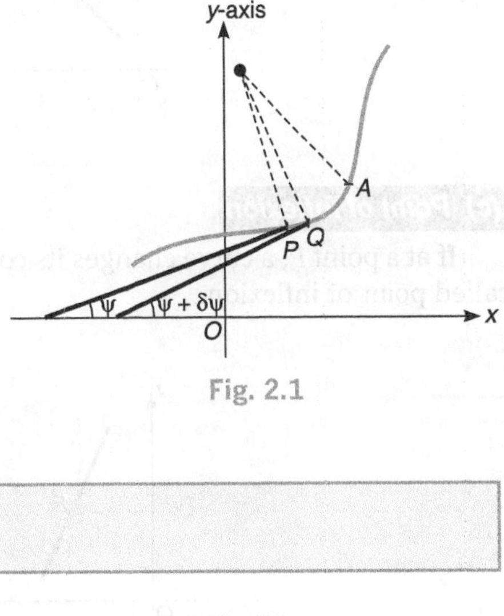

Fig. 2.1

Note To study curvature we shall define $\dfrac{d\psi}{ds}$ or $\dfrac{d^2y}{dx^2}$.

2.2 CONCAVITY, CONVEXITY AND POINTS OF INFLEXION

(a) Concave upwards

If in the neighbourhood of a point P on a curve is above the tangent at P, it is said to be concave upwards.

Mathematically

$\dfrac{dy}{dx}$ increases as x increases. $\Rightarrow$ $\dfrac{d^2y}{dx^2} > 0$

Geometrically

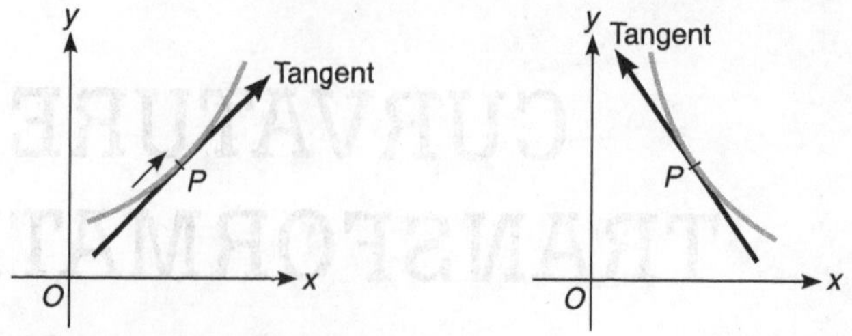

Fig. 2.2

(b) Convex upwards or Concave downwards

If the curve is below the tangent at P, it is said to be convex upward or concave downward.

Mathematically

$\frac{dy}{dx}$ decreases as x decreases. $\Rightarrow$ $\frac{d^2y}{dx^2} < 0$

Geometrically

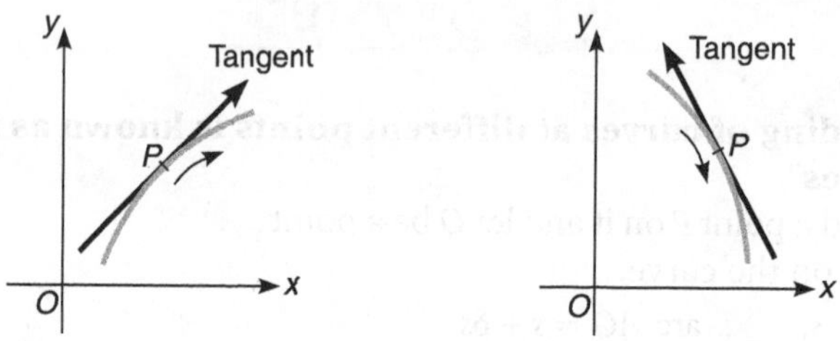

Fig. 2.3

(c) Point of Inflexion

If at a point P, a curve changes its concavity from upwards to downwards or vice versa. Then P is called point of inflexion.

Geometrically

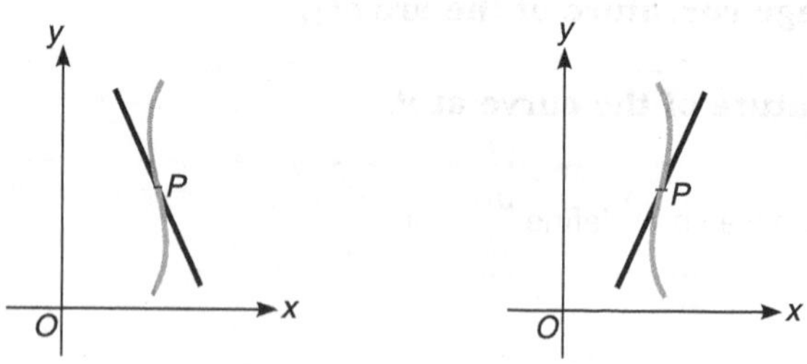

Fig. 2.4

Mathematically

(i) $\frac{d^2y}{dx^2} = 0$ at the point.

(ii) $\frac{d^2y}{dx^2}$ changes its sign as x increases through the value at which $\frac{d^2y}{dx^2} = 0$; *i.e.*, $\frac{d^3y}{dx^3} \neq 0$.

Note In general we can represent concavity as;.

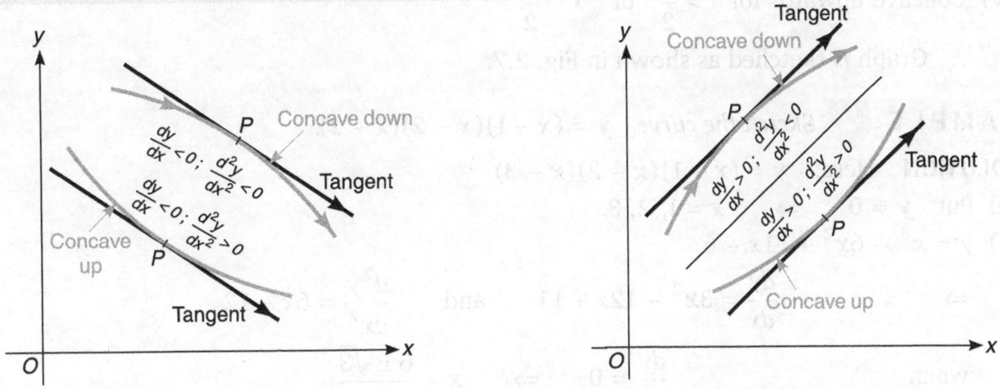

Here; $\frac{dy}{dx} < 0$ for both curves. Here; $\frac{dy}{dx} > 0$ for both curves.

Fig. 2.5

2.3 PLOTTING OF ALGEBRAIC CURVES USING CONCAVITY

Here; if $y = f(x) = (x - \alpha)(x - \beta)$. $0 < \alpha < \beta$. Then we know it has roots α and β and would be plotted as shown in Fig. 2.6.

From above discussion it becomes clear that to plot curves we require;

(i) Point of intersection on x-axis. (*i.e.*, $y = 0$)

(ii) Point of maximum and minimum value.

(iii) Interval for which function increases or decreases.

(iv) Point at which concave up, concave down and point of inflexion.

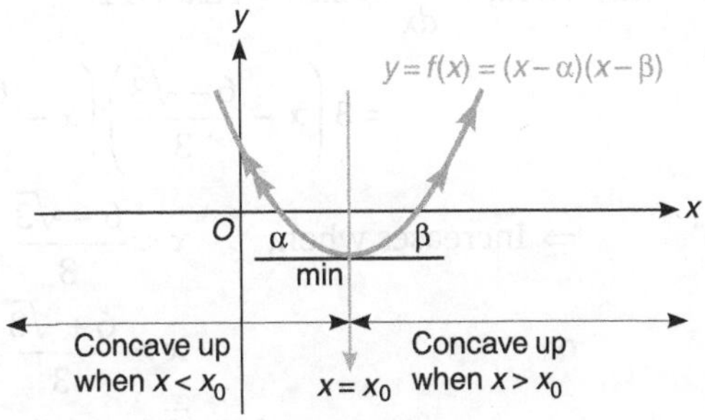

Fig. 2.6

EXAMPLE 1 *Sketch* $y = (x - 1)(x - 2)$.

SOLUTION Here; $y = (x - 1)(x - 2)$

(i) Put $y = 0 \Rightarrow x = 1, 2$. (for point of intersection on x-axis.)

(ii) $y = x^2 - 3x + 2$

$$\Rightarrow \frac{dy}{dx} = 2x - 3 \quad \text{and} \quad \frac{d^2y}{dx^2} = 2$$

$$\therefore \text{ minimum at } x = \frac{3}{2} \quad \left(\text{as } \frac{d^2y}{dx^2} > 0\right)$$

(iii) Increases when $x > \frac{3}{2}$ and decreases when $x < \frac{3}{2}$.

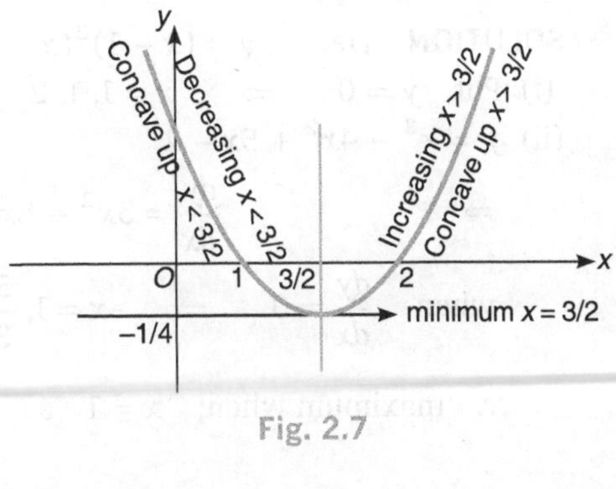

Fig. 2.7

(iv) Concave upwards for $x > \frac{3}{2}$ or $x < \frac{3}{2}$.

$\therefore$ Graph is sketched as shown in Fig. 2.7.

EXAMPLE 2 *Sketch the curve* $y = (x-1)(x-2)(x-3)$.

SOLUTION Here; $y = (x-1)(x-2)(x-3)$

(i) Put $y = 0 \Rightarrow x = 1, 2, 3.$

(ii) $y = x^3 - 6x^2 + 11x - 6$

$$\Rightarrow \quad \frac{dy}{dx} = 3x^2 - 12x + 11 \quad \text{and} \quad \frac{d^2y}{dx^2} = 6x - 12$$

$$\text{when} \quad \frac{dy}{dx} = 0 \quad \Rightarrow \quad x = \frac{6 \pm \sqrt{3}}{3}$$

$$\therefore \text{ maximum when; } \quad x = \frac{6-\sqrt{3}}{3} \quad \text{as} \quad \frac{d^2y}{dx^2} = -2\sqrt{3}$$

$$\text{minimum when; } \quad x = \frac{6+\sqrt{3}}{3} \quad \text{as} \quad \frac{d^2y}{dx^2} = 2\sqrt{3}$$

(iii) Here; $\frac{dy}{dx} = 3x^2 - 12x + 11$

$$= 3\left(x - \frac{6-\sqrt{3}}{3}\right)\left(x - \frac{6+\sqrt{3}}{3}\right)$$

$$\Rightarrow \text{Increases when; } \quad x < \frac{6-\sqrt{3}}{3}$$

$$\text{or} \quad x > \frac{6+\sqrt{3}}{3}$$

$$\text{decreases when; } \quad \frac{6-\sqrt{3}}{3} < x < \frac{6+\sqrt{3}}{3}$$

(iv) Concave upwards when $x > 2$ and concave down when $x < 2$.

$\therefore$ Graph is sketched as shown in Fig. 2.8.

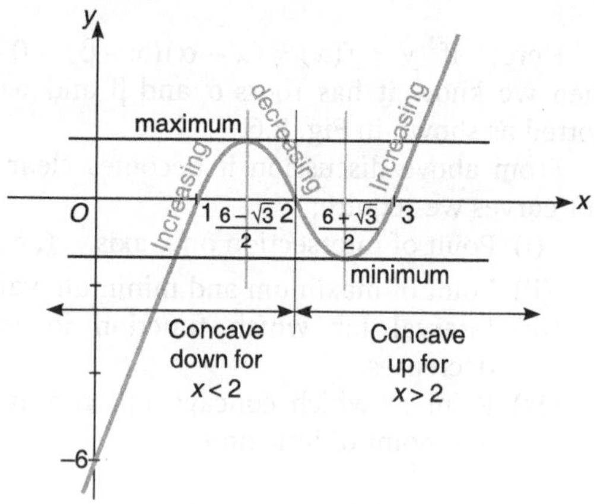

Fig. 2.8

EXAMPLE 3 *Sketch the curve* $y = (x-1)^2(x-2)$.

SOLUTION Here; $y = (x-1)^2(x-2)$

(i) Put $y = 0 \Rightarrow x = 1, 1, 2$

(ii) $y = x^3 - 4x^2 + 5x - 2$

$$\Rightarrow \quad \frac{dy}{dx} = 3x^2 - 8x + 5 \quad \text{and} \quad \frac{d^2y}{dx^2} = 6x - 8$$

$$\text{when} \quad \frac{dy}{dx} = 0 \quad \Rightarrow \quad x = 1, \frac{5}{3}$$

$$\therefore \text{ maximum when; } \quad x = 1 \quad \text{as} \quad \frac{d^2y}{dx^2} = -2.$$

minimum when; $x = \frac{5}{3}$ as $\frac{d^2y}{dx^2} = 2$.

(iii) Here; $\frac{dy}{dx} = 3x^2 - 8x + 5 = 3(x-1)\left(x - \frac{5}{3}\right)$

$\Rightarrow$ Increases when; $x < 1$ or $x > \frac{5}{3}$.

Decreases when; $1 < x < \frac{5}{3}$.

(iv) Concave up when $x > \frac{4}{3}$ and

concave down when $x < \frac{4}{3}$.

$\therefore$ Graph is sketched as shown in Fig. 2.9.

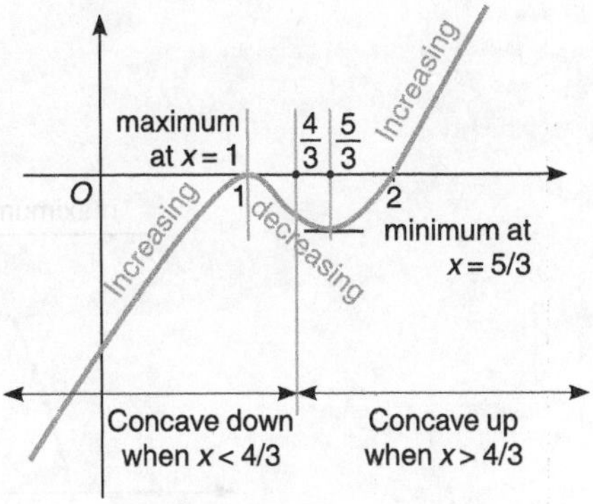

Fig. 2.9

EXAMPLE 4 *Sketch the curve $y = 3x^2 - 2x^3$.*

SOLUTION Here; $y = 3x^2 - 2x^3$

$\Rightarrow$ $x = 0, \frac{3}{2}$ when $y = 0$ ...(i)

also $\frac{dy}{dx} = 6x - 6x^2 = 6x(1-x)$

and $\frac{d^2y}{dx^2} = 6 - 12x = 6(1-2x)$

$\Rightarrow$ maximum when; $x = 1$ as $\frac{d^2y}{dx^2} = -6$

and minimum when; $x = 0$ as $\frac{d^2y}{dx^2} = 6$ ...(ii)

y increases when; $0 < x < 1$.

y decreases when; $x < 0$ and $x > 1$ ...(iii)

Concave up when; $x < \frac{1}{2}$

Concave down when; $x > \frac{1}{2}$ ...(iv)

$\therefore$ Graph is sketched as shown in Fig. 2.10.

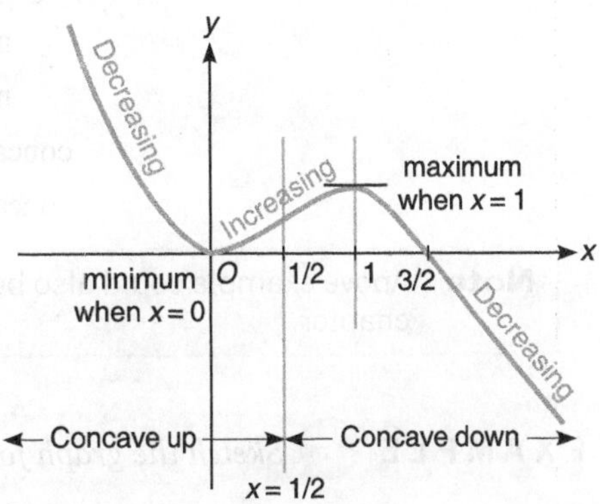

Fig. 2.10

EXAMPLE 5 *Sketch the graph for the function: $f(x) = |x+3|(x+1)$.*

SOLUTION Here; $y = |x+3|(x+1) = \begin{cases} (x+3)(x+1); & x \geq -3 \\ -(x+3)(x+1); & x < -3 \end{cases}$

$\Rightarrow$ $x = -1, -3$ when $y = 0$ ...(i)

also; $\frac{dy}{dx} = \begin{cases} +2x+4; & x \geq -3 \\ -2x-4; & x < -3 \end{cases}$ and $\frac{d^2y}{dx^2} = \begin{cases} 2; & x \geq -3 \\ -2; & x < -3 \end{cases}$

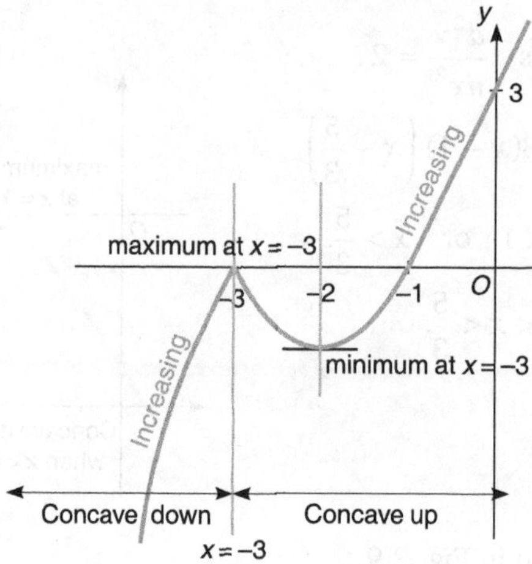

Fig. 2.11

$$\Rightarrow \quad \left.\begin{array}{ll}\text{Increasing when} & x < -3 \text{ or } x > -2 \\ \text{decreasing when} & -3 < x < -2\end{array}\right\} \quad \ldots\text{(ii)}$$

$$\left.\begin{array}{l}\text{maximum at } x = -3 \\ \text{minimum at } x = -2\end{array}\right\} \quad \ldots\text{(iii)}$$

$$\left.\begin{array}{l}\text{concave up when } x < -3 \\ \text{concave down when } x > -3\end{array}\right\} \quad \ldots\text{(iv)}$$

> **Note** Above example could also be solved by using transformations discussed in later part of chapter.

EXAMPLE 6 *Sketch the graph for:* $f(x) = \dfrac{x+1}{x^2+3}$

SOLUTION Here; $y = \dfrac{x+1}{x^2+3} \quad \Rightarrow \quad \left.\begin{array}{ll} x = -1 & \text{when } y = 0 \\ y = \dfrac{1}{3} & \text{when } x = 0\end{array}\right\} \quad \ldots\text{(i)}$

$$\frac{dy}{dx} = \frac{-x^2 - 2x + 3}{(x^2+3)^2} = \frac{-(x+3)(x-1)}{(x^2+3)^2} \Rightarrow \left.\begin{array}{l}\text{Increasing when; } -3 < x < 1 \\ \text{Decreasing when; } x < -3 \text{or } x > 1\end{array}\right\} \quad \ldots\text{(ii)}$$

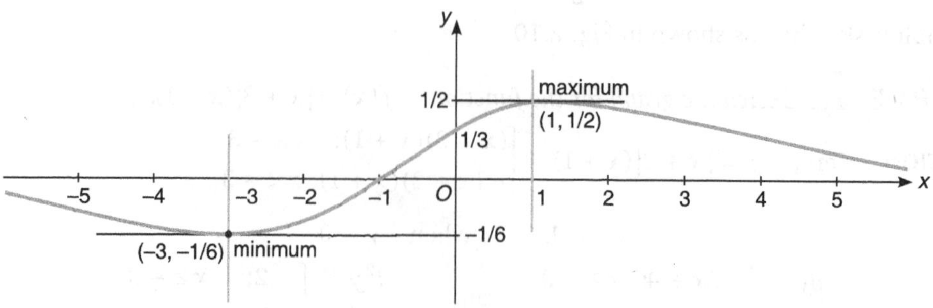

Fig. 2.12

{Using number line rule for $\frac{dy}{dx}$, $\xleftarrow{\quad - \quad}$ dec^n -3 $\xrightarrow{+}$ Inc^n 1 dec^n $\xrightarrow{-}$ }

also; $$\frac{d^2y}{dx^2} = \frac{2(x^3 + 3x^2 - 9x - 3)}{(x^2 + 3)^3}$$

$$\left.\begin{array}{llll} \text{minimum at} & x = -3 & \text{as;} & \dfrac{d^2y}{dx^2} = +\dfrac{1}{36} > 0 \\ \text{maximum at} & x = 1 & \text{as;} & \dfrac{d^2y}{dx^2} = -\dfrac{1}{4} < 0 \end{array}\right\} \quad \ldots\text{(iii)}$$

> **Note** In above curve x-axis works as asymptote, *i.e.*, the curve would never meet x-axis. For detail refer chapter 3.

2.4 GRAPHICAL TRANSFORMATIONS

Here, we shall discuss the transformations as;

(i) $f(x)$ transforms to $f(x) \pm a$
(ii) $f(x)$ transforms to $f(x \pm a)$.
(iii) $f(x)$ transforms to $(a\,f(x))$
(iv) $f(x)$ transforms to $f(ax)$.
(v) $f(x)$ transforms to $f(-x)$.
(vi) $f(x)$ transforms to $-f(x)$.
(vii) $f(x)$ transforms to $-f(-x)$

(viii) $f(x)$ transforms to $|f(x)|$.
(ix) $f(x)$ transforms to $f(|x|)$.
(x) $f(x)$ transforms to $|f(|x|)|$.
(xi) $y = f(x)$ transforms to $|y| = f(x)$.
(xii) $y = f(x)$ transforms to $|y| = |f(x)|$.
(xiii) $y = f(x)$ transforms to $|y| = |f(|x|)|$.

Where $|\bullet|$ means modulus or absolute value function.

(xiv) $y = f(x)$ transforms to $y = [f(x)]$.
(xv) $y = f(x)$ transforms to $y = f([x])$.
(xvi) $y = f(x)$ transforms to $y = [f([x])]$.
(xvii) $y = f(x)$ transforms to $[y] = f(x)$.
(xviii) $y = f(x)$ transforms to $[y] = [f(x)]$.

Where $[\bullet]$ denotes greatest integer less than or equal to x.

(xix) $y = f(x)$ transforms to $y = f(\{x\})$.
(xx) $y = f(x)$ transforms to $y = \{f(x)\}$.
(xxi) $y = f(x)$ transforms to $y = \{f(\{x\})\}$.
(xxii) $y = f(x)$ transforms to $\{y\} = f(x)$.
(xxiii) $y = f(x)$ transforms to $\{y\} = \{f(x)\}$.

Where $\{\bullet\}$ denotes fractional part of x.

(xxiv) $y = f(x)$ transforms to $y = f^{-1}(x)$, $f^{-1}(x)$ represents inverse of $f(x)$.

Now, we shall study the following cases as;

(i) When $f(x)$, transforms to $f(x) \pm a$. (where a is + ve)

i.e., $$f(x) \longrightarrow f(x) + a$$

shift the given graph of $f(x)$ **upward** through 'a' units

again, $$f(x) \longrightarrow f(x) - a$$

shift the given graph of $f(x)$ **downward** through 'a' units.

Graphically it could be stated as:

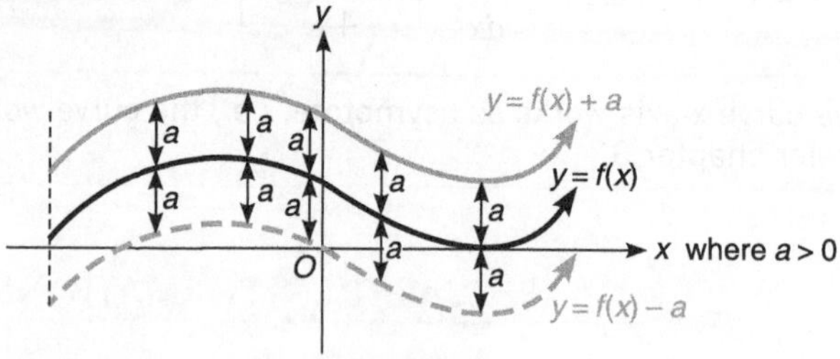

Fig. 2.13

EXAMPLE 1 *Plot* $y = e^x + 1$; $y = e^x - 1$, *with the help of* $y = e^x$.

SOLUTION We know; $y = e^x$ (exponential function) could be plotted as;

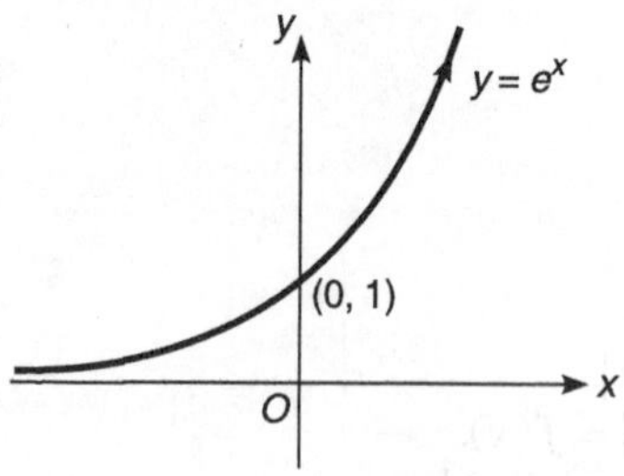

Fig. 2.14

$\Rightarrow$ $y = e^x + 1$, is shifted **upwards** by 1 unit, shown as

Also $y = e^x - 1$, is shifted downwards by 1 unit, shown as

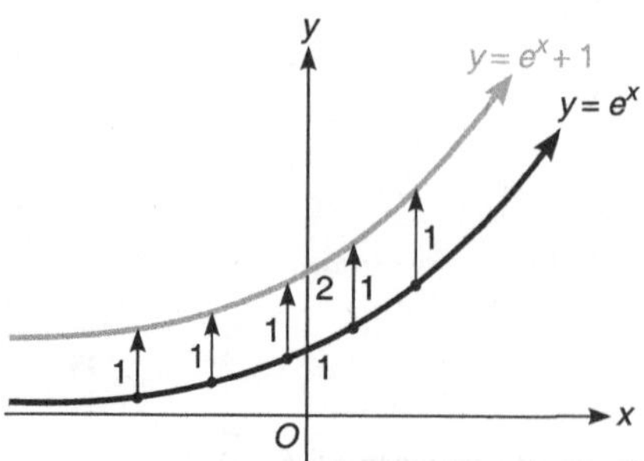

Fig. 2.15

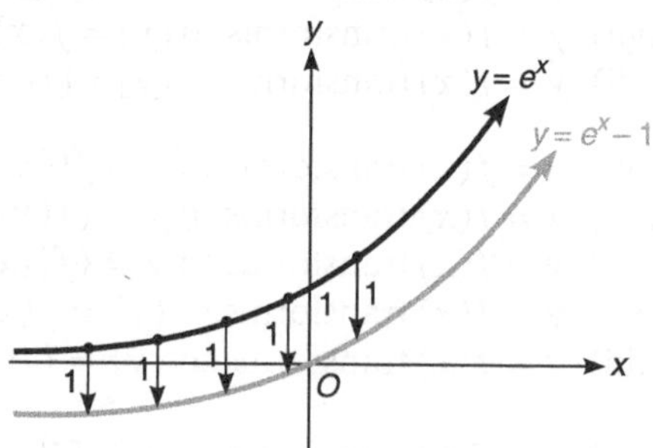

Fig. 2.16

EXAMPLE 2 *Plot* $y = |x| + 2$ *and* $y = |x| - 2$, *with the help of* $y = |x|$.

SOLUTION We know; $y = |x|$ (modulus function) could be plotted as;

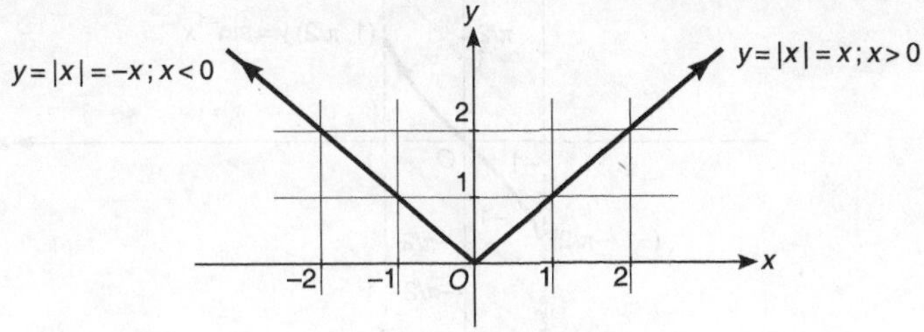

Fig. 2.17

$\Rightarrow$ $y = |x| + 2$ is shifted **upwards** by 2 units.

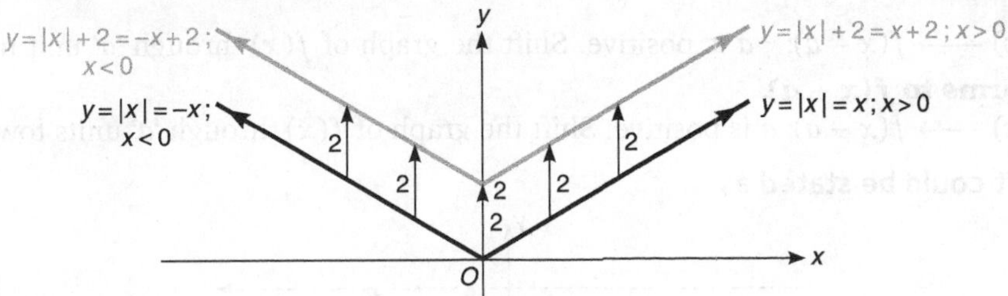

Fig. 2.18

also $y = |x| - 2$ is shifted **downwards** by 2 units.

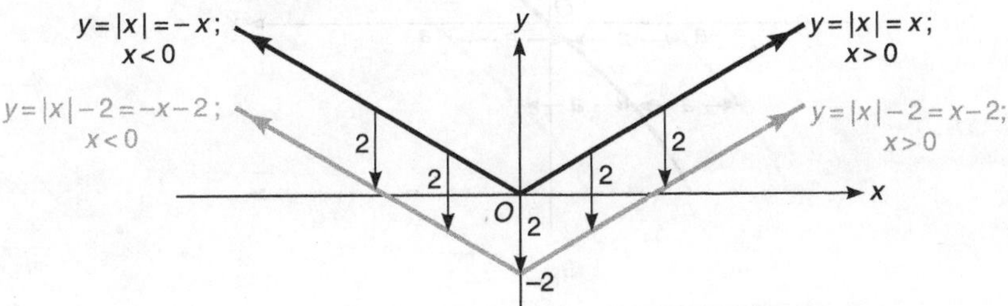

Fig. 2.19

EXAMPLE 3 *Plot* $y = \sin^{-1} x$; $y = (\sin^{-1} x) + 1$ *and* $y = (\sin^{-1} x) - 1$.

SOLUTION We know, $y = \sin^{-1} x$ (Inverse trigonometric) could be plotted as;

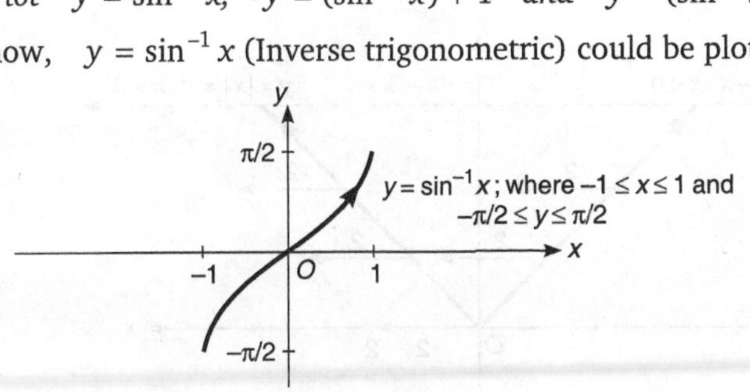

Fig. 2.20

$\Rightarrow$ $y = \sin^{-1} x + 1$, is shifted **upwards** by 1 unit.

and $y = \sin^{-1} x - 1$, is shifted **downwards** by 1 unit.

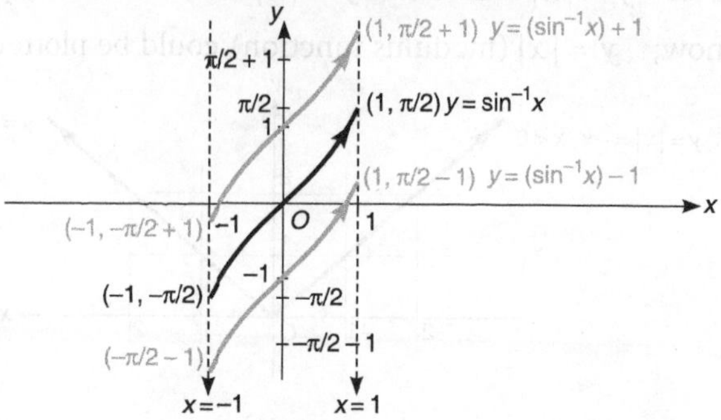

Fig. 2.21

(ii) $f(x)$ transforms to $f(x - a)$

i.e., $f(x) \longrightarrow f(x - a)$; a is positive. Shift the graph of $f(x)$ through 'a' unit towards right

$f(x)$ transforms to $f(x + a)$.

i.e., $f(x) \longrightarrow f(x + a)$; a is positive. Shift the graph of $f(x)$ through 'a' units towards left.

Graphically it could be stated as

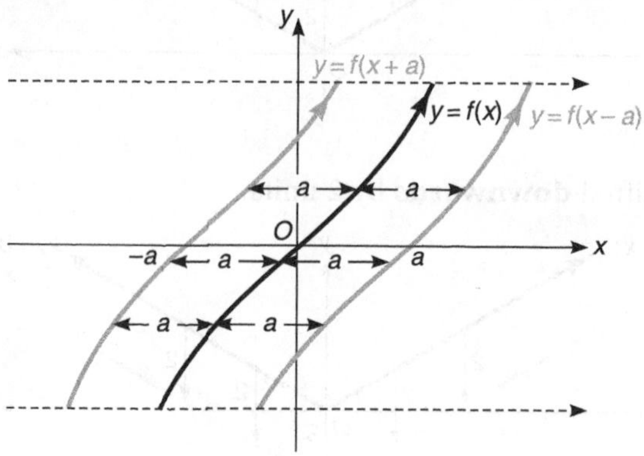

Fig. 2.22

EXAMPLE 1 *Plot* $y = |x|$, $y = |x - 2|$ *and* $y = |x + 2|$.

SOLUTION As discussed $f(x) \longrightarrow f(x - a)$; shift towards right.

$\Rightarrow$ $y = |x - 2|$ is shifted '2' units **towards right**.

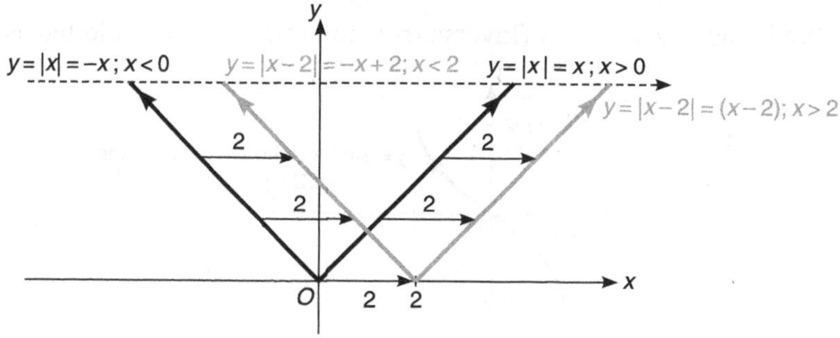

Fig. 2.23

also $\quad y = |x + 2|$ is shifted '2' units **towards left**.

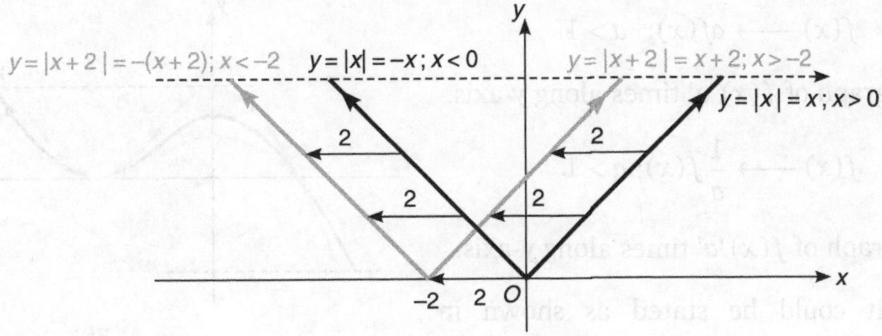

Fig. 2.24

EXAMPLE 2 *Plot* $y = \sin\left(x - \dfrac{\pi}{2}\right)$ *and* $y = \sin\left(x + \dfrac{\pi}{2}\right)$.

SOLUTION As we know; $y = \sin x$ could be plotted as;

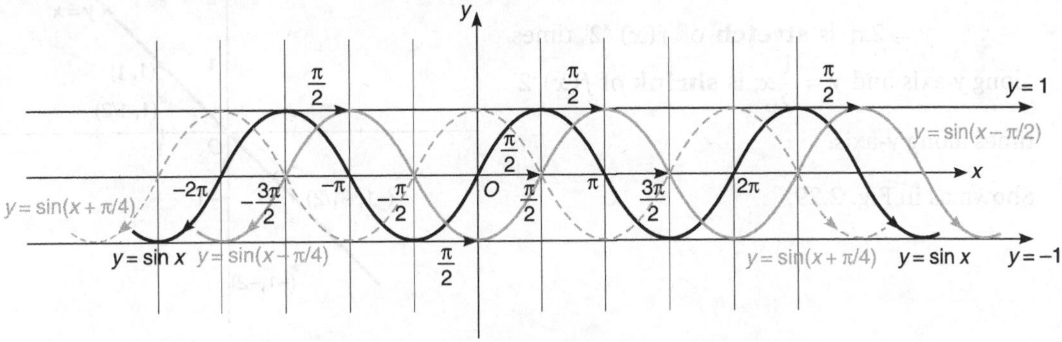

Fig. 2.25

EXAMPLE 3 *Plot* $y = \sin^{-1} x$; $\quad y = \sin^{-1}(x - 1)$ *and* $y = \sin^{-1}(x + 1)$.

SOLUTION We know;

$y = \sin^{-1} x$ could be plotted as shown in Fig. 2.26.

$\Rightarrow \quad y = \sin^{-1}(x - 1)$ is shifted '1' unit **towards right**.

and $\quad y = \sin^{-1}(x + 1)$ is shifted '1' unit **towards left**.

Shown as in Fig. 2.27.

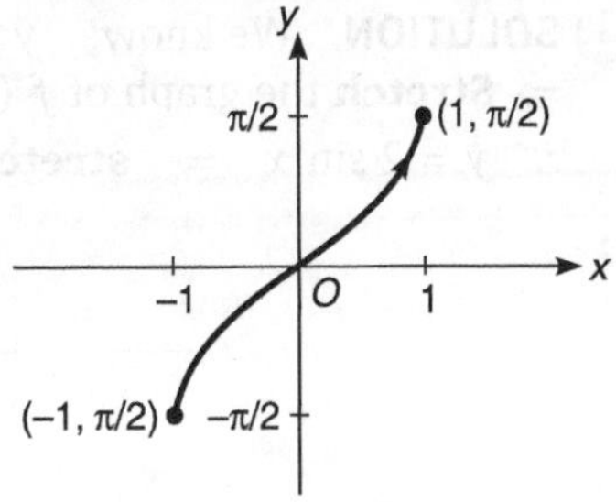

Fig. 2.26

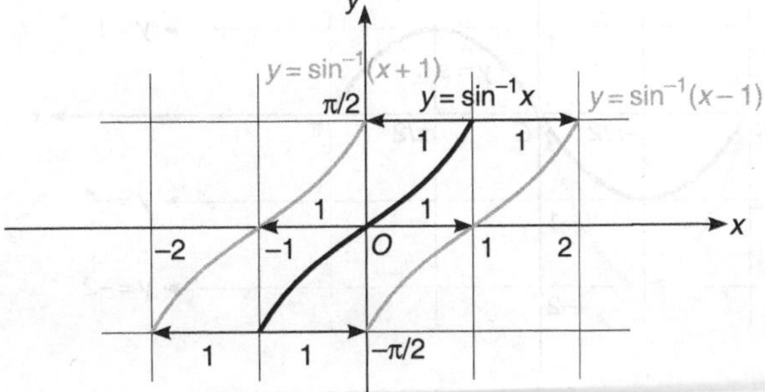

Fig. 2.27

(iii) $f(x)$ transforms to $a\,f(x)$

i.e., $f(x) \longrightarrow af(x);\ a > 1$

Stretch the graph of $f(x)$ 'a' times along y-axis.

$$f(x) \longrightarrow \frac{1}{a}f(x);\ a > 1.$$

Shrink the graph of $f(x)$ 'a' times along y-axis.

Graphically it could be stated as shown in Fig. 2.28.

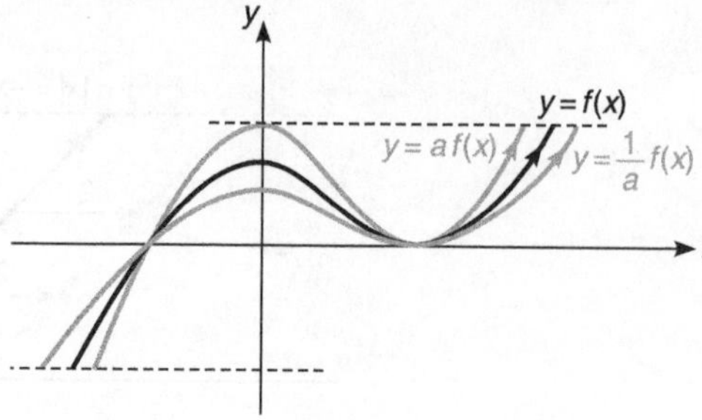

Fig. 2.28

EXAMPLE 1 *Plot* $y = x;\ y = 2x$ *and* $y = \frac{1}{2}x$.

SOLUTION As we know graph for $y = 2x$.

$\therefore\quad y = 2x$; is **stretch** of $f(x)$ '2' times along y-axis and $y = \frac{1}{2}x$; is **shrink** of $f(x)$ '2' times along y-axis.

Shown as in Fig. 2.29.

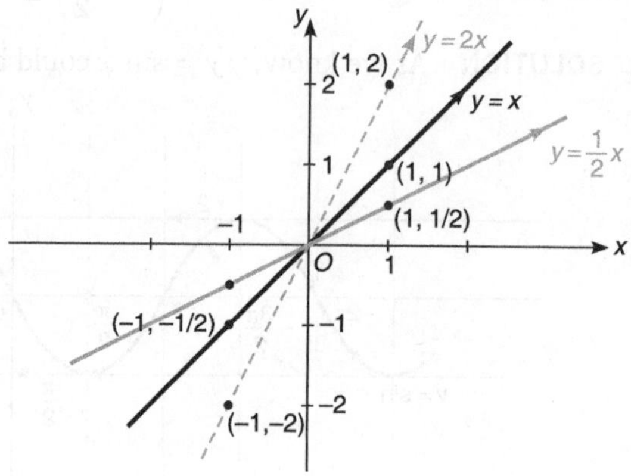

Fig. 2.29

EXAMPLE 2 *Plot* $y = \sin x$ *and* $y = 2\sin x$.

SOLUTION We know; $y = \sin x$ and $f(x) \to a\,f(x)$

$\Rightarrow$ **Stretch** the graph of $f(x)$ 'a' times along y-axis.

$\therefore\ y = 2\sin x \ \Rightarrow$ **stretch** the graph of $\sin x$ '2' times along y-axis.

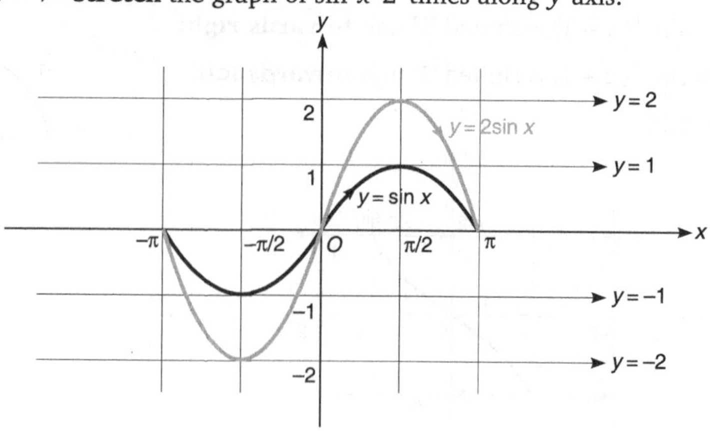

Fig. 2.30

Above curve is plotted for the interval $[-\pi, \pi]$ as periodic with period 2π.

EXAMPLE 3 *Plot* $y = \sin x$ *and* $y = \frac{1}{2}\sin x$.

SOLUTION As we know;

$$y = \frac{1}{a} f(x)$$

$\Rightarrow$ **shrink** the graph of $f(x)$ 'a' times along y-axis.

$$\therefore \quad y = \frac{1}{2} \sin x$$

$\Rightarrow$ **shrink** the graph of $f(x)$ '2' times along y-axis.

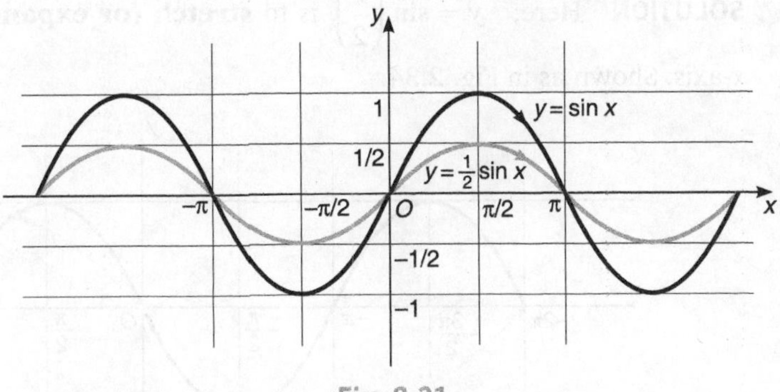

Fig. 2.31

(iv) $f(x)$ transforms to $f(ax)$

i.e., $\quad f(x) \longrightarrow f(ax); \quad a > 1$

Shrink (or contract) the graph of $f(x)$ 'a' times along x-axis.

again $\quad f(x) \longrightarrow f\left(\frac{1}{a}x\right); \quad a > 1$

Stretch (or expand) the graph of $f(x)$ 'a' times along x-axis.

Graphically it could be stated as shown in Fig. 2.32.

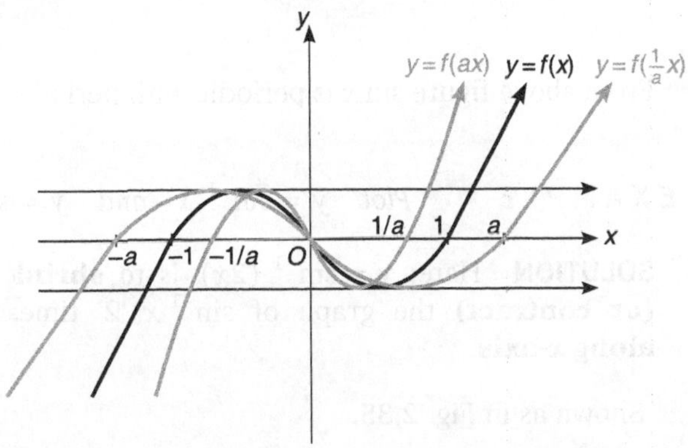

Fig. 2.32

EXAMPLE 1 *Plot* $y = \sin x$ *and* $y = \sin 2x$.

SOLUTION Here; $y = \sin 2x$, is to **shrink (or contract)** the graph of $\sin x$ by '2' units along x-axis. Shown as in Fig. 2.33.

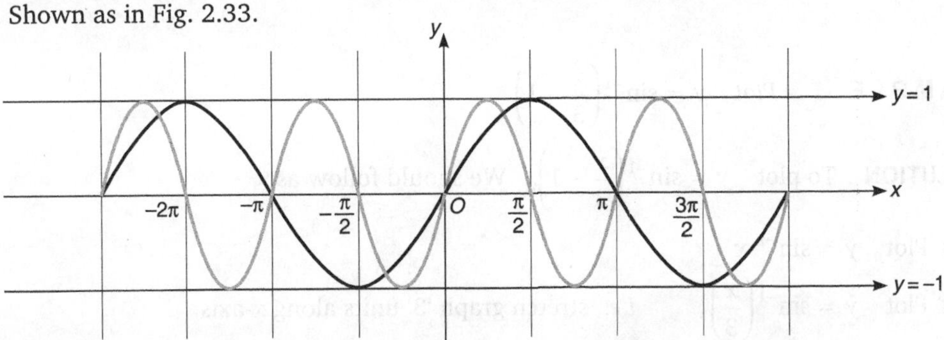

Fig. 2.33

From above figure $\sin x$ is periodic with period 2π and $\sin 2x$ with period π.

EXAMPLE 2 *Plot* $y = \sin x$ *and* $y = \sin\frac{x}{2}$.

SOLUTION Here; $y = \sin\left(\frac{x}{2}\right)$, is to **stretch (or expand)** the graph of $\sin x$ '2' times along x-axis. Shown as in Fig. 2.34.

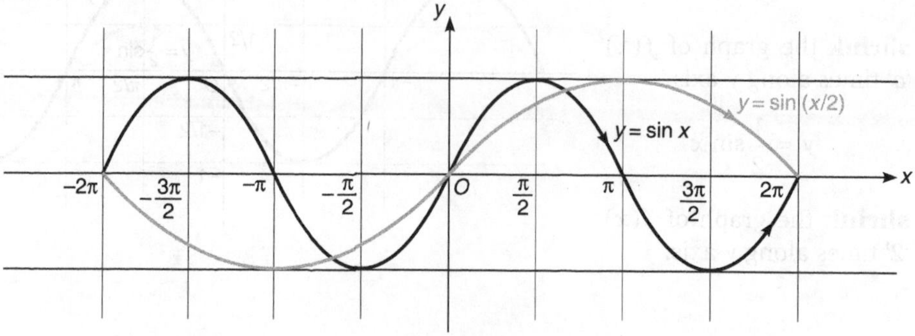

Fig. 2.34

From above figure $\sin x$ is periodic with period 2π and $\sin\left(\frac{x}{2}\right)$ is periodic with period 4π.

EXAMPLE 3 *Plot* $y = \sin^{-1} x$ *and* $y = \sin^{-1}(2x)$.

SOLUTION Here; $y = \sin^{-1}(2x)$, is to **shrink (or contract)** the graph of $\sin^{-1} x$ '2' times **along x-axis**.

Shown as in Fig. 2.35.

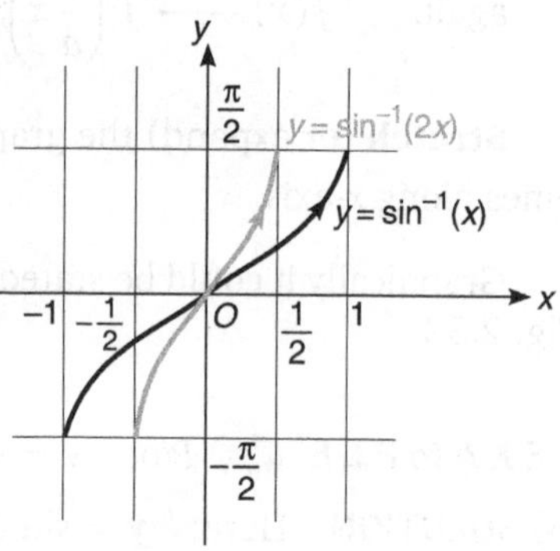

Fig. 2.35

EXAMPLE 4 *Plot* $y = \sin^{-1}\left(\frac{x}{3} - 1\right)$.

SOLUTION To plot $y = \sin^{-1}\left(\frac{x}{3} - 1\right)$. We should follow as;

(i) Plot $y = \sin^{-1} x$

(ii) Plot $y = \sin^{-1}\left(\frac{x}{3}\right)$, *i.e.*, stretch graph '3' units along x-axis.

(iii) Plot $y = \sin^{-1}\left(\frac{x}{3} - 1\right)$, *i.e.*, shift the graph (ii) by '3' unit towards right.

(i) Plotting $y = \sin^{-1} x$:

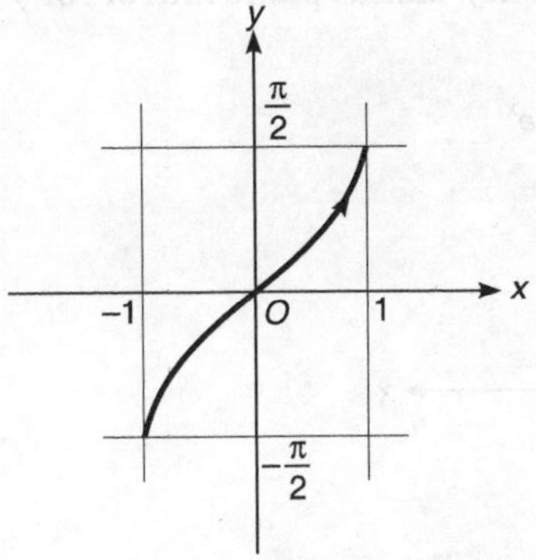

Fig. 2.36

(ii) Plot $y = \sin^{-1}\left(\dfrac{x}{3}\right)$:

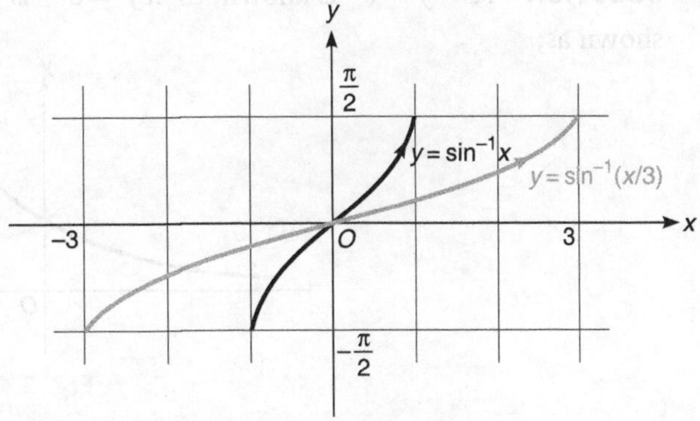

Fig. 2.37

(iii) Plot $y = \sin^{-1}\left(\dfrac{x}{3} - 1\right)$:

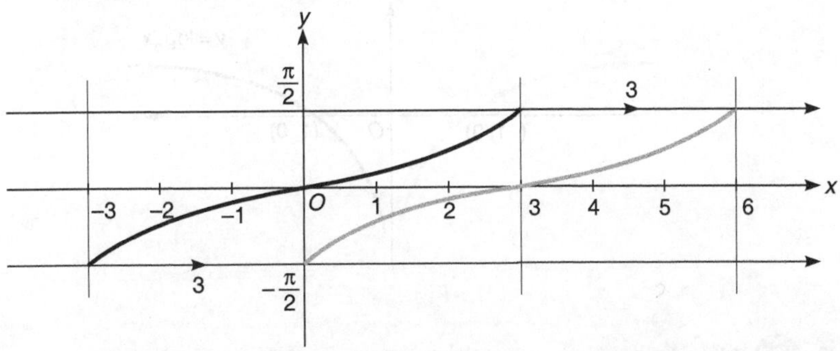

Fig. 2.38

(v) $f(x)$ transforms to $f(-x)$

i.e.,

$$f(x) \longrightarrow f(-x)$$

To draw $y = f(-x)$, take the image of the curve $y = f(x)$ in y-axis as plane mirror.

OR

"Turn the graph of $f(x)$ by 180° about y-axis."

Graphically it is stated as;

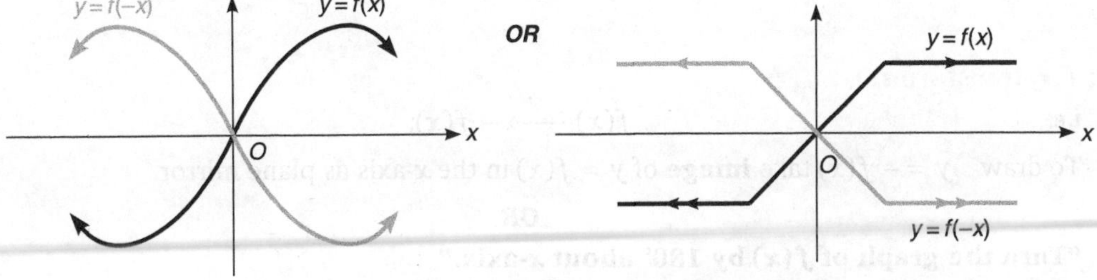

Fig. 2.39

EXAMPLE 1 *Plot* $y = e^{-x}$.

SOLUTION As $y = e^x$ is known; then $y = e^{-x}$ is the image in y-axis as plane mirror for $y = e^x$; shown as;

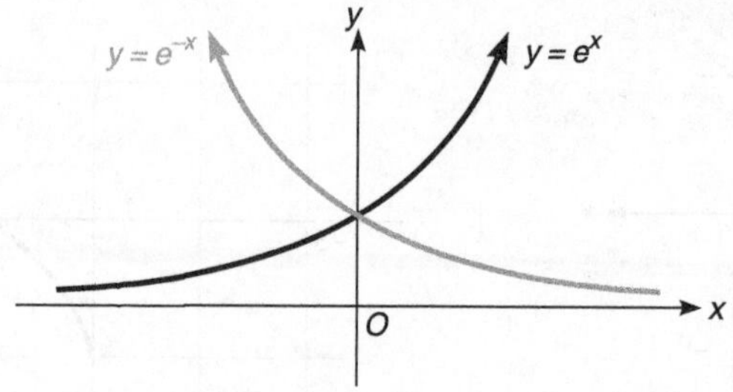

Fig. 2.40

EXAMPLE 2 *Plot the curve* $y = \log_e(-x)$.

SOLUTION Here; $y = \log_e(-x)$; is to take **mirror image** of $y = \log_e x$ about y-axis. Shown as;

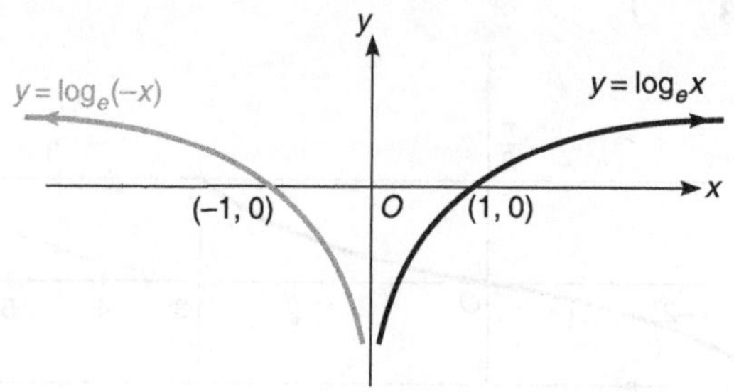

Fig. 2.41

EXAMPLE 3 *Plot the curve* $y = \sin^{-1}(-x)$.

SOLUTION Here; $y = \sin^{-1}(-x)$; is the **mirror image** of $y = \sin^{-1}(x)$ about y-axis. Shown as in Fig. 2.42.

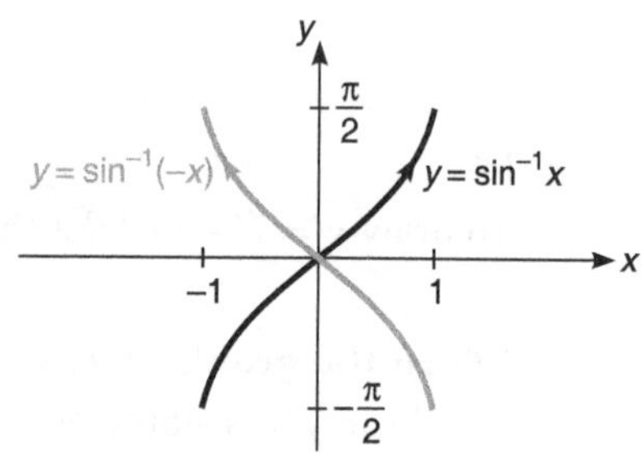

Fig. 2.42

(vi) $f(x)$ transforms to $-f(x)$

i.e., $$f(x) \longrightarrow -f(x);$$

To draw $y = -f(x)$ take **image** of $y = f(x)$ in the x-axis as plane mirror.

OR

"Turn the graph of $f(x)$ by 180° about x-axis."

EXAMPLE 1 *Plot the curve* $y = -e^x$.

SOLUTION As $y = e^x$ is known;

$\therefore$ $y = -e^x$ take image of $y = e^x$ in the **x-axis as plane mirror**.

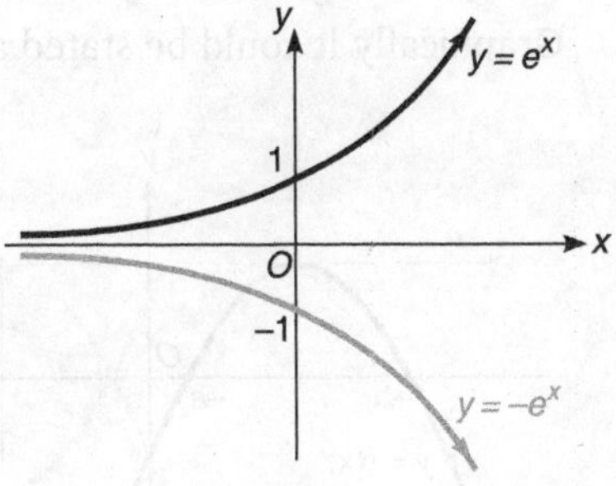

Fig. 2.43

EXAMPLE 2 *Plot the curve* $y = -(\log x)$.

SOLUTION As $y = \log x$ is given then $y = -\log x$ is the image of $y = \log x$ in the **x-axis as plane mirror.**

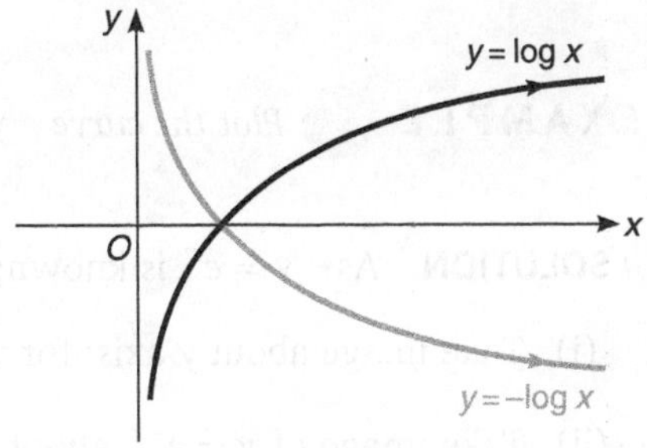

Fig. 2.44

EXAMPLE 3 *Plot the curve* $y = -\{x\}$; *where* $\{\cdot\}$ *denotes the fractional part of x.*

SOLUTION As $y = \{x\}$ is known;

$\therefore$ $y = -\{x\}$ is the image of $y = \{x\}$ about **x-axis as plane mirror**.

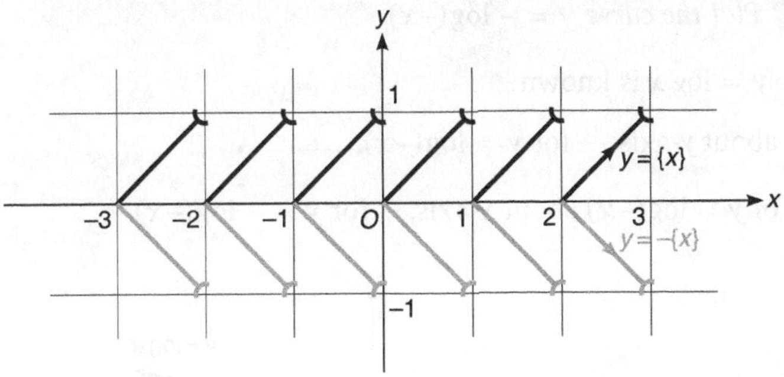

Fig. 2.45

(vii) $f(x)$ transforms to $-f(-x)$

i.e.,
$$f(x) \longrightarrow -f(-x);$$

to draw $y = -f(-x)$ take image of $f(x)$ about y-axis to obtain $f(-x)$ and then take image of $f(-x)$ about x-axis to obtain $-f(-x)$.

$\therefore$
$$f(x) \longrightarrow -f(-x)$$

⇒ (i) **Image about *y*-axis.** (ii) **Image about *x*-axis.**

Graphically it could be stated as;

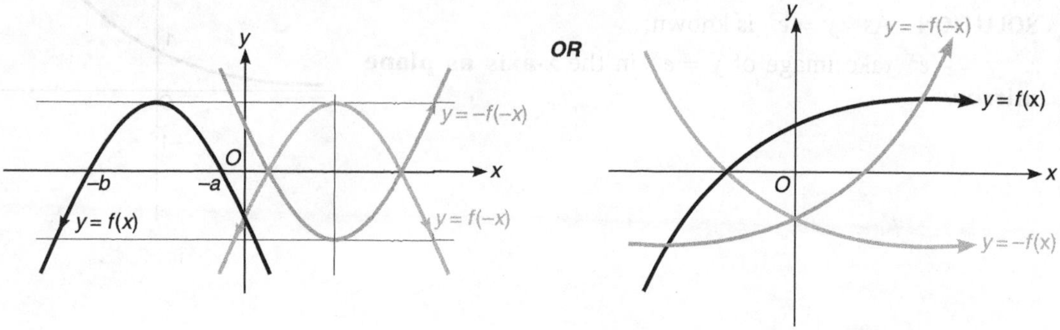

Fig. 2.46

EXAMPLE 1 *Plot the curve* $y = -e^{-x}$.

SOLUTION As $y = e^x$ is known;

(i) Take image about *y*-axis; for $y = e^{-x}$.

(ii) Take image of $y = e^{-x}$ about *x*-axis; for $y = -e^{-x}$. Shown as in Fig. 2.47.

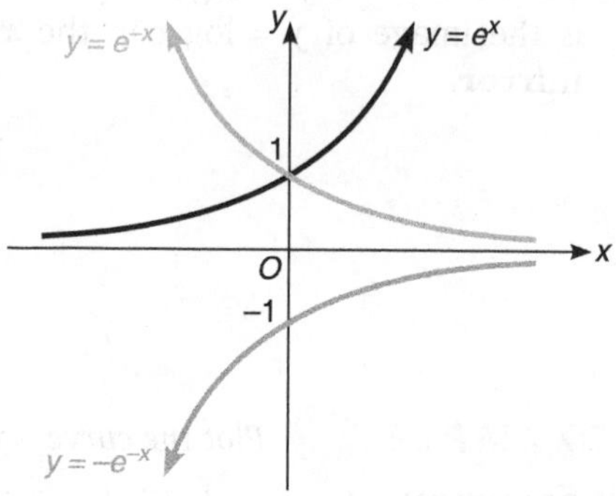

Fig. 2.47

EXAMPLE 2 *Plot the curve* $y = -\log(-x)$.

SOLUTION As $y = \log x$ is known;

(i) Take image about *y*-axis, for $y = \log(-x)$.

(ii) Take image of $y = \log(-x)$ about *x*-axis, for $y = -\log(-x)$.

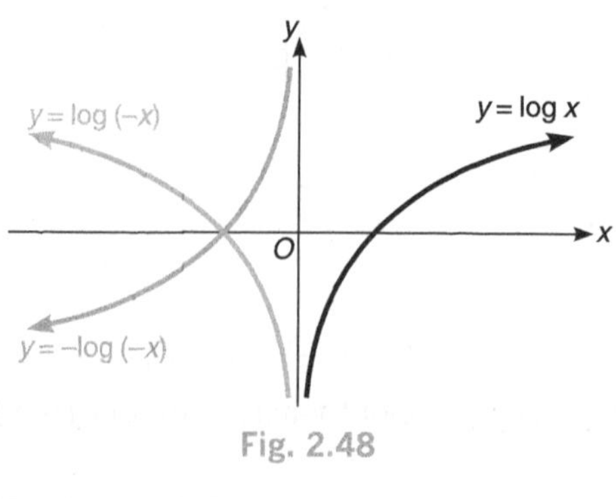

Fig. 2.48

EXAMPLE 3 *Plot the curve* $y = -\{-x\}$. *(where* $\{\cdot\}$ *denote fractional part).*

SOLUTION As we know the curve for $y = \{x\}$.

$\therefore$ **To plot y = – {–x}**

(i) Take image about *x*-axis.

(ii) Take image about *y*-axis.

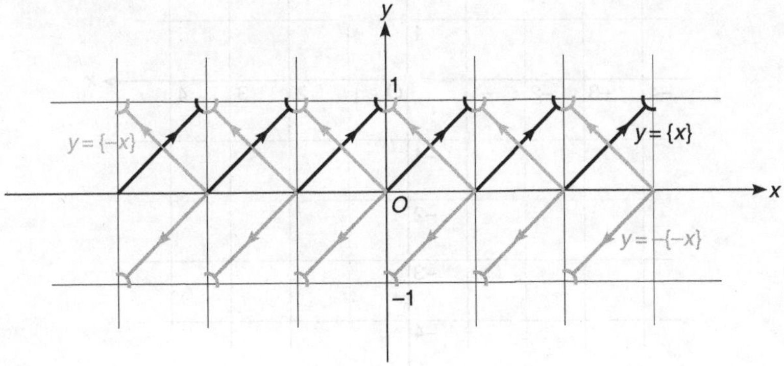

Fig. 2.49

EXAMPLE 4 *Plot the curve for* $y = -[-x]$. *(where* $[\cdot]$ *denotes the greatest integer function.)*

SOLUTION As we know the curve for $y = [x]$.

$\therefore$ to plot $y = -[-x]$

(i) Take image about *x*-axis.

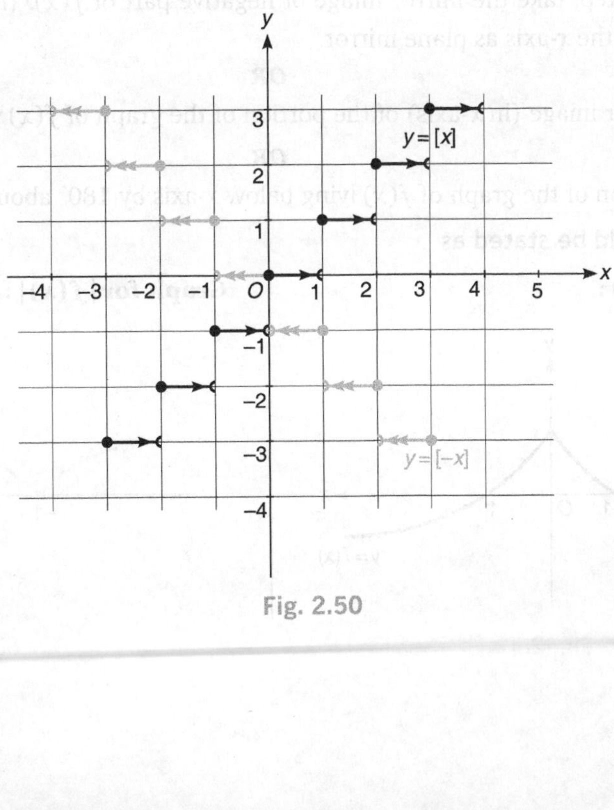

Fig. 2.50

(ii) Take image about y-axis.

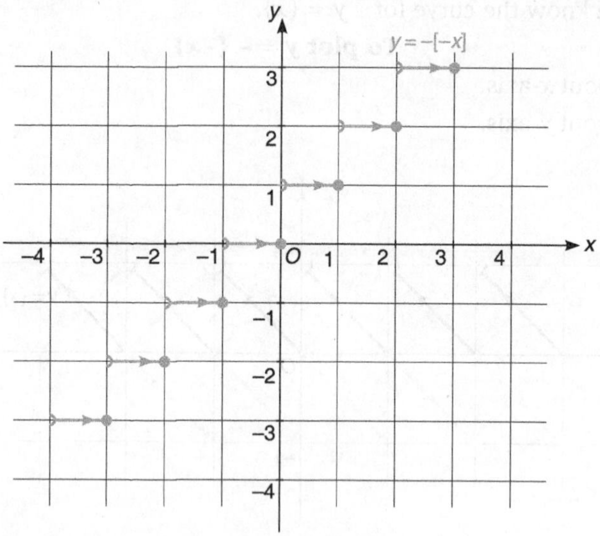

Fig. 2.51

(viii) $f(x)$ transforms to $|f(x)|$; (where $|\bullet|$ represents modulus function)

i.e.,
$$f(x) \longrightarrow |f(x)|$$

Here; $y = |f(x)|$ is drawn in two steps.

(a) In the I step, leave the positive part of $f(x)$, {*i.e.*, the part of $f(x)$ above x-axis) as it is.

(b) In the II step, take the mirror image of negative part of $f(x)$. {*i.e.*, the part of $f(x)$ below x-axis} in the x-axis as plane mirror.

OR

Take the mirror image (in x-axis) of the portion of the graph of $f(x)$ which lies below x-axis.

OR

Turn the portion of the graph of $f(x)$ lying below x-axis by 180° about x-axis.

Graphically it could be stated as

Graph of $f(x)$:

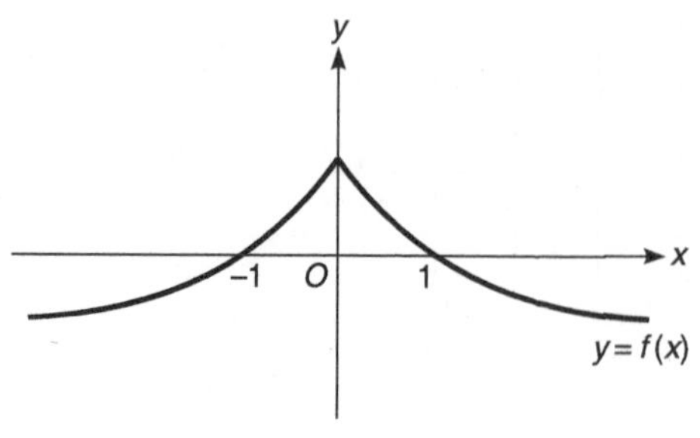

Fig. 2.52

Graph for $|f(x)|$:

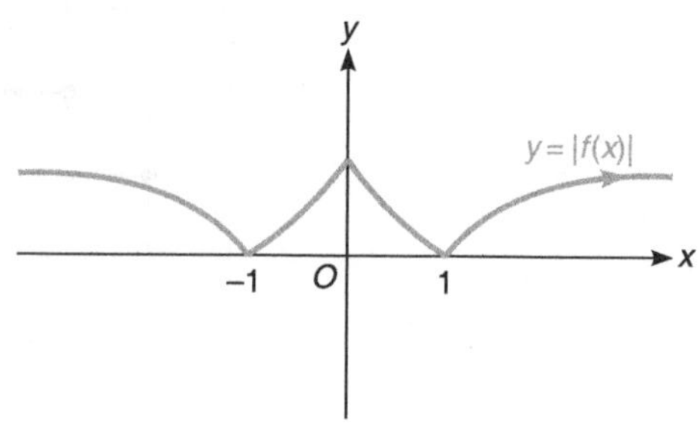

Fig. 2.53

> **Note** Above transformation of graph is very important as to discuss differentiability of $f(x)$.
> As from above example we could say $y = f(x)$ is differentiable for all $x \in R - \{0\}$.
> But; $y = |f(x)|$ is differentiable for all $x \in R - \{-1, 0, 1\}$ as, "at sharp edges function is not differentiable."

EXAMPLE 1 *Draw the graph for $y = |\log x|$.*

SOLUTION To draw graph for $y = |\log x|$ we have to follow two steps:

(i) Leave the (+ ve) part of $y = \log x$, as it is

(ii) Take images of (–ve) part of $y = \log x$, *i.e.*, the part below x-axis in the x-axis as plane mirror.

Shown as:

Graph for $y = \log x$:

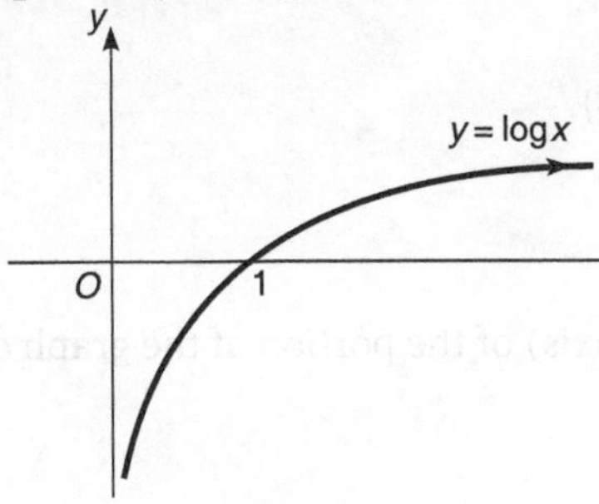

Fig. 2.54

which is differentiable for all $x \in (0, \infty)$

Graph for $y = |\log x|$:

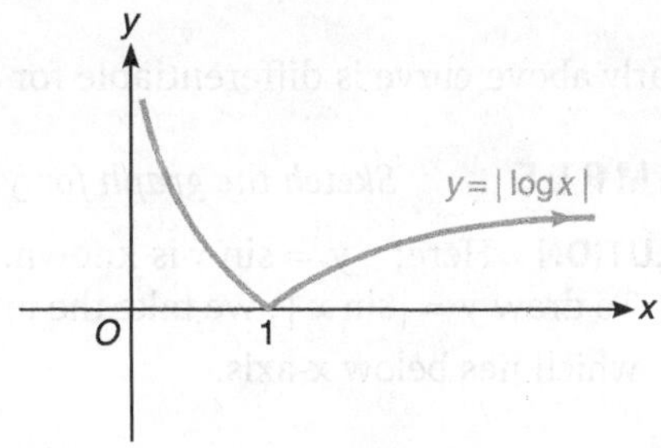

Fig. 2.55

which is clearly differentiable for all $x \in (0, \infty) - \{1\}$. **"as at $x = 1$ their is a sharp edge".**

EXAMPLE 2 *Draw the graph for $y = |x^2 - 2x - 3|$.*

SOLUTION As we know the graph for $y = x^2 - 2x - 3 = (x - 3)(x + 1)$ is a parabola; so to sketch $y = |x^2 - 2x - 3|$ we have to follow two steps.

(i) Leave the positive part of $y = x^2 - 2x - 3$, as it is.

(ii) Take the **image of negative part** of $y = x^2 - 2x - 3$, *i.e.*, the part below x-axis in the **x-axis as plane mirror** shown as in Fig. 2.56.

Graph for $y = x^2 - 2x - 3 = (x - 3)(x + 1)$:

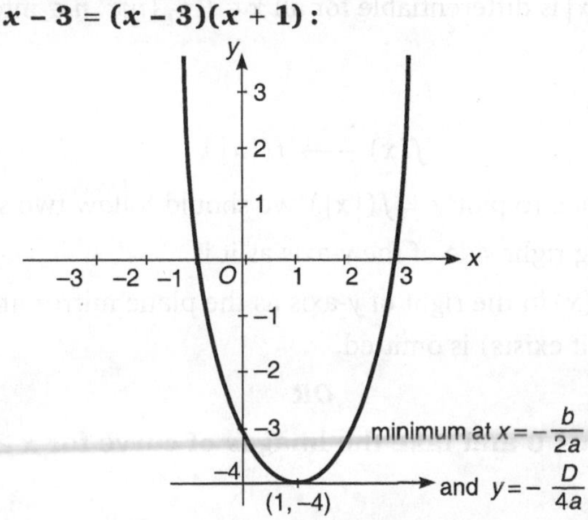

Fig. 2.56

Graph for $y = |x^2 - 2x - 3|$:

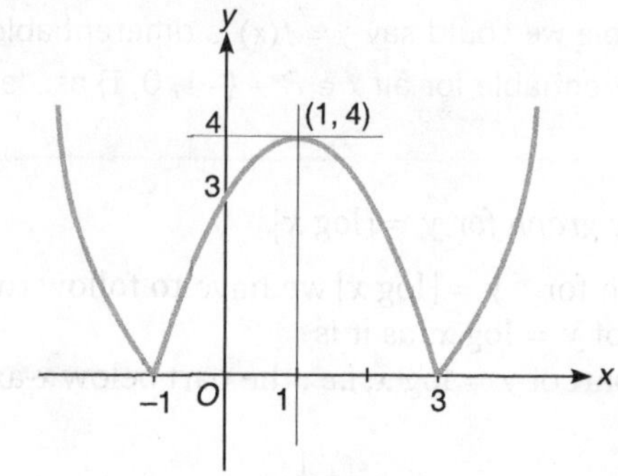

Fig. 2.57

Clearly above curve is differentiable for all $x \in R - \{-1, 3\}$.

EXAMPLE 3 *Sketch the graph for $y = |\sin x|$.*

SOLUTION Here; $y = \sin x$ is known.

$\therefore$ To draw $y = |\sin x|$, we take the mirror image (in x-axis) of the portion of the graph of $\sin x$ which lies below x-axis.

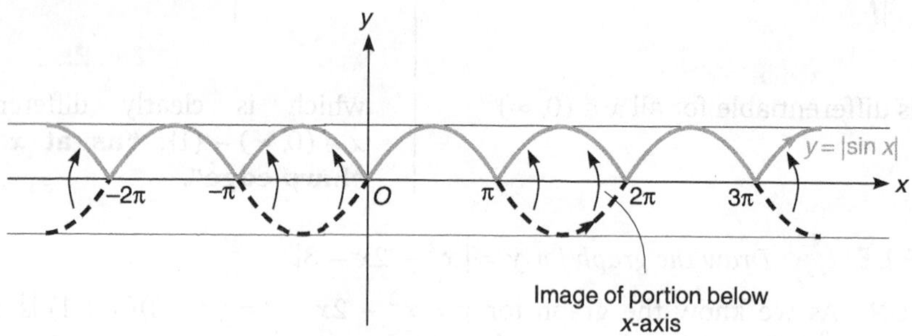

Fig. 2.58

From above figure it is clear;

$$y = |\sin x| \text{ is differentiable for all } x \in R - \{n\pi;\ n \in \text{integer}\}.$$

(ix) $f(x)$ transforms to $f(|x|)$

i.e.,
$$f(x) \longrightarrow f(|x|).$$

If we know $y = f(x)$, then to plot $y = f(|x|)$, we should follow two steps:

(i) Leave the graph lying right side of the y-axis as it is.

(ii) Take the image of $f(x)$ in the right of y-axis as the plane mirror and the graph of $f(x)$ lying leftward of the y-axis (if it exists) is omitted.

OR

Neglect the curve for $x < 0$ and take the images of curve for $x \geq 0$ about y-axis.

Graphically shown as;

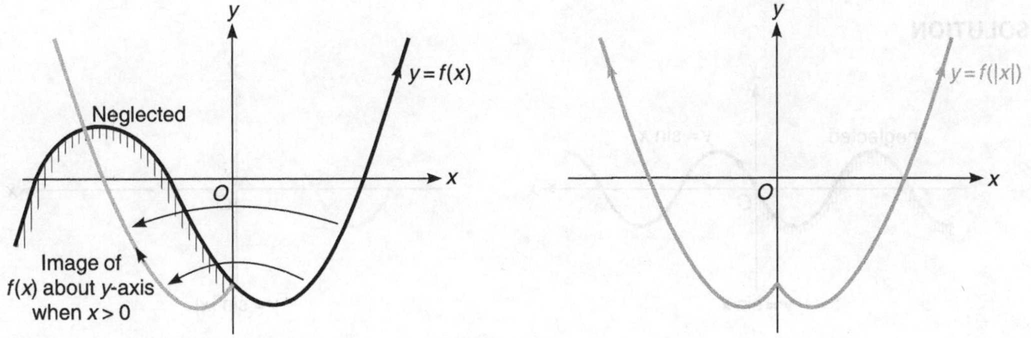

Fig. 2.59

EXAMPLE 1 *Sketch the curve* $y = \log|x|$.

SOLUTION As we know, the curve $y = \log x$.

$\therefore$ $y = \log|x|$ could be drawn in two steps:

(i) Leave the graph lying right side of y-axis as it is.

(ii) Take the image of $f(x)$ in the y-axis as plane mirror.

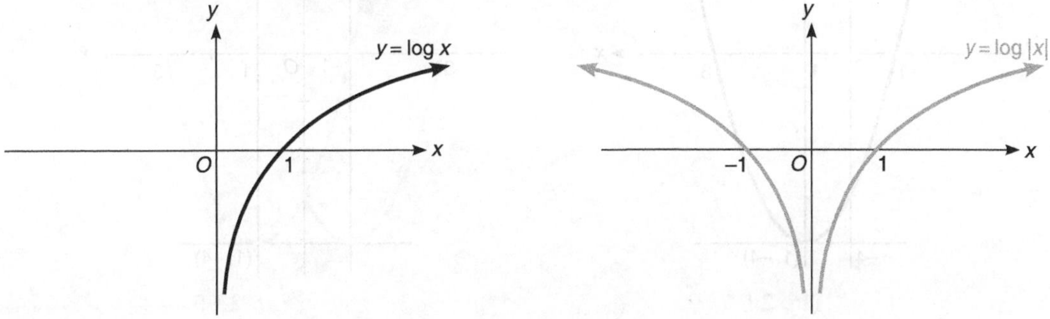

Fig. 2.60

EXAMPLE 2 *Plot the curve* $y = e^{|x|}$.

SOLUTION As we know the curve for $y = e^x$.

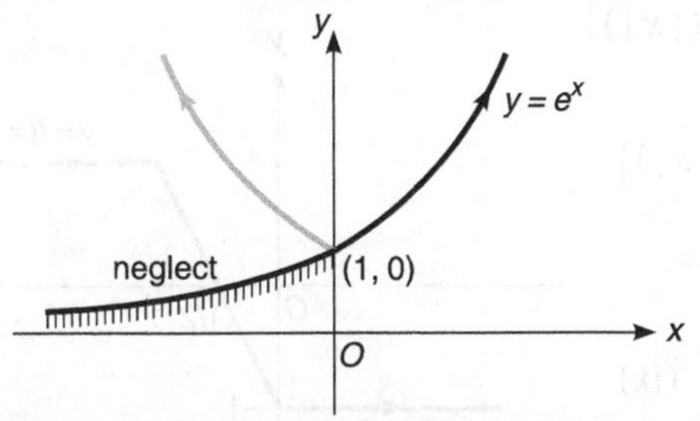

Fig. 2.61

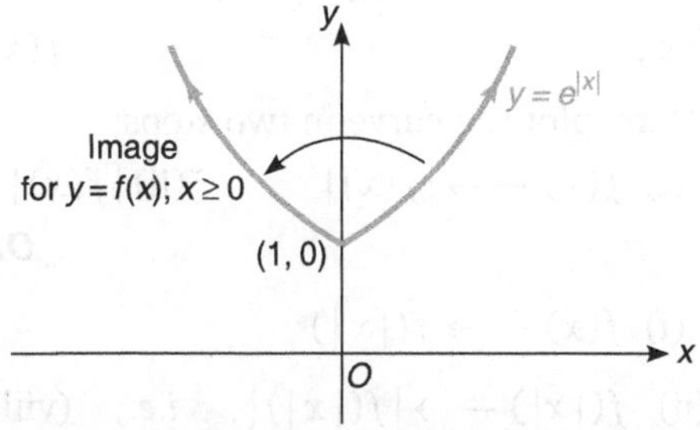

Fig 2.62

$\therefore$ To plot $y = e^{|x|}$, neglect the curve for $x < 0$ and take image about y-axis for $x \geq 0$. Shown as in Fig. 2.62.

EXAMPLE 3 *Plot the curve* $y = \sin|x|$.

SOLUTION

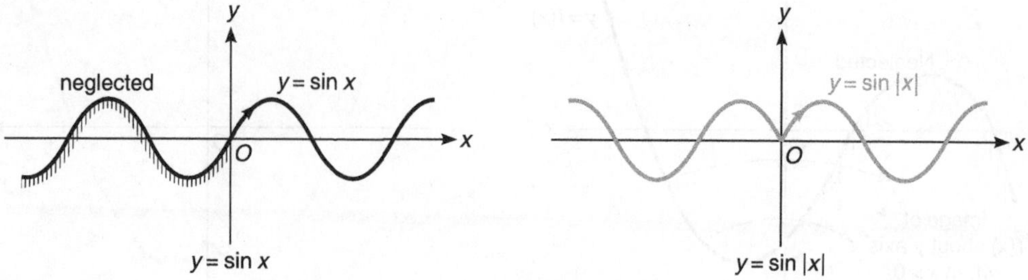

Fig. 2.63

EXAMPLE 4 *Plot the curve* $y = |x|^2 - 2|x| - 3$.

SOLUTION As we know, the curve for $y = x^2 - 2x - 3$ is plotted as shown in Fig. 2.64.

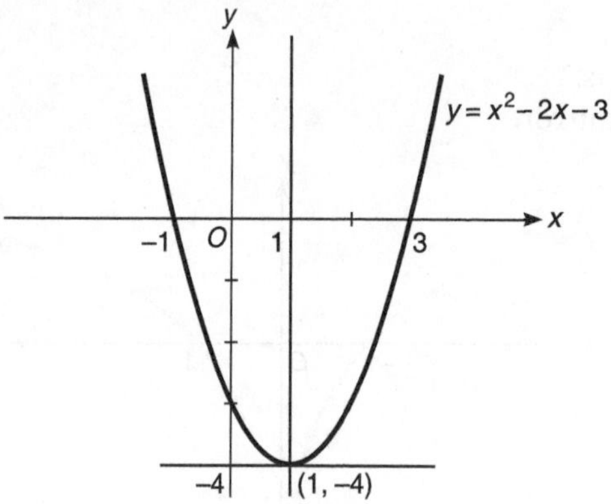

Fig. 2.64

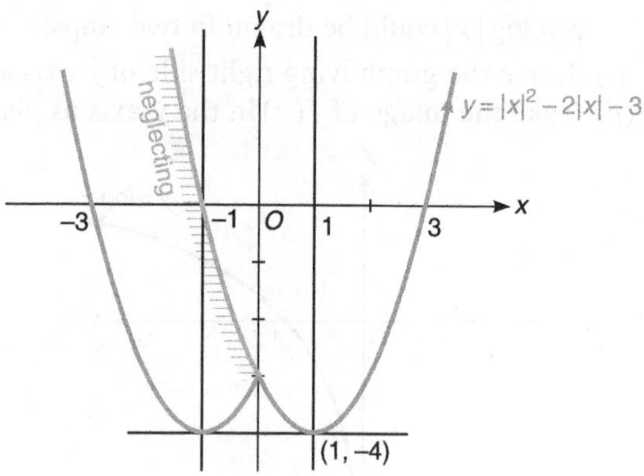

Fig. 2.65

$\therefore$ $y = f(|x|)$, *i.e.*, $y = |x|^2 - 2|x| - 3$ is to be plotted as shown in Fig. 2.65.

which shows $y = |x|^2 - 2|x| - 3$ is differentiable for all $x \in R - \{0\}$.

(x) $f(x)$ transforms to $|f(|x|)|$

i.e., $$f(x) \longrightarrow |f(|x|)|$$

Here, plot the curve in two steps;

(i) $f(x) \longrightarrow |f(x)|$ (ii) $|f(x)| \longrightarrow |f(|x|)|$

OR

(i) $f(x) \longrightarrow f(|x|)$

(ii) $f(|x|) \longrightarrow |f(|x|)|$., *i.e.*, (viii) and (ix) transformations.

Graphically it could be stated as shown in Fig. 2.66.

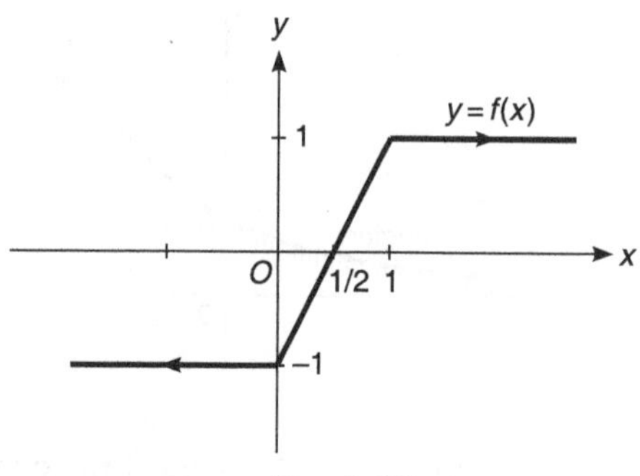

Fig. 2.66

(i) $y = |f(x)|$:

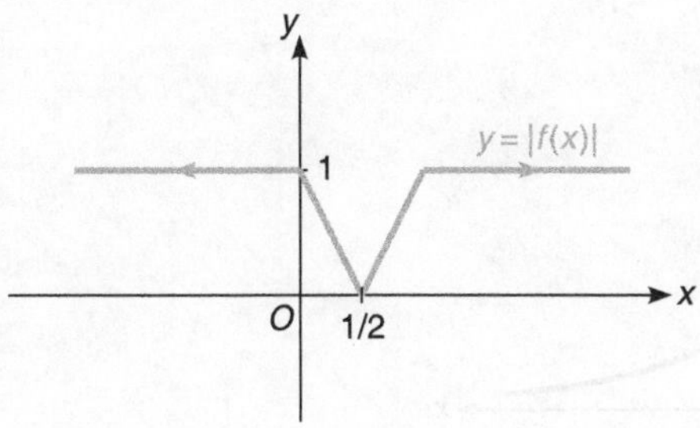

Fig. 2.67

(ii) $y = |f(|x|)|$:

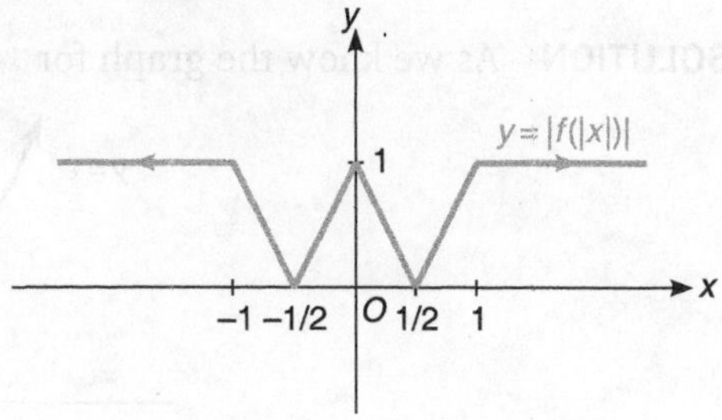

Fig. 2.68

EXAMPLE 1 *Sketch the curve for* $y = ||x|^2 - 2|x| - 3|$.

SOLUTION As we know the graph for $y = x^2 - 2x - 3$, shown as;

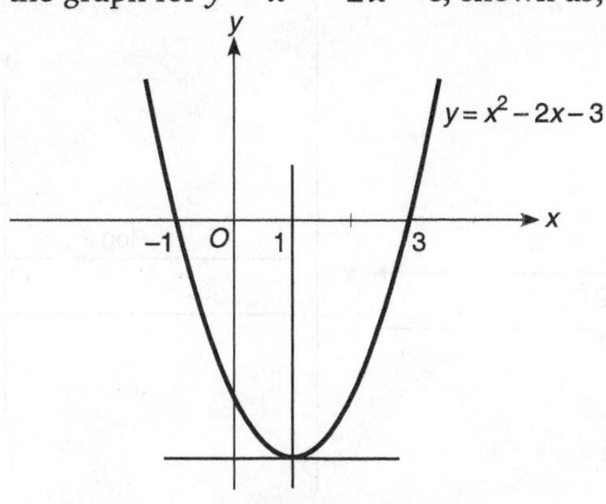

Fig. 2.69

(i) $y = x^2 - 2x - 3 \rightarrow y = |x|^2 - 2|x| - 3.$

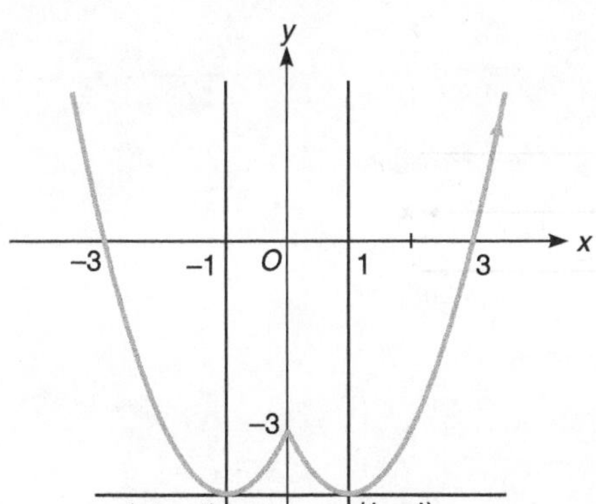

Fig. 2.70

(ii) $y = |x|^2 - 2|x| - 3 \rightarrow y = ||x|^2 - 2|x| - 3|$

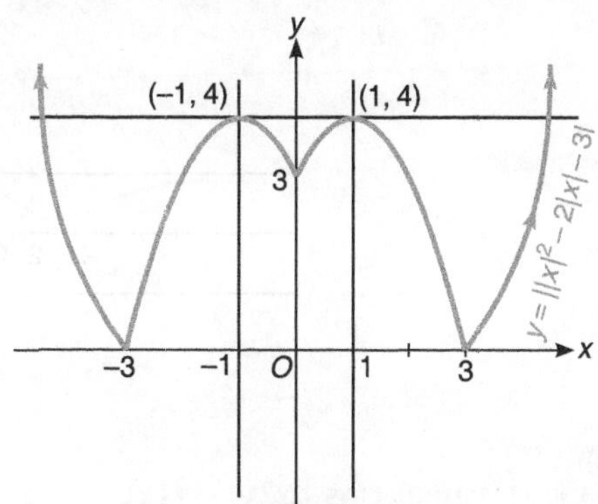

Fig. 2.71

Clearly, above figure is differentiable for all $x \in R - \{-3, 0, 3\}$.

EXAMPLE 2 *Sketch the graph for* $y = \left| e^{-|x|} - \frac{1}{2} \right|$.

SOLUTION As we know the graph for $y = e^{-x}$.

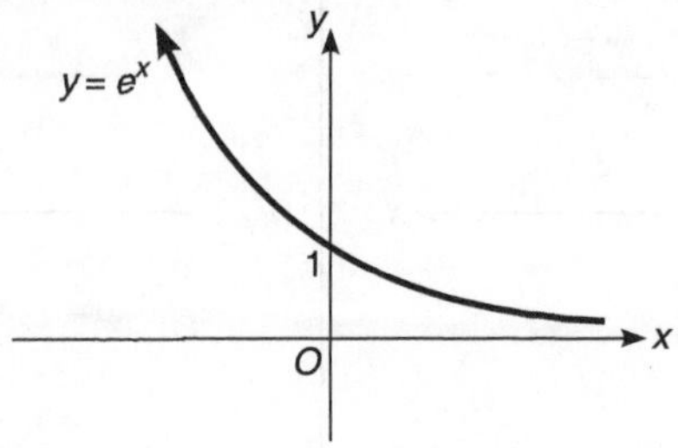

Fig. 2.72

(i) $y = e^{-x} \longrightarrow y = e^{-x} - \frac{1}{2}$.

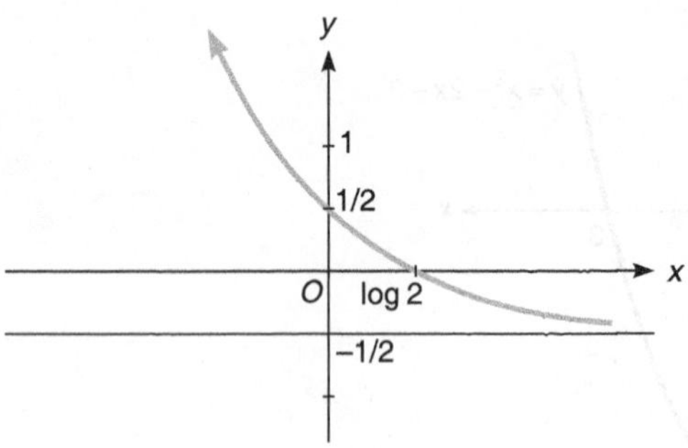

Fig. 2.73

(ii) $y = e^{-x} - \frac{1}{2} \longrightarrow y = e^{-|x|} - \frac{1}{2}$.

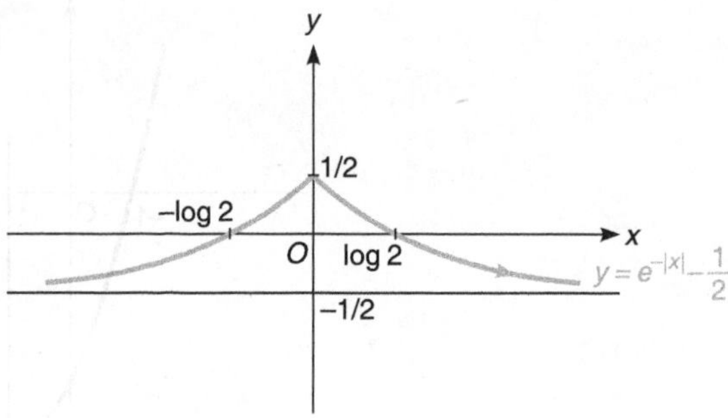

Fig. 2.74

(iii) $y = \left| e^{-|x|} - \frac{1}{2} \right|$

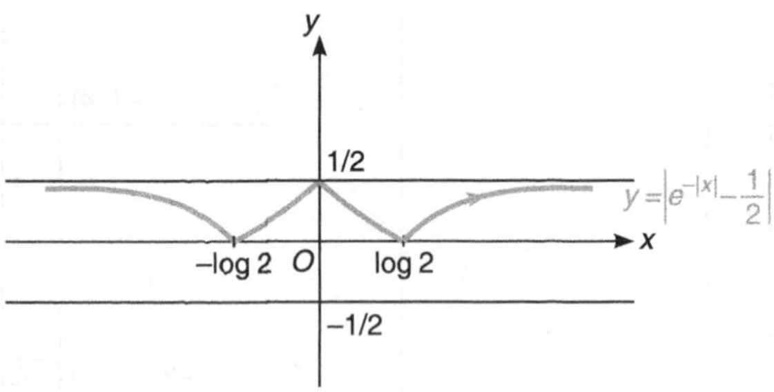

Fig. 2.75

(xi) $y = f(x)$ transforms to $|y| = f(x)$

Clearly $|y| \geq 0 \Rightarrow$ if $f(x) < 0$; graph of $|y| = f(x)$ would not exist.

if $f(x) \geq 0$; $|y| = f(x)$ would be given as $y = \pm f(x)$.

Hence, the graph of $|y| = f(x)$ exists only in the regions where $f(x)$ is non-negative and will be reflected about x-axis only when $f(x) \geq 0$. "Region where $f(x) < 0$ is neglected".

OR

(i) Remove (or neglect) the portion of the graph which lies below x-axis.
(ii) Plot the remaining portion of the graph, and also its mirror image in the x-axis.
Graphically it could be stated as shown in Fig. 2.76.

Graph for $y = f(x)$:

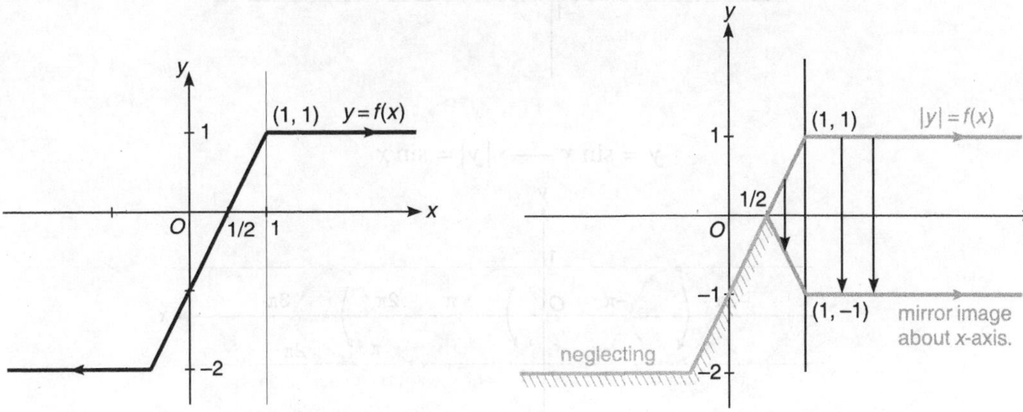

Fig. 2.76

EXAMPLE 1 *Sketch the curve* $|y| = (x - 1)(x - 2)$.

SOLUTION

As we know the graph for $y = (x - 1)(x - 2)$, is shown in Fig. 2.77.

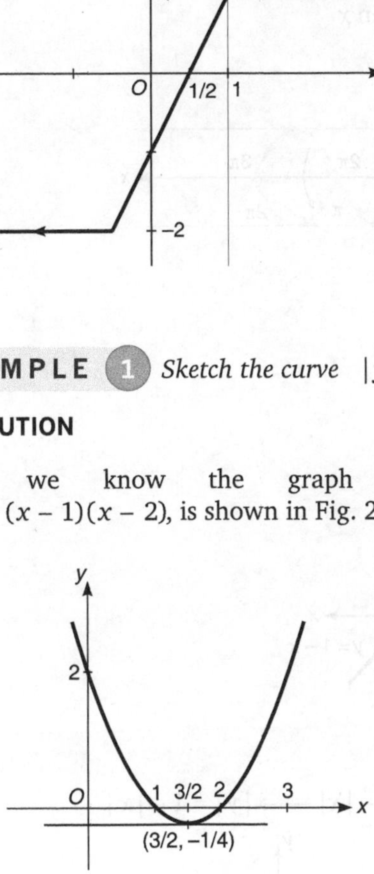

Fig. 2.77

$\Rightarrow y = (x - 1)(x - 2) \longrightarrow |y| = (x - 1)(x - 2)$, as shown in Fig. 2.78.

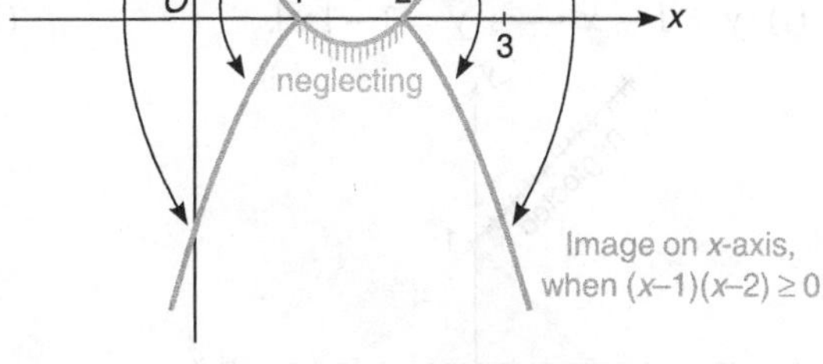

Fig. 2.78

EXAMPLE 2 *Plot the curve* $|y| = \sin x$.

SOLUTION Here, we know the curve for $y = \sin x$.

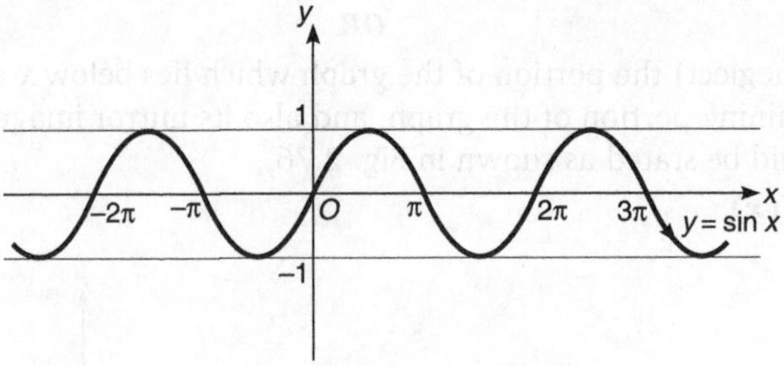

Fig. 2.79

$$\therefore \quad y = \sin x \longrightarrow |y| = \sin x$$

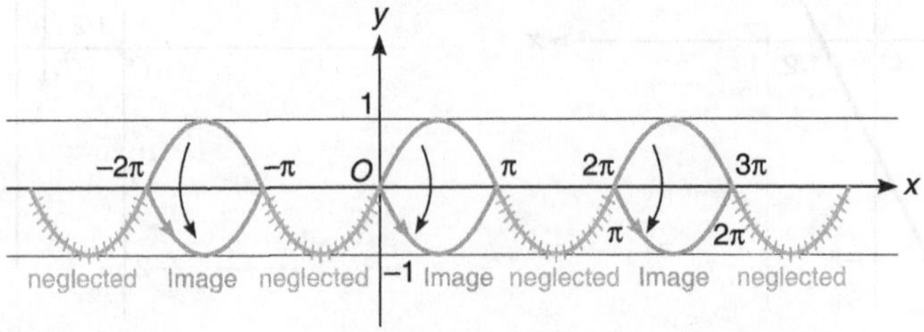

Fig. 2.80

EXAMPLE 3 *Sketch the curve* $|x| + |y| = 1$.

SOLUTION As the graph for $y = 1 - x$ is;

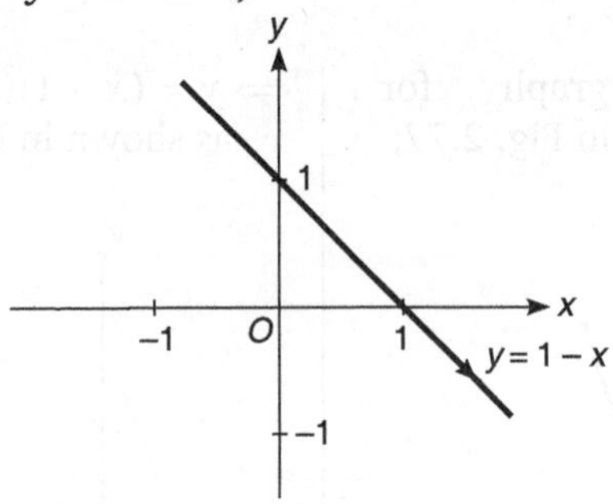

Fig. 2.81

(i) $y = 1 - x \longrightarrow y = 1 - |x|$.

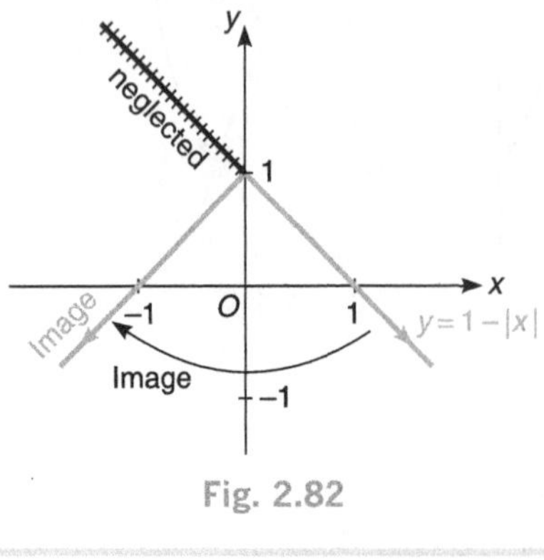

Fig. 2.82

(ii) $y = 1 - |x| \longrightarrow |y| = 1 - |x|$.

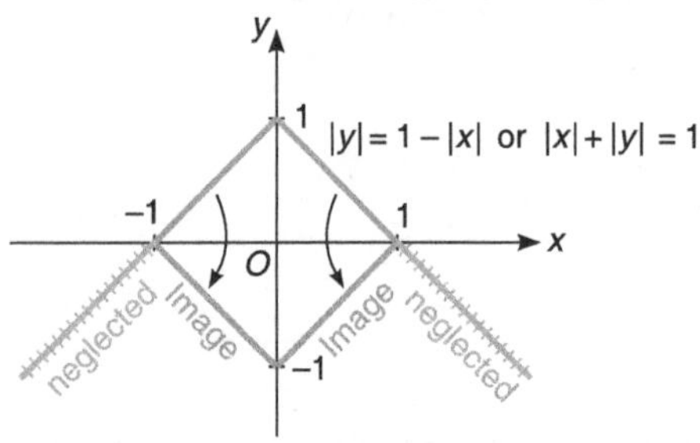

Fig. 2.83

Clearly above figure represents a square.

EXAMPLE 4 *Sketch the curve* $|x| - |y| = 1$.

SOLUTION As the graph for $y = x - 1$ is known;

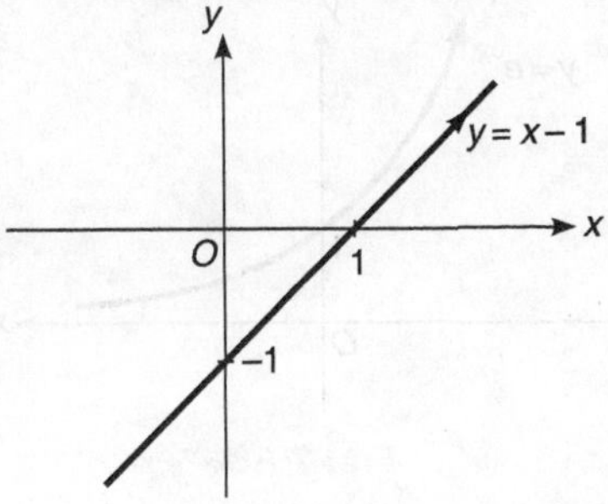

Fit. 2.84

(i) $y = x - 1 \longrightarrow y = |x| - 1$

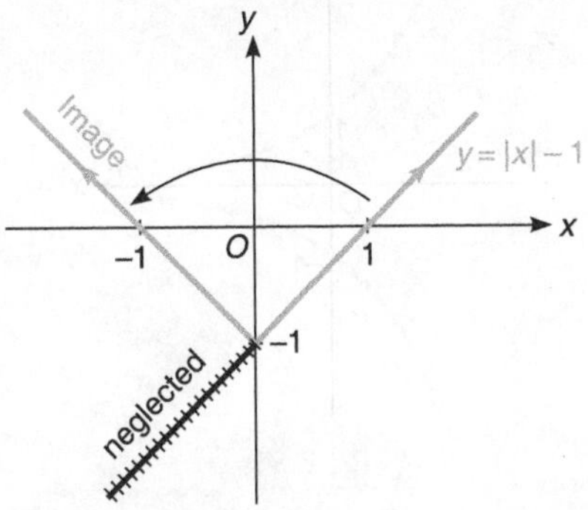

Fig. 2.85

(ii) $y = |x| - 1 \longrightarrow |y| = |x| - 1$

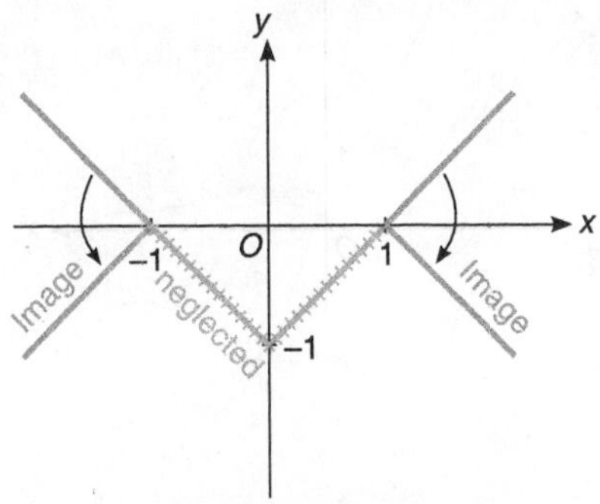

Fig. 2.86

(xii) $y = f(x)$ transforms to $|y| = |f(x)|$

i.e., $y = f(x) \longrightarrow |y| = |f(x)|$; is plotted in two steps.

(i) $y = f(x) \longrightarrow y = |f(x)|$

(ii) $y = |f(x)| \longrightarrow |y| = |f(x)|$

Graphically it could be stated as;

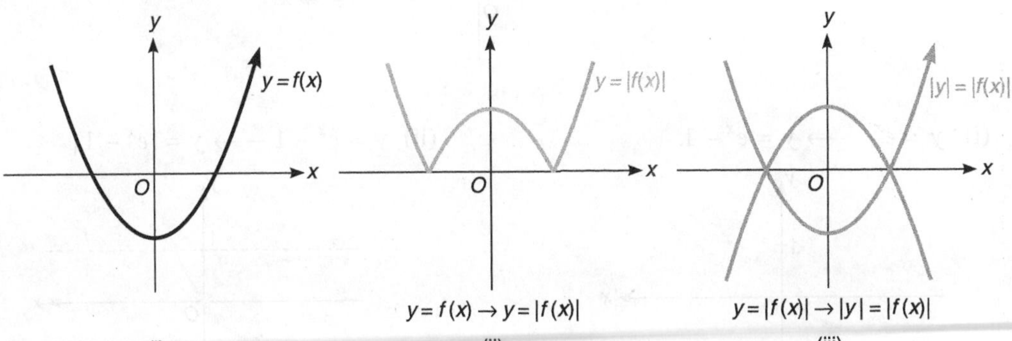

Fig. 2.87

Curvature and Transformations

EXAMPLE 1 *Plot the curve for* $|y| = |e^{-x}|$.

SOLUTION Here; curve for $y = e^{-x}$ is shown as;

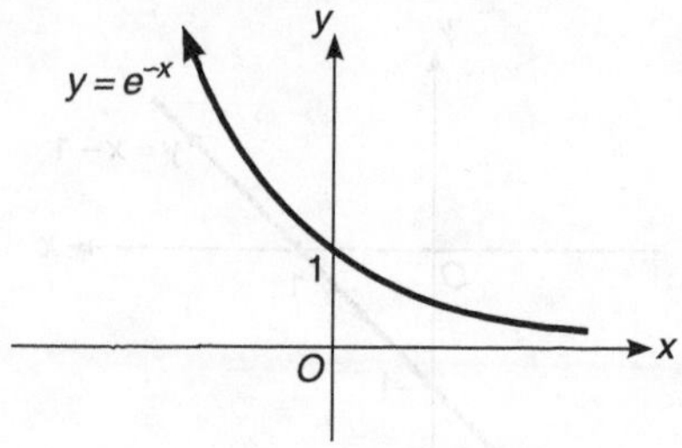

Fig. 2.88

(i) $y = e^{-x} \longrightarrow y = |e^{-x}|$

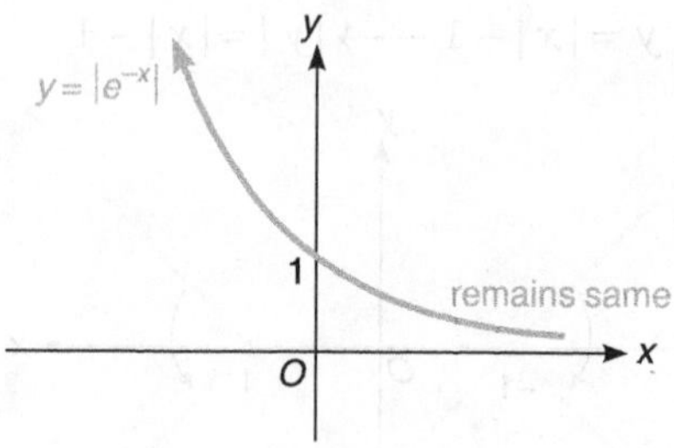

Fig. 2.89

(ii) $y = |e^{-x}| \longrightarrow |y| = |e^{-x}|$.

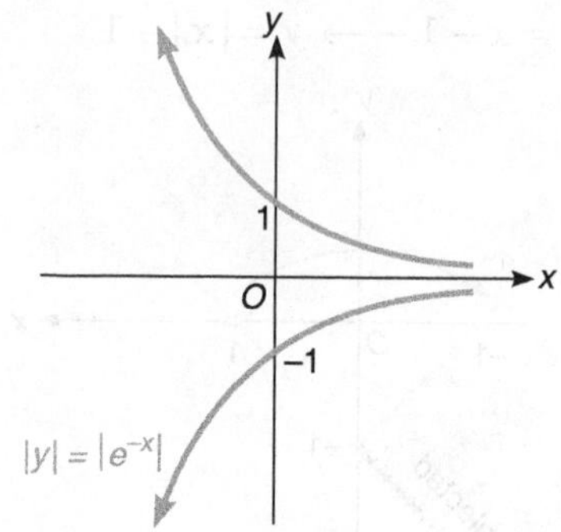

Fig. 2.90

EXAMPLE 2 *Plot the curve* $|y| = |e^x - 1|$.

SOLUTION As we know the curve for $y = e^x$ is shown as;

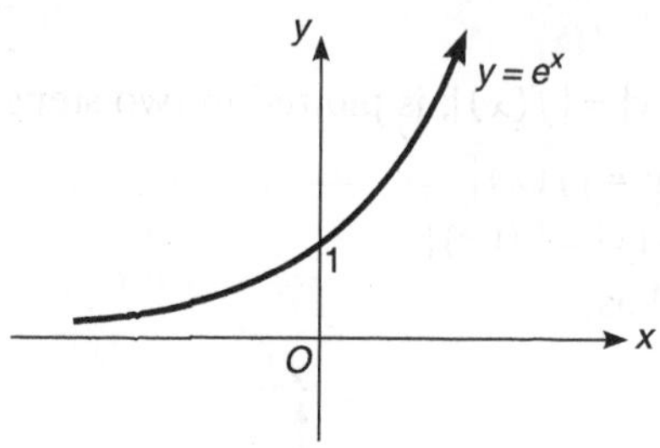

Fig. 2.91

(i) $y = e^x \longrightarrow y = e^x - 1$.

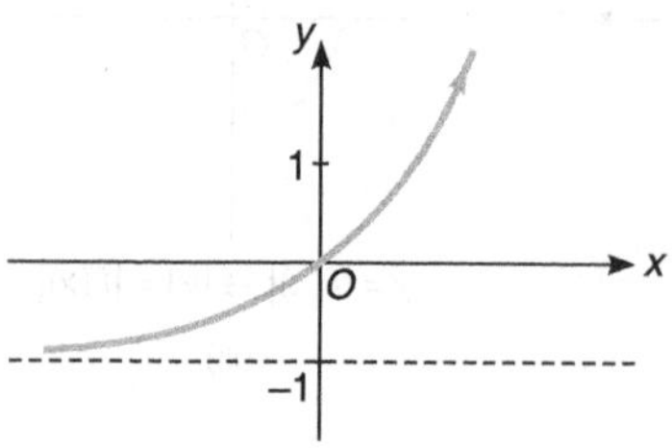

Fig. 2.92

(ii) $y = e^x - 1 \longrightarrow y = |e^x - 1|$.

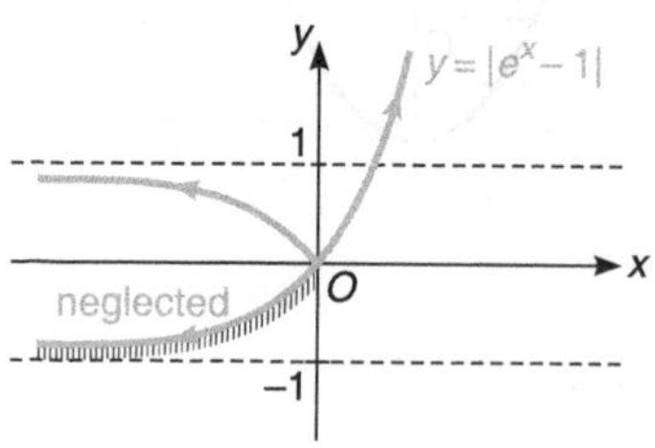

Fig. 2.93

(iii) $y = |e^x - 1| \longrightarrow |y| = |e^x - 1|$.

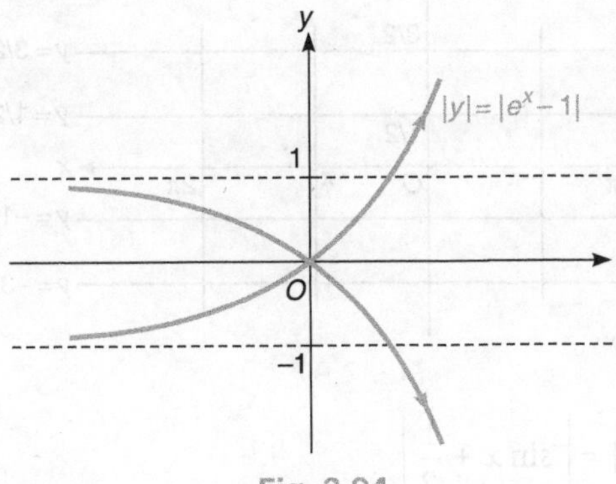

Fig. 2.94

EXAMPLE 3 *Plot the curve* $|y| = \left|\sin x + \dfrac{1}{2}\right|$.

SOLUTION Here; we know the graph for $y = \sin x$, is shown as

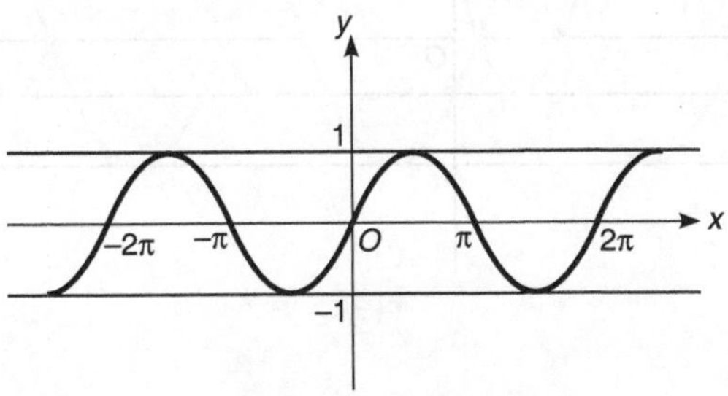

Fig. 2.95

(i) $y = \sin x \longrightarrow y = \sin x + \dfrac{1}{2}$.

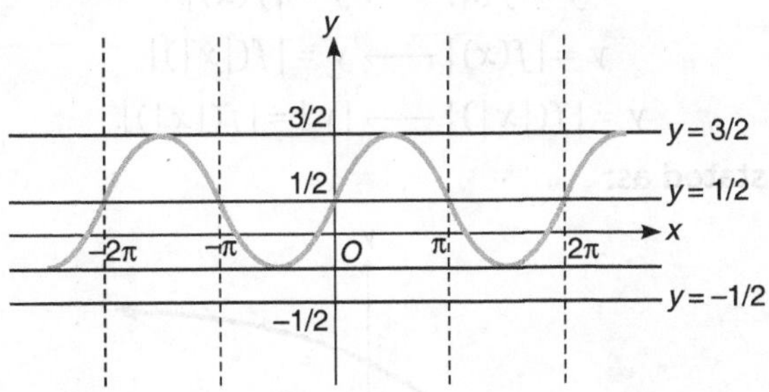

Fig. 2.96

(ii) $y = \sin x + \dfrac{1}{2} \longrightarrow y = \left|\sin x + \dfrac{1}{2}\right|$

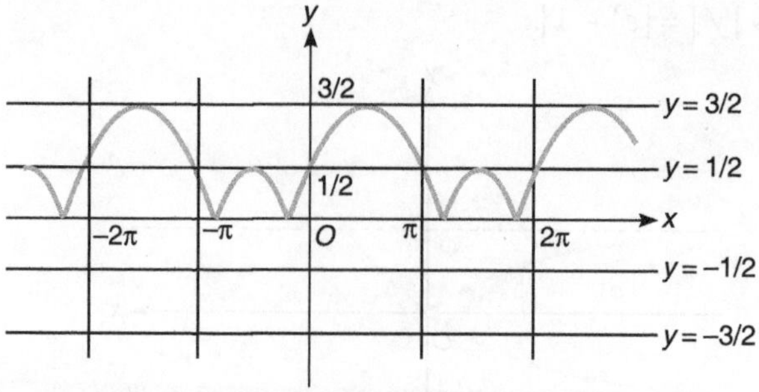

Fig. 2.97

(iii) $y = \left|\sin x + \frac{1}{2}\right| \longrightarrow |y| = \left|\sin x + \frac{1}{2}\right|$

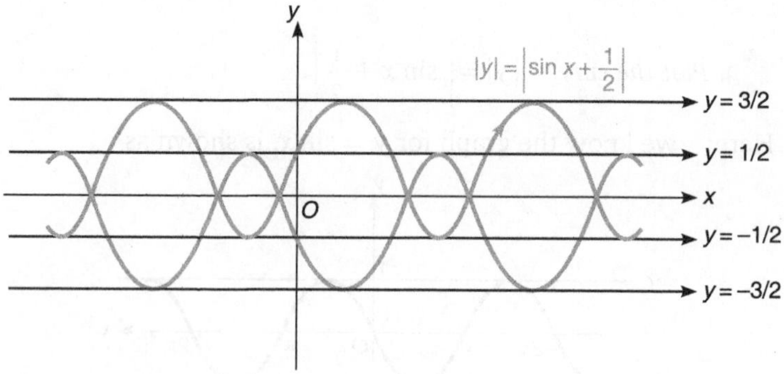

Fig. 2.98

(xiii) $y = f(x)$ transforms to $|y| = |f(|x|)|$

i.e., $$y = f(x) \longrightarrow |y| = |f(x)|.$$

The steps followed are:

(i) $$y = f(x) \longrightarrow y = |f(x)|.$$

(ii) $$y = |f(x)| \longrightarrow y = |f(|x|)|$$

(iii) $$y = |f(|x|)| \longrightarrow |y| = |f(|x|)|.$$

Graphically it could be stated as:

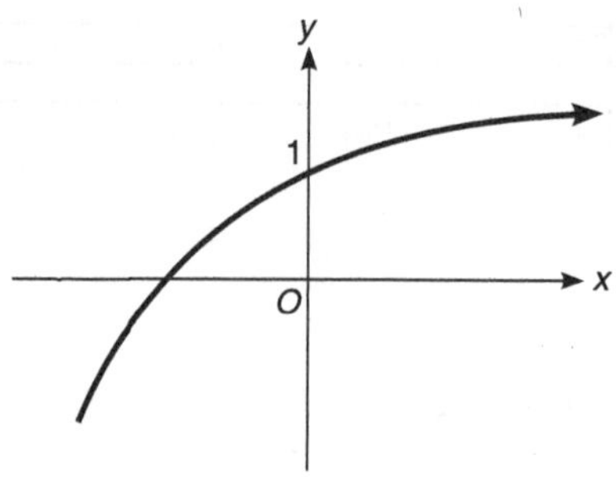

Fig. 2.99

(i) $y = f(x) \longrightarrow y = |f(x)|$

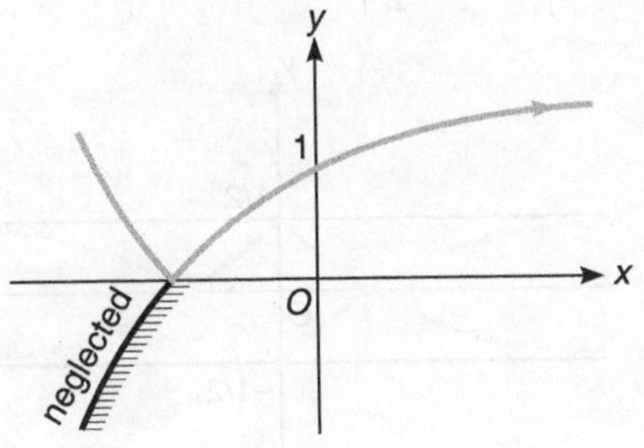

Fig. 2.100

(ii) $y = |f(x)| \longrightarrow y = |f(|x|)|$

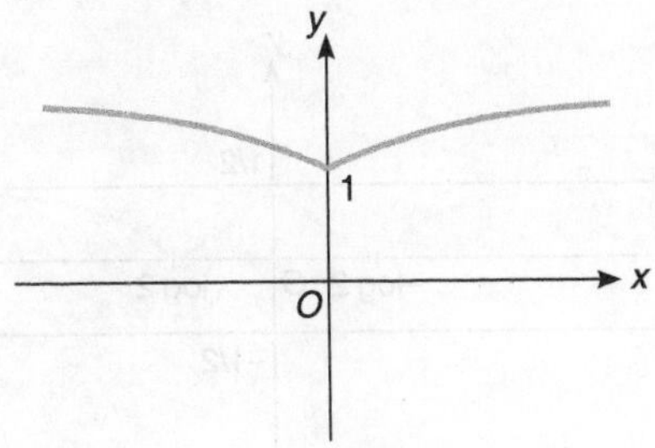

Fig. 2.101

(iii) $y = |f(|x|)| \longrightarrow |y| = |f(|x|)|$.

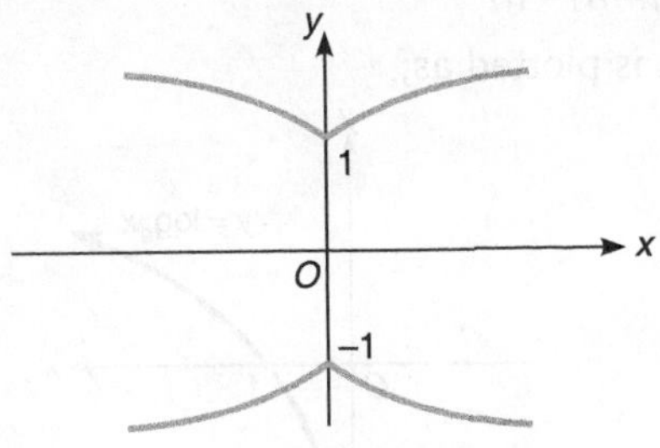

Fig. 2.102

EXAMPLE 1 *Plot the* $|y| = \left| e^{-|x|} - \dfrac{1}{2} \right|$.

SOLUTION Here; we know the graph for $y = e^{-x}$.

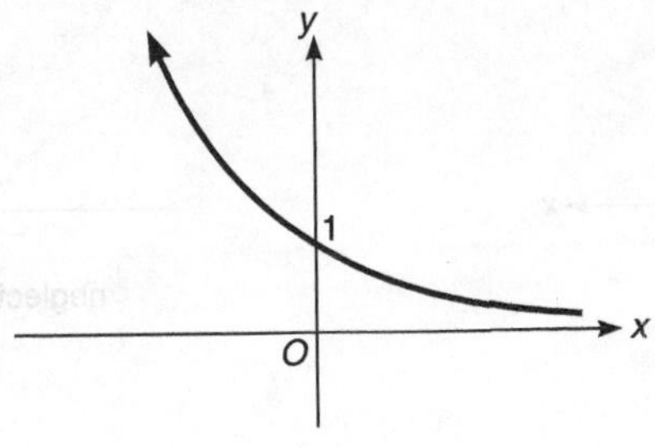

Fig. 2.103

(i) $y = e^{-x} \longrightarrow y = e^{-x} - \dfrac{1}{2}$

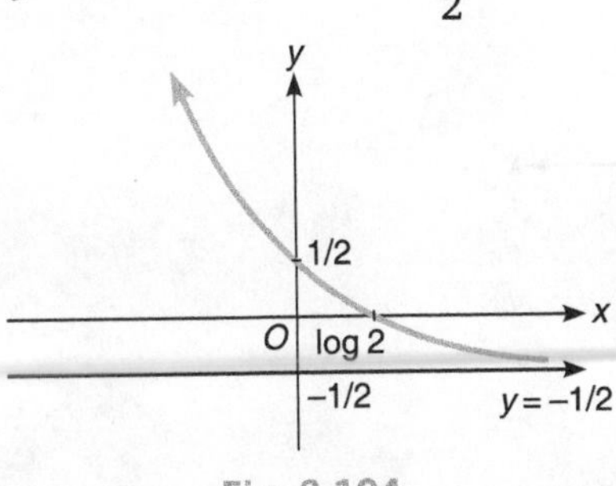

Fig. 2.104

(ii) $y = e^{-x} - \dfrac{1}{2} \longrightarrow y = e^{-|x|} - \dfrac{1}{2}$.

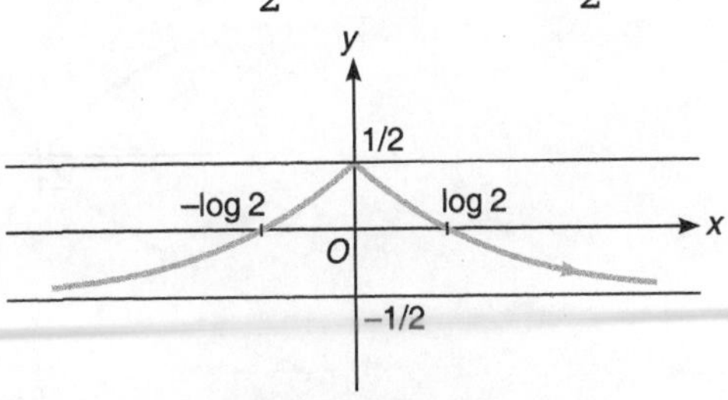

Fig. 2.105

(iii) $y = e^{-|x|} - \dfrac{1}{2} \longrightarrow y = \left| e^{-|x|} - \dfrac{1}{2} \right|$.

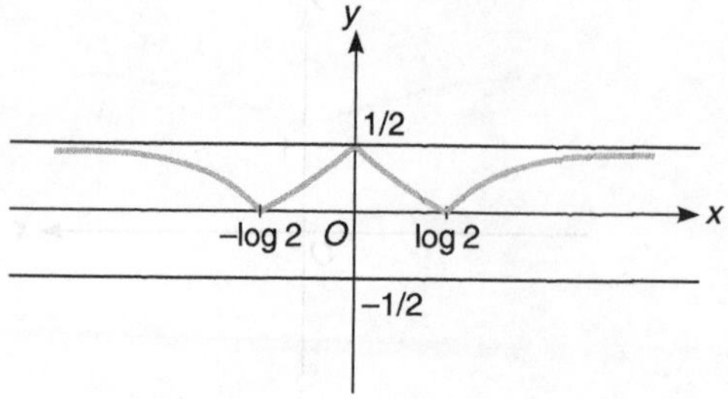

Fig. 2.106

(iv) $y = \left| e^{-|x|} - \dfrac{1}{2} \right| \longrightarrow |y| = \left| e^{-|x|} - \dfrac{1}{2} \right|$.

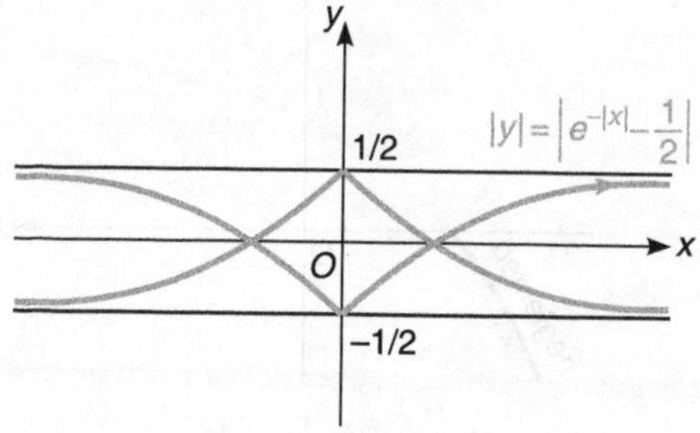

Fig. 2.107

EXAMPLE 2 *Plot* $|y| = |\log|x||$.

SOLUTION Here; $y = \log x$ is plotted as;

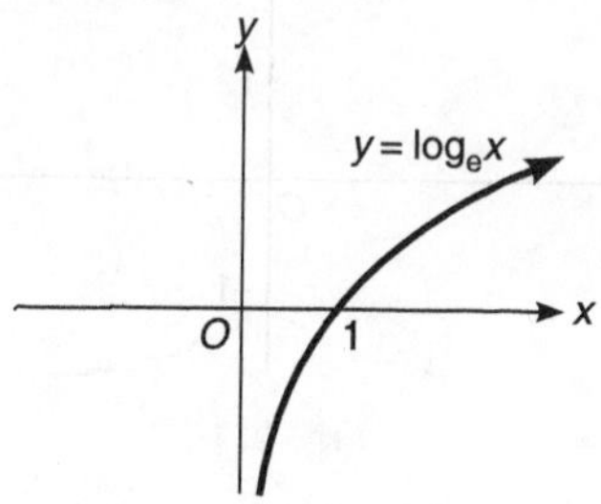

Fig. 2.108

(i) $y = \log x \longrightarrow y = \log|x|$

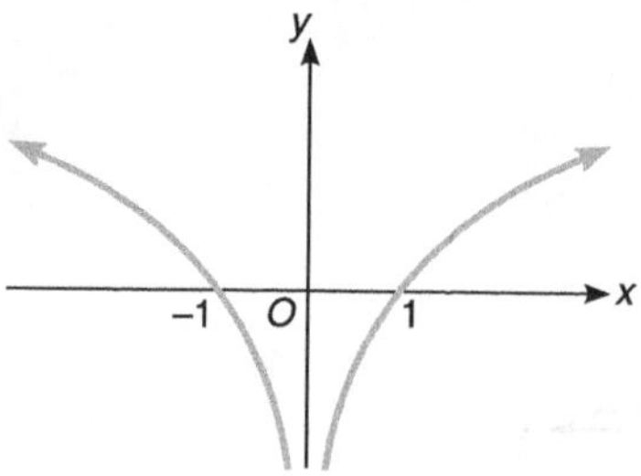

Fig. 2.109

(ii) $y = \log|x| \longrightarrow y = |\log|x||$

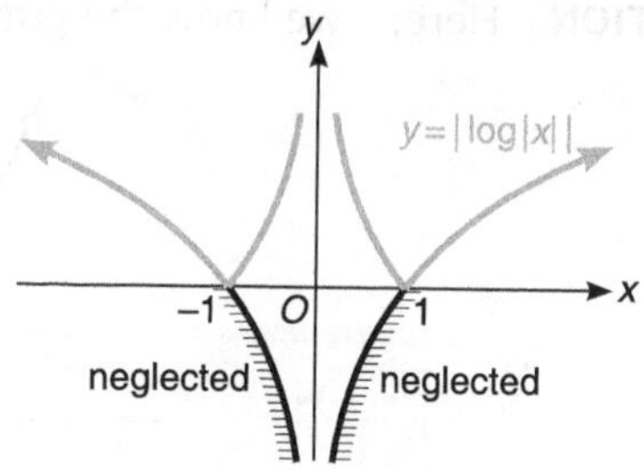

Fig. 2.110

(iii) $y = |\log|x|| \longrightarrow |y| = |\log|x||$

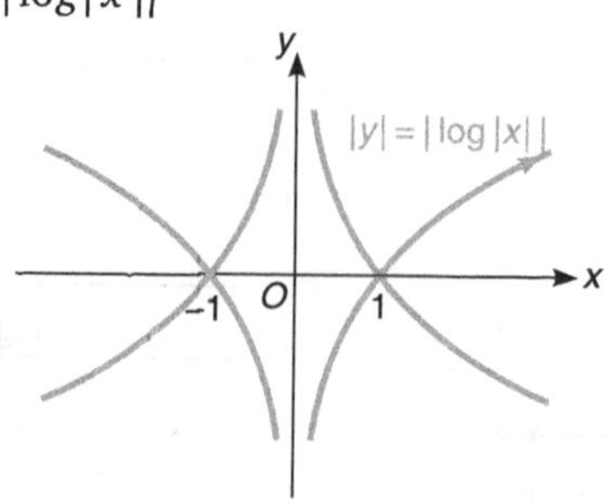

Fig. 2.111

EXAMPLE 3 *Sketch the curve* $|y| = ||x|^2 - 3|x| - 2|$.

SOLUTION As we know the graph for $y = x^2 - 3x - 2$.

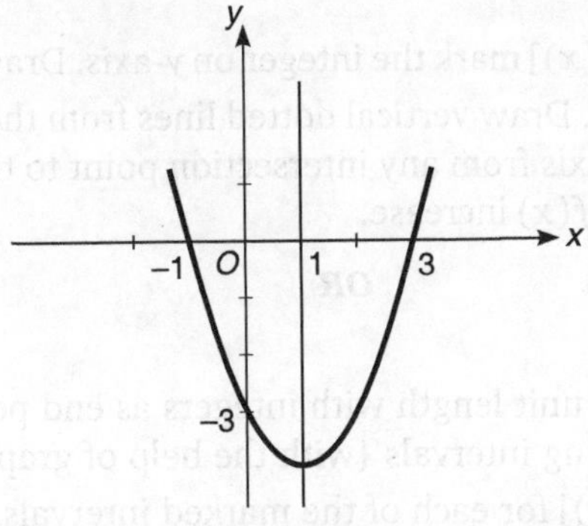

Fig. 2.112

(i) $y = x^2 - 3x - 2 \rightarrow y = |x|^2 - 3|x| - 2$ (ii) $y = |x|^2 - 3|x| - 2 \rightarrow y = ||x|^2 - 3|x| - 2|$.

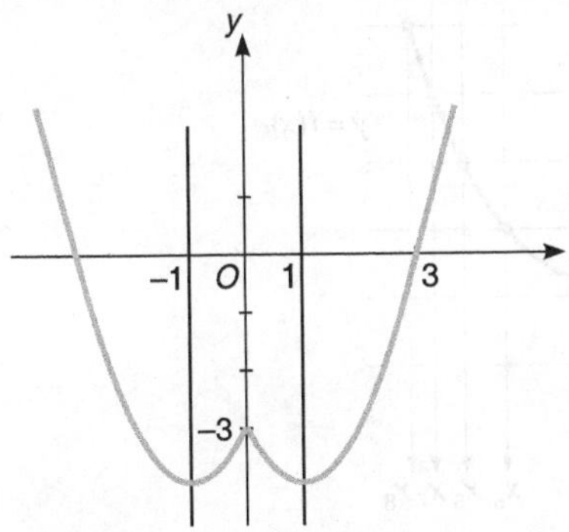

Fig. 2.113

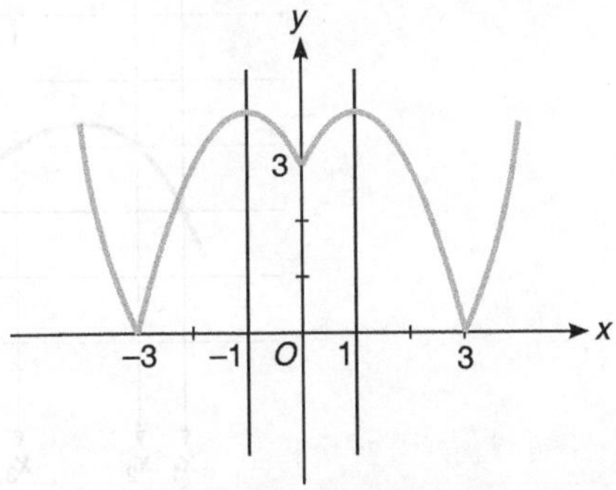

Fig. 2.114

(iii) $|y| = ||x|^2 - 3|x| - 2|$

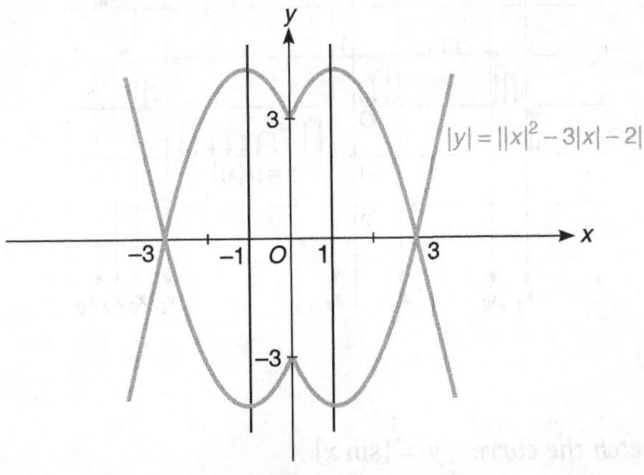

Fig. 2.115

(xiv) $y = f(x)$ transforms to $y = [f(x)]$; (where [•] denotes the greatest integer function)

i.e., $$f(x) \longrightarrow [f(x)]$$

Here; in order to draw $y = [f(x)]$ mark the integer on y-axis. Draw the horizontal lines through integers till they intersect the graph. Draw vertical dotted lines from these intersection points; finally draw horizontal lines parallel to x-axis from any intersection point to the nearest vertical dotted line with blank dot at right end in case $f(x)$ increase.

OR

Step 1. Plot $f(x)$.

Step 2. Mark the intervals of unit length with integers as end points on y-axis.

Step 3. Mark the corresponding intervals {with the help of graph of $f(x)$} on x-axis.

Step 4. Plot the value of $[f(x)]$ for each of the marked intervals.

Graphically it could be shown as:

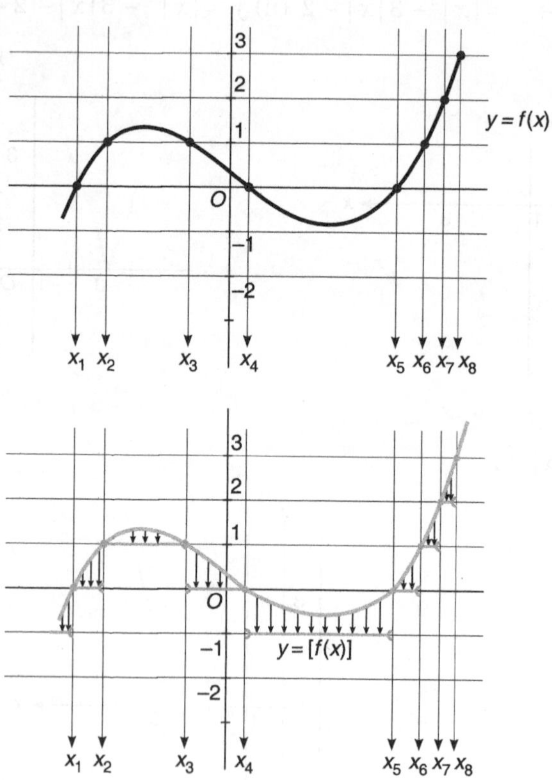

Fig. 2.116

EXAMPLE 1 *Sketch the curve* $y = [\sin x]$.

SOLUTION Here, sketch for $y = \sin x$ is shown as in Fig. 2.117.

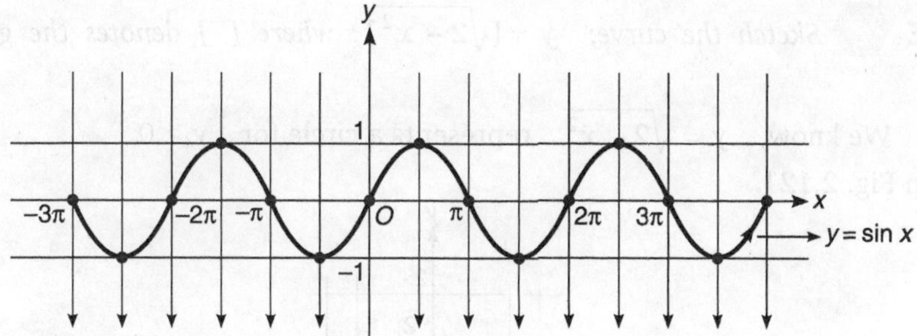

Fig. 2.117

$$y = \sin x \rightarrow y = [\sin x]$$

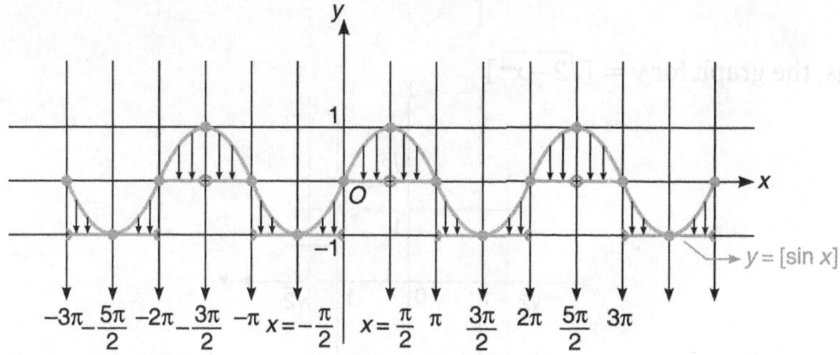

Fig. 2.118

EXAMPLE 2 *Sketch the curve $y = [x^2 - 1]$. (where $[\cdot]$ denotes greatest integer function). When $-2 \le x \le 2$.*

SOLUTION Here $y = x^2 - 1$ could be plotted as shown in Fig. 2.119.

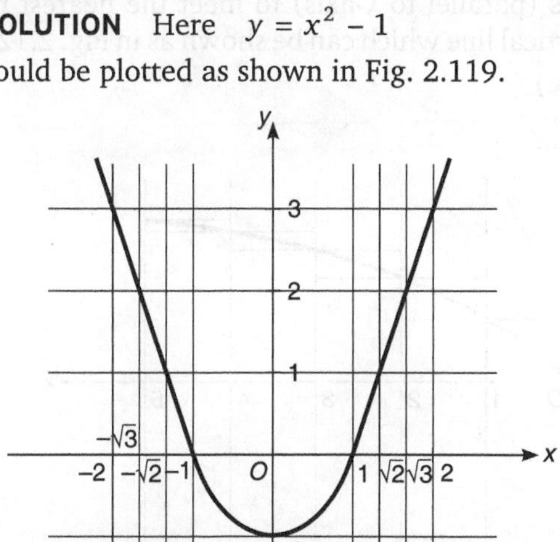

Fig. 2.119

(i) $y = x^2 - 1 \longrightarrow y = [x^2 - 1]$.

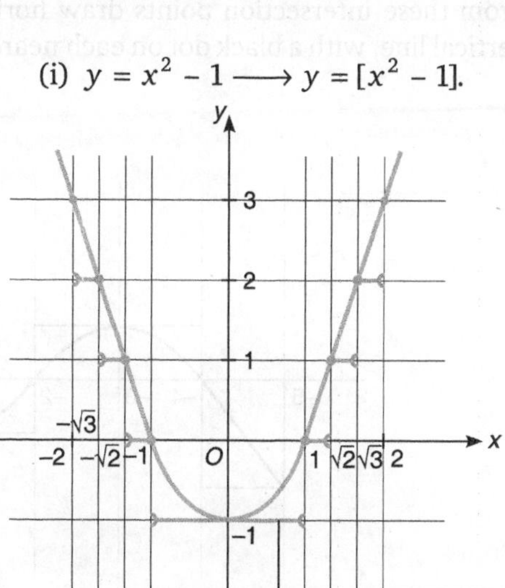

Fig. 2.120

EXAMPLE 3 *Sketch the curve;* $y = [\sqrt{2 - x^2}]$; *where* $[\cdot]$ *denotes the greatest integer function.*

SOLUTION We know, $y = \sqrt{2 - x^2}$ represents a circle for $y \geq 0$.
Shown as in Fig. 2.121.

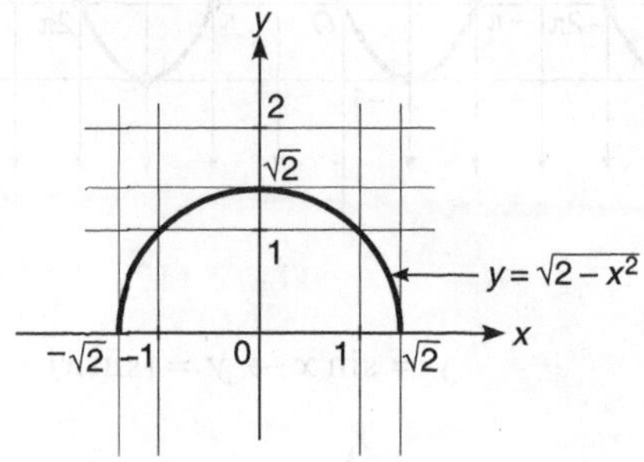

Thus, the graph for $y = [\sqrt{2 - x^2}]$

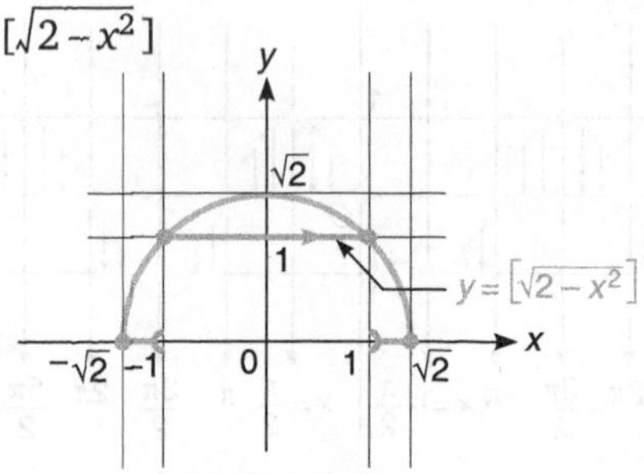

Fig. 2.121

(xv) $y = f(x)$ transforms to $y = f([x])$

Here, mark the integers on the x-axis. Draw vertical lines till they intersect the graph of $f(x)$. From these intersection points draw horizontal lines (parallel to x-axis) to meet the nearest right vertical line, with a black dot on each nearest right vertical line which can be shown as in Fig. 2.122.

$y = f(x)$

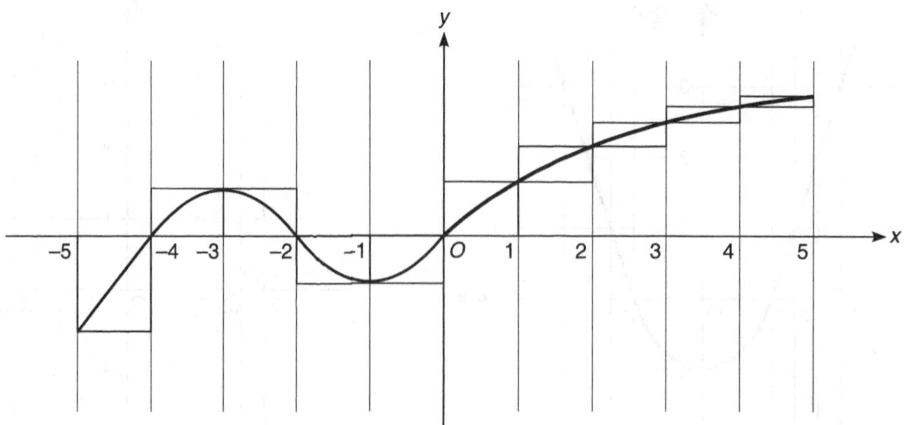

Fig. 2.122

$y = f([x])$

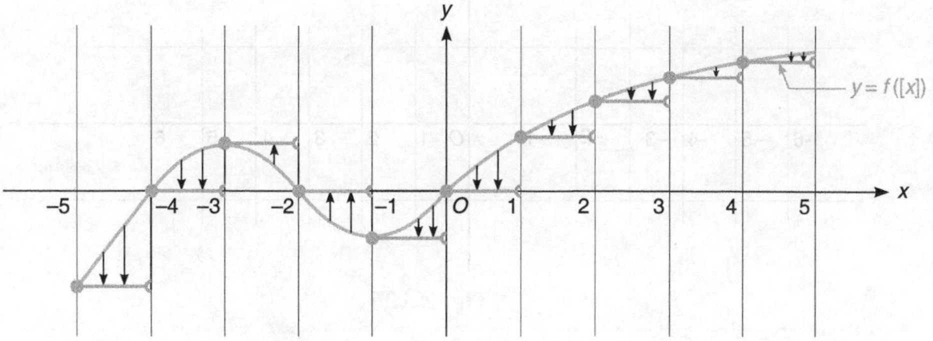

Fig. 2.123

OR

$$y = f(x) \longrightarrow y = f([x])$$

Step 1. Plot the straight lines parallel to y-axis for integral values of x (say – 3, – 2, – 1, 0, 1, 2, 3, ...)

Step 2. Now mark the points at which $x = -3$, $x = -2$, $x = -1$, $x = 0$, $x = 1, \ldots$ on the curve.

Step 3. Take the lower marked point for x say if $n < x < n + 1$ then take the point at $x = n$ and draw a horizontal line to the nearest vertical line formed by $x = n + 1$, proceeding in this way we get required curve.

EXAMPLE 1 *Plot the curve* $y = e^{[x]}$.

SOLUTION Here the graph for $y = e^{[x]}$ is shown as;

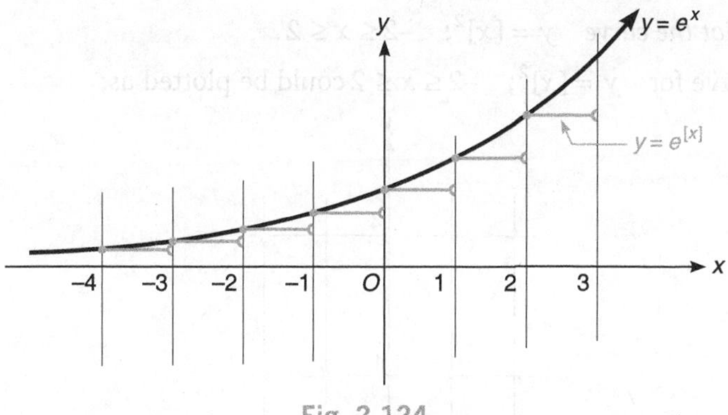

Fig. 2.124

EXAMPLE 2 *Sketch the curve* $y = \sin[x]$ *when* $-2\pi \le x \le 2\pi$.

SOLUTION The curve for

$y = \sin[x]$; could be plotted as shown in Fig. 2.125.

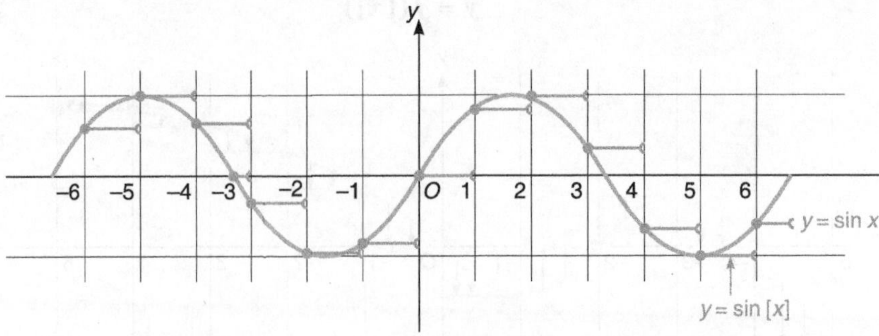

Fig. 2.125

EXAMPLE 3 *Sketch the curve for* $y = \cos[x]$; $-\pi \le x \le \pi$.

SOLUTION The curve for $y = \cos[x]$ could be plotted as;

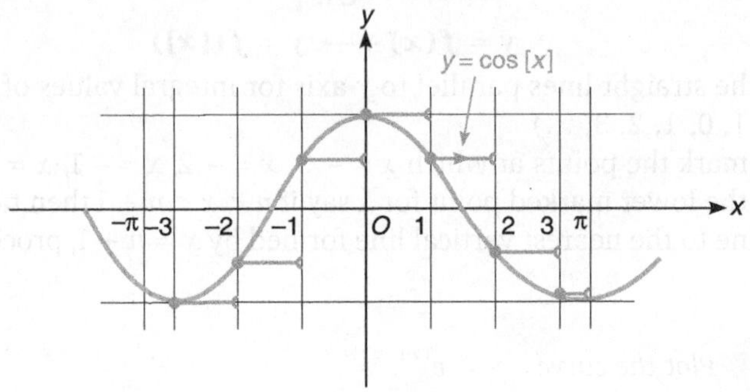

Fig. 2.126

EXAMPLE 4 *Plot the curve* $y = [x]^2$; $-2 \le x \le 2$.

SOLUTION The curve for $y = [x]^2$; $-2 \le x \le 2$ could be plotted as;

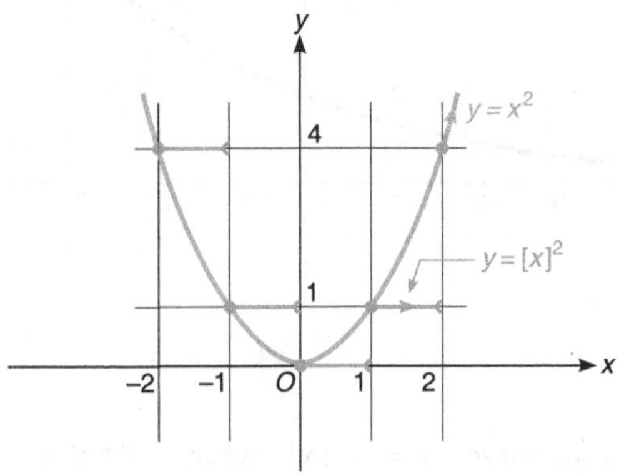

Fig. 2.127

(xvi) $y = f(x)$ transforms to $y = [f([x])]$

Here, we should follow two steps;

(i) $y = f(x) \longrightarrow y = f([x])$

(ii) $y = f([x]) \longrightarrow y = [f([x])]$

EXAMPLE 1 *Sketch the curve $y = [\sin[x]]$; where $[\cdot]$ denotes the greatest integral function when $0 \le x \le \pi$.*

SOLUTION Here; first we shall plot the curve for $y = \sin[x]$, when $x \in [0, \pi]$.

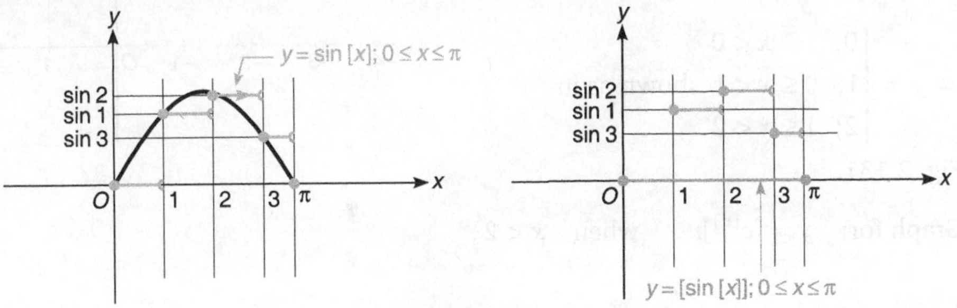

Fig. 2.128

From above figure we conclude that;

when $\quad 0 \le x \le \pi \;\Rightarrow\; y = \sin[x] \in [0, 1).$

$\Rightarrow \quad y = [\sin[x]] \longrightarrow 0 \quad \text{for all} \quad 0 \le x \le \pi.$

EXAMPLE 2 *Plot the curve $y = [e^{[x]}]$; when $-4 < x < 2$.*

SOLUTION Here to sketch $y = [e^{[x]}]$, we should follow the steps as;

(i) $y = e^x$

(ii) $y = e^{[x]}$

(iii) $y = [e^{[x]}]$

(i) $y = e^x$ **(ii) $y = e^{[x]}$**

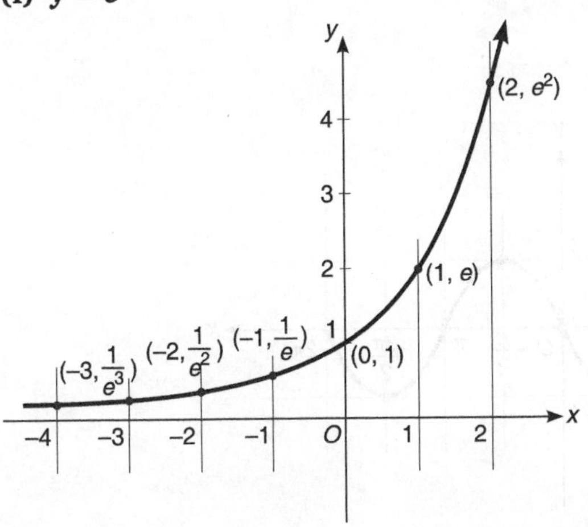

Fig. 2.129

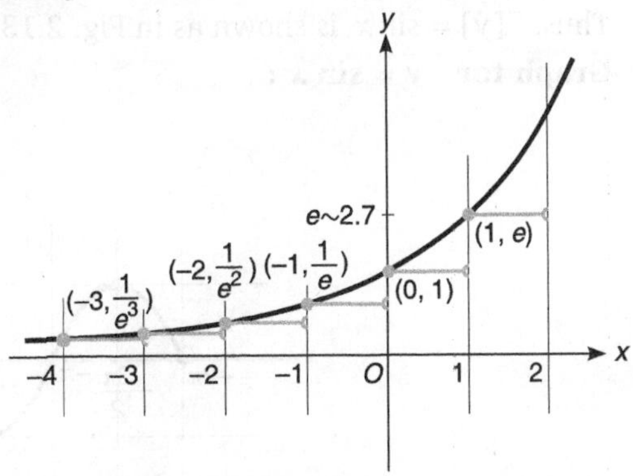

Fig. 2.130

(iii) $y = [e^{[x]}]$; from Fig. 2.130.

$$e^{[x]} = \begin{cases} 0 < e^{[x]} < 1; & x < 0 \\ e^{[x]} = 1 & ; \ 0 \le x < 1 \\ e^{[x]} = 2.7 & ; \ 1 \le x < 2 \end{cases}$$

Thus; $y = [e^{[x]}]$

$$\Rightarrow y = \begin{cases} 0; & x < 0 \\ 1; & 0 \le x < 1 \\ 2; & 1 \le x < 2 \end{cases}$$ shown as in

Fig. 2.131.

Graph for; $y = [e^{[x]}]$; when $x < 2$.

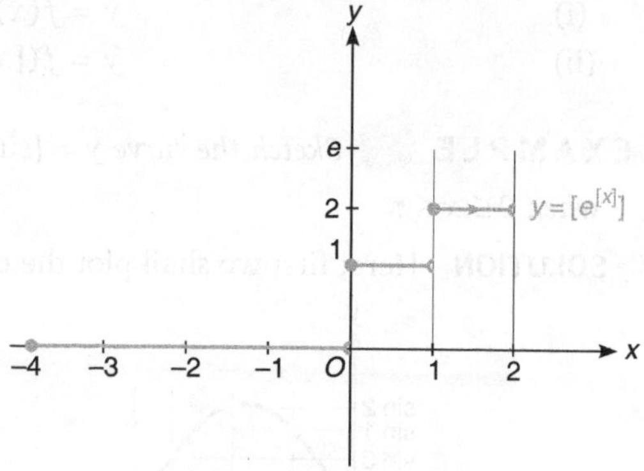

Fig. 2.131

(xvii) $y = f(x)$ transforms to $[y] = f(x)$ / Graph of Candles

Here, to plot $[y] = f(x)$; we check only those points for which $f(x) \in$ integers, as $[y] \in$ integers for all x.

Thus; $[y] = f(x)$ represents only integral values of y. Here, domain of $f(x)$ are set of values of x for which $f(x) \in$ integers.

EXAMPLE 1 *Sketch the curve;*

$$[y] = \sin x.$$

SOLUTION As we know; $-1 \le \sin x \le 1$ but since; $[y] = \sin x$.

$\Rightarrow$ $\sin x = -1, 0, 1$ are only solutions;

or $$x = \frac{n\pi}{2} \in \text{Domain of } [y] = \sin x.$$

Thus, $[y] = \sin x$ is shown as in Fig. 2.132.

Graph for $y = \sin x$:

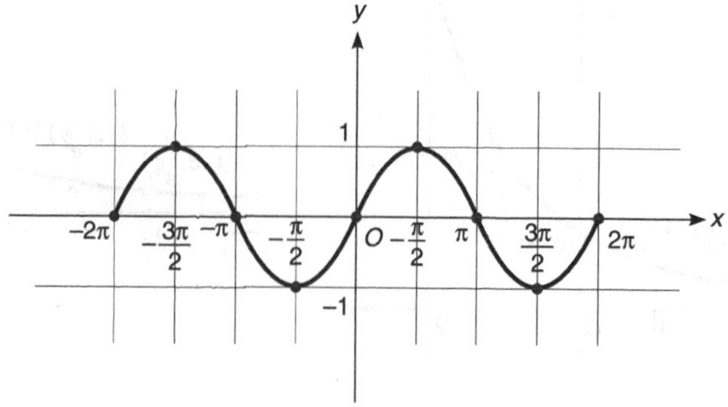

Fig. 2.132

Graph for [y] = sin x :

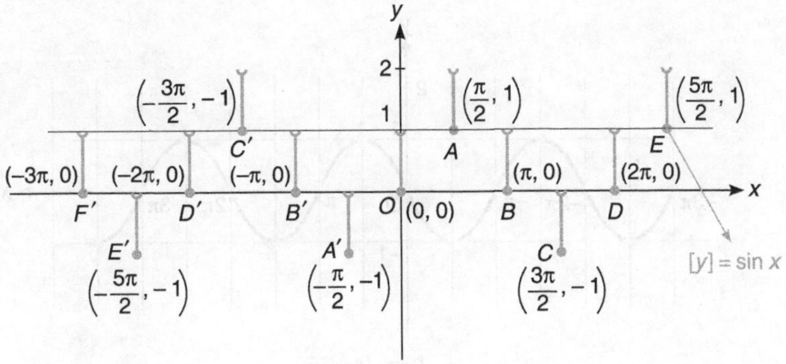

Fig. 2.133

From above figure, (the points marked $O, A, B, C, D, E, A', B', C', D', E', \ldots$ is the graph for candles), or graph for $[y] = \sin x$.

EXAMPLE 2 *Sketch the curve* $[y] = \sin^{-1} x$.

SOLUTION As we know the graph for $y = \sin^{-1} x$; shown as in Fig. 2.134 and 2.135.

Graph for $y = \sin^{-1} x$;

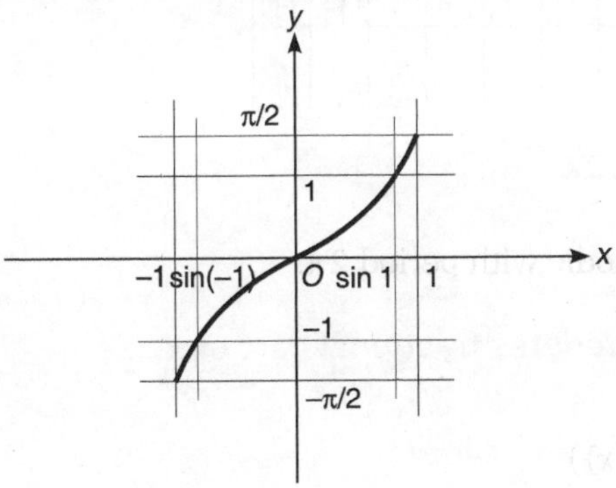

Fig. 2.134

Graph for $[y] = \sin^{-1} x$;

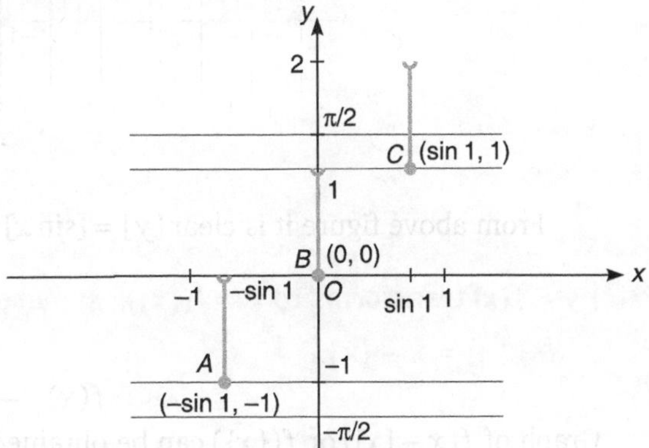

Fig. 2.135

From above figure, $[y] = \sin x$.

(xviii) $y = f(x)$ transforms to $[y] = [f(x)]$

As we have earlier discussed $y = [f(x)]$, *i.e.*, **transformation (xiv)**, and we know $[y]$ implies only those values of x for which $f(x) \in$ integer.

EXAMPLE 1 *Sketch the curve;* $[y] = [\sin x]$.

SOLUTION To sketch the curve $[y] = [\sin x]$ we first plot $y = [\sin x]$.

(i) Graph for $y = [\sin x]$

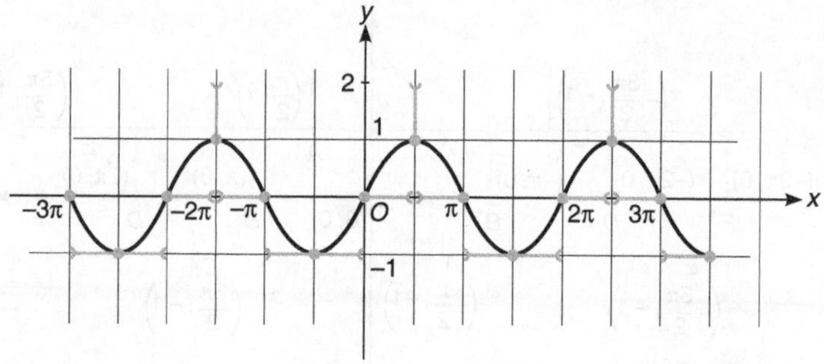

Fig. 2.136

(ii) Graph for $[y] = [\sin x]$

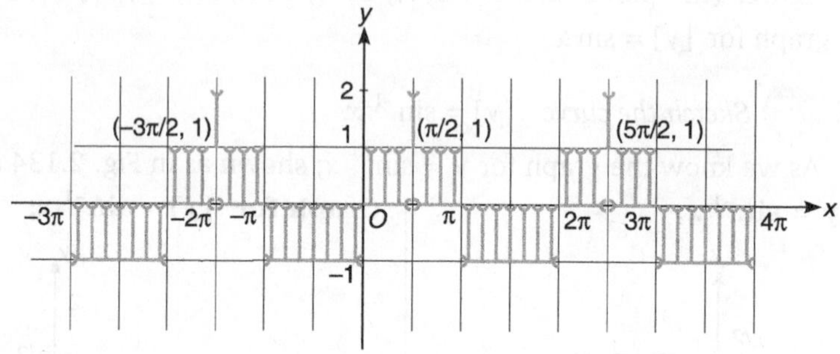

Fig. 2.137

From above figure it is clear $[y] = [\sin x]$ is periodic with period 2π.

(xix) $y = f(x)$ transforms to $y = f(\{x\})$; (where $\{\cdot\}$ denotes fractional part of x, *i.e.*, $\{x\} = x - [x]$)

$$f(x) \longrightarrow f(\{x\})$$

Graph of $f(x - [x])$ or $f(\{x\})$ can be obtained from the graph of $f(x)$ by following rule.

"Retain the graph of $f(x)$ for values of x lying between interval $[0, 1)$. Now it can be repeated for rest of the points. (taking periodicity 1).

New obtained function is graph for $y = f(\{x\})$".

Graphically it could be stated as;

Graph for $y = f(x)$

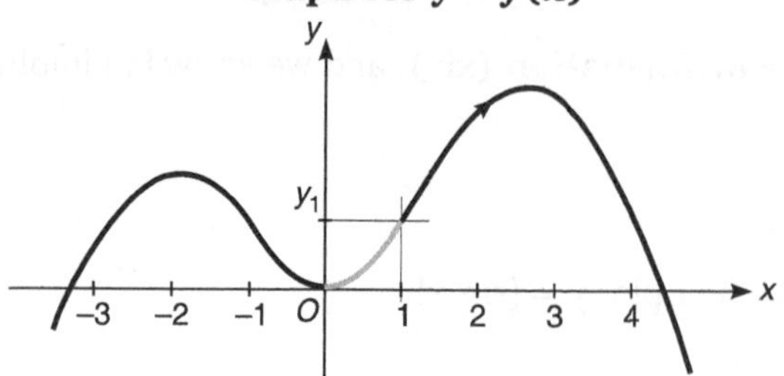

Fig. 2.138

Graph for $y = f(\{x\})$

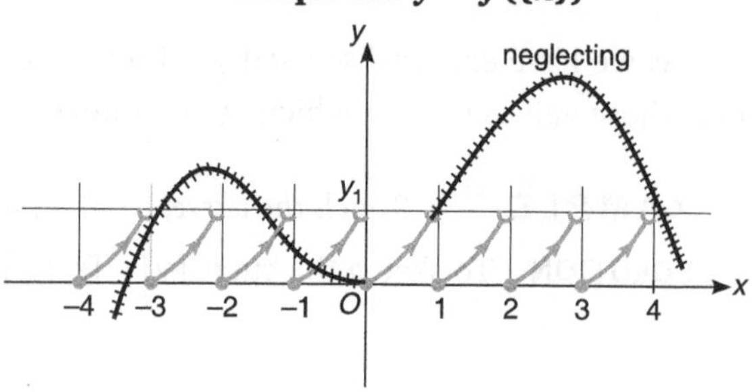

Fig. 2.139

EXAMPLE 1 *Sketch the curve* $y = (\{x\} - 1)^2$.

SOLUTION Here, we know the curve for $y = (x - 1)^2$ shown as;

Graph for $y = (x - 1)^2$:

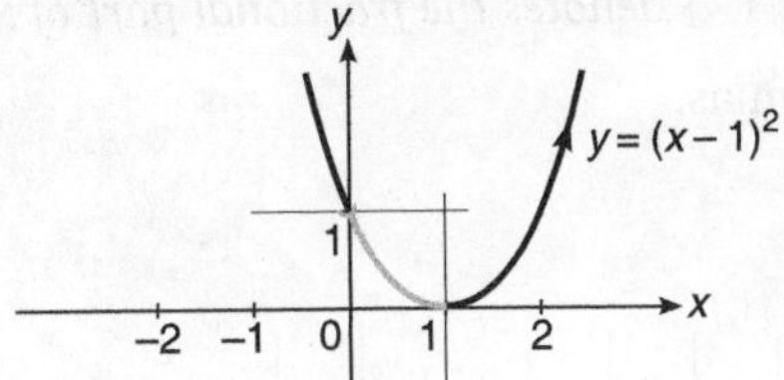

Fig. 2.140

Now; to plot $y = (\{x\} - 1)^2$ retain the graph for the interval $x \in [0, 1)$ and repeat for length 'one'.

Graph for $y = (\{x\} - 1)^2$:

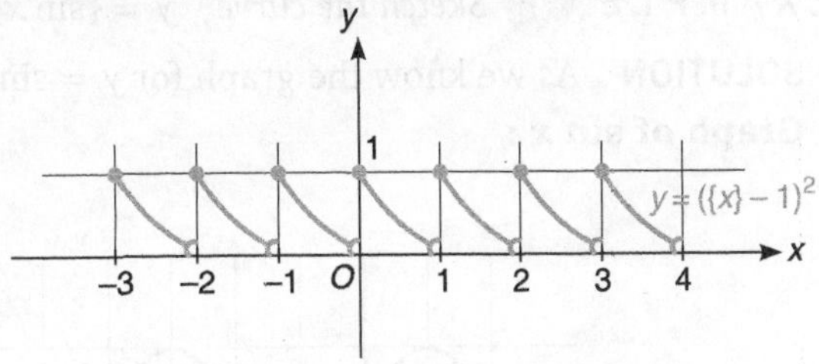

Fig. 2.141

EXAMPLE 2 *Sketch the curve* $|y| = (\{x\} - 1)^2$.

SOLUTION As discussed in above example $y = (\{x\} - 1)^2$.

Thus, $|y| = (\{x\} - 1)^2$ is image of $(\{x\} - 1)^2$ on x-axis whenever $(\{x\} - 1)^2$ is positive.

Graph of $|y| = (\{x\} - 1)^2$:

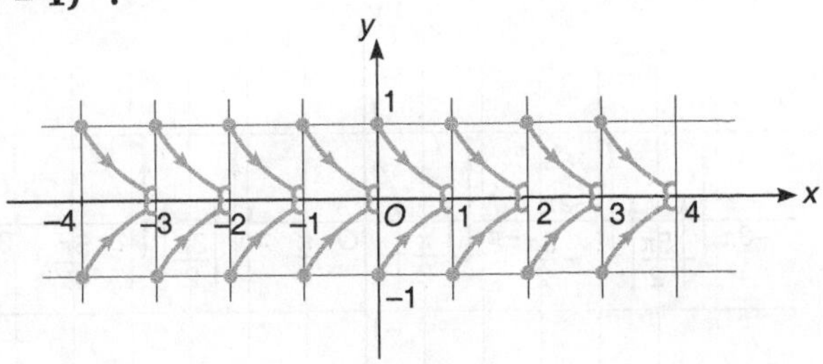

Fig. 2.142

EXAMPLE 3 *Sketch the graph of* $y = \dfrac{2^x}{2^{[x]}}$.

SOLUTION As we know 2^x is exponential function and we want to transform it to $2^{x-[x]}$, it retain the graph for $x \in [0, 1)$ and repeat for rest points.

Graph for $y = 2^x$

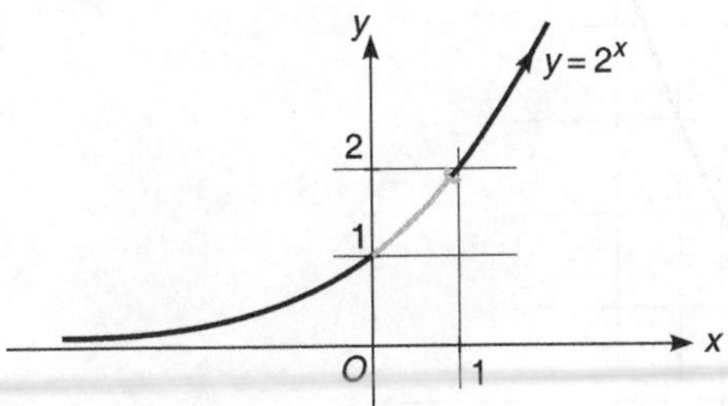

Fig. 2.143

Graph for $y = 2^{\{x\}}$

To retain graph between $x \in [0, 1)$.

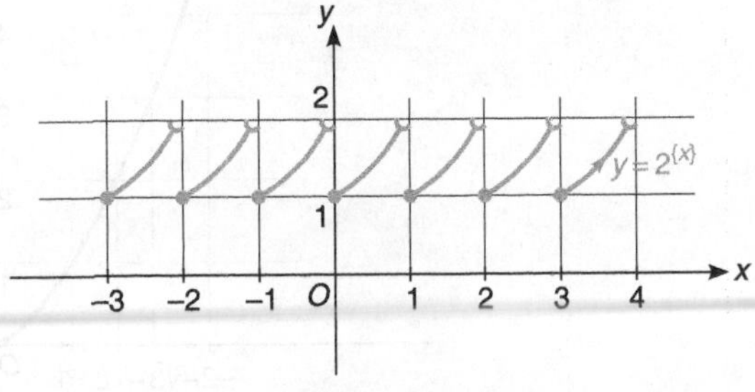

Fig. 2.144

(xx) $y = f(x)$ transforms to $y = \{f(x)\}$

Here, plot the horizontal lines for all integral values of y and for the point of intersection on $y = f(x)$ plot draw vertical lines and translate the graph for boundary $y = 0$ and $y = 1$.

EXAMPLE 1 *Sketch the curve* $y = \{\sin x\}$. *(where* $\{\cdot\}$ *denotes the fractional part of* x*).*

SOLUTION As we know the graph for $y = \sin x$. Shown as.

Graph of sin *x* :

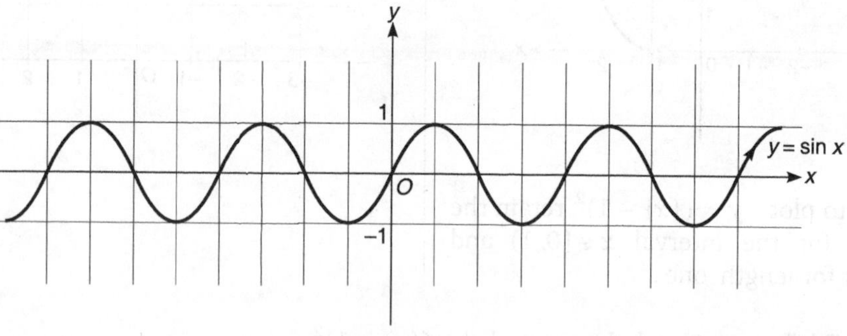

Fig. 2.145

As to retain the curve when $0 \le y < 1$; and shift other sections of graph between $y = 0$ to $y = 1$.

Graph for $y = \{\sin x\}$:

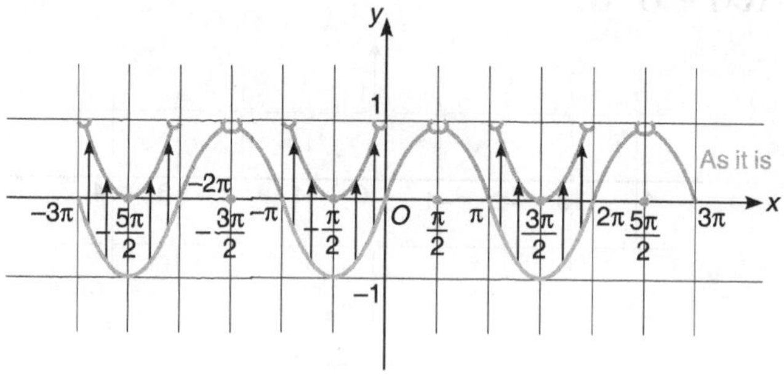

Fig. 2.146

EXAMPLE 2 *Sketch the curve* $y = \{x^2\}$.

SOLUTION As we know the curve $y = x^2$, is shown as:

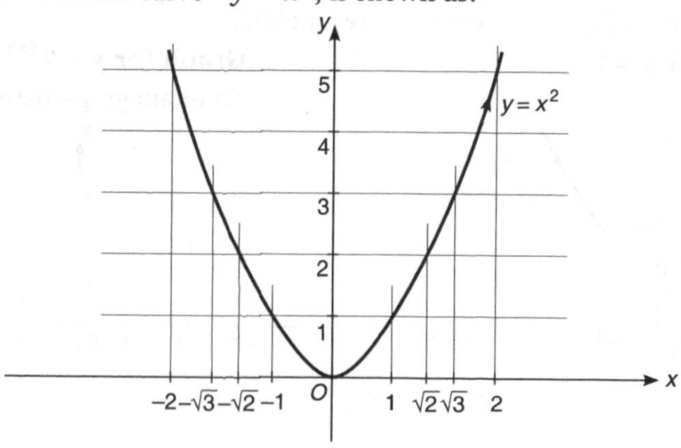

Fig. 2.147

Now to sketch $y = \{x^2\}$; retain the graph for $0 \le y < 1$ and for other intervals transform the graph between $0 \le y < 1$.

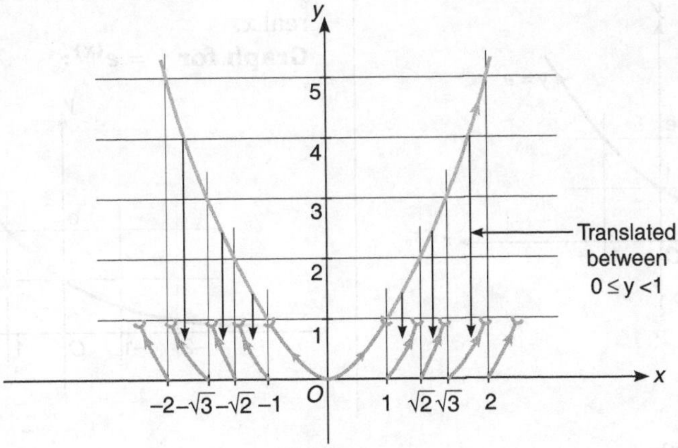

Fig. 2.148

EXAMPLE 3 *Sketch the curve* $y = \{e^x\}$.

SOLUTION As we know the curve $y = e^x$;

shown as:

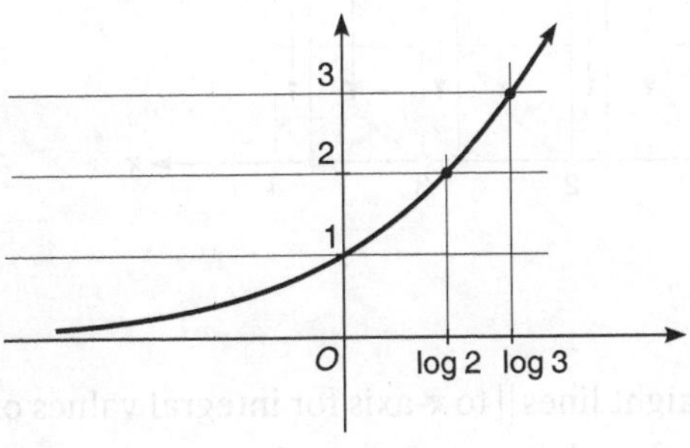

Fig. 2.149

Graph for $y = \{e^x\}$:

To retain the graph for $0 \le y < 1$ and transform the others to $0 \le y < 1$.

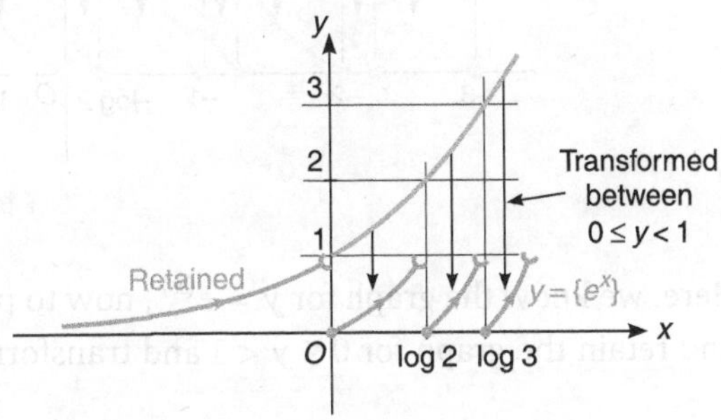

Fig. 2.150

(xxi) $y = f(x)$ transforms to $y = \{f(\{x\})\}$

Here, we have to follow two steps :

(i) Draw the graph for $y = f(\{x\})$.

(ii) $f(\{x\}) \longrightarrow \{f(\{x\})\}$

EXAMPLE 1 *Sketch the curve* $y = \{e^{\{x\}}\}$.

SOLUTION As we know the curve for $y = e^x$, is plotted as shown in Fig. 2.151.

Graph for $y = e^x$:

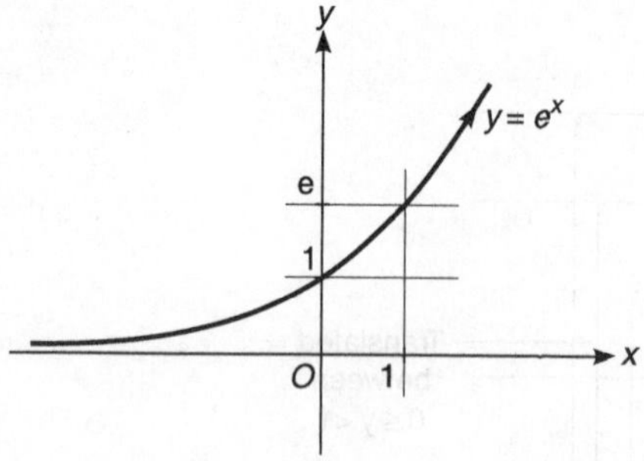

Fig. 2.151

Now to sketch $y = e^{\{x\}}$ retain the graph for $y = e^x$ between $0 \le x < 1$ and repeat for entire real x.

Graph for $y = e^{\{x\}}$:

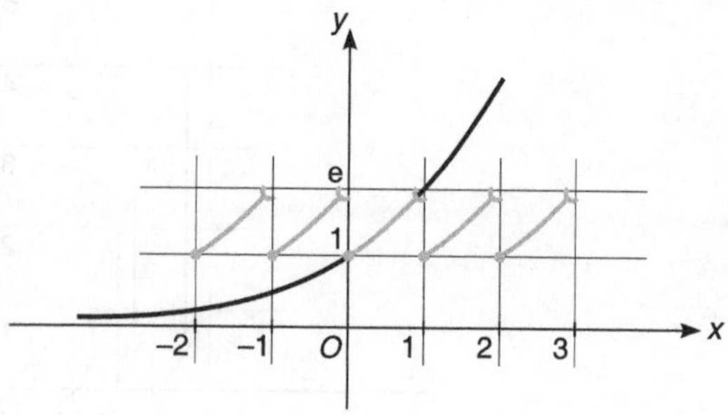

Fig. 2.152

Graph for $y = \{e^{\{x\}}\}$:

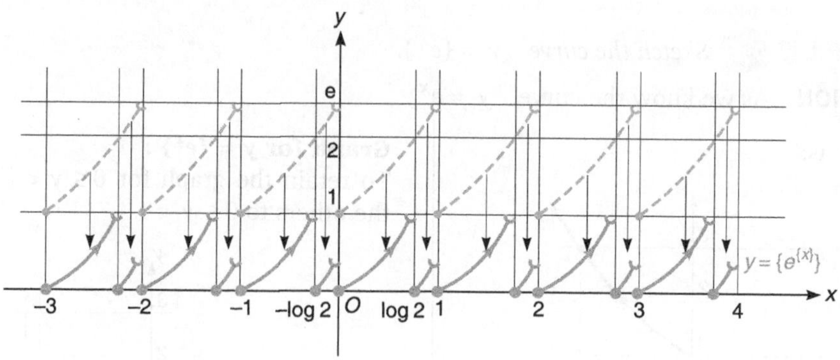

Fig. 2.153

Here, we know the graph for $y = e^{\{x\}}$, now to plot straight lines || to x-axis for integral values of y and retain the graph for $0 \le y < 1$ and transform the others between $0 \le y < 1$.

EXAMPLE 2 *Sketch the curve* $y = \{\sin\{x\}\}$.

SOLUTION As we know the curve $y = \sin x$; is plotted as;

Graph for $y = \sin x$:

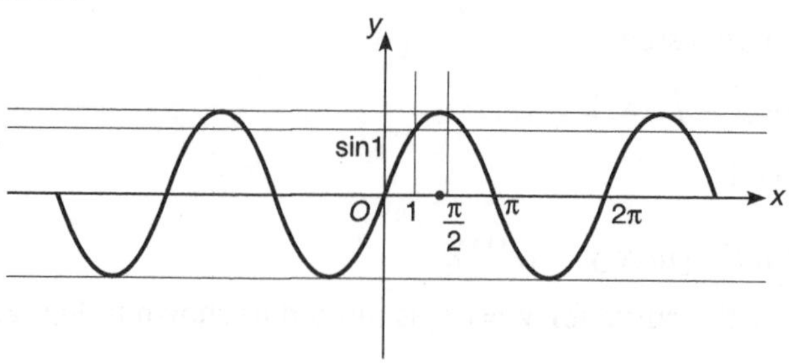

Fig. 2.154

Graph for $y = \sin\{x\}$:

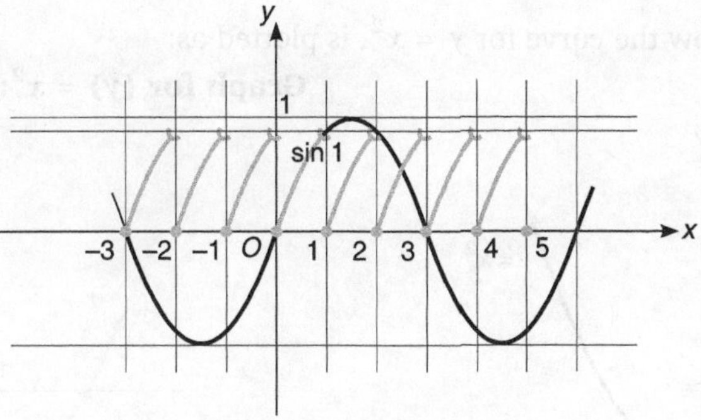

Fig. 2.155

Graph for $y = \{\sin\{x\}\}$:

From above figure

$$y = \sin\{x\} \Rightarrow 0 \le y < 1.$$

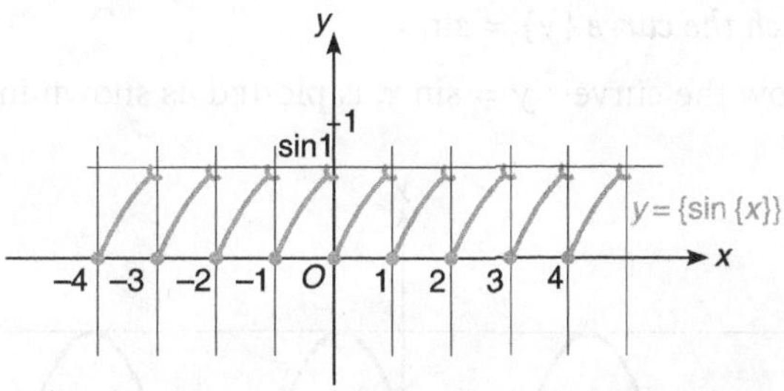

Fig. 2.156

So, the graph of $y = \sin\{x\}$ and $y = \{\sin\{x\}\}$ are same.

(xxii) $y = f(x)$ transforms to $\{y\} = f(x)$

Here; retain the graph of $y = f(x)$ only when $y = f(x)$ lies between $y \in [0, 1)$ and neglect the graph for other values.

Graphically it could be stated as;

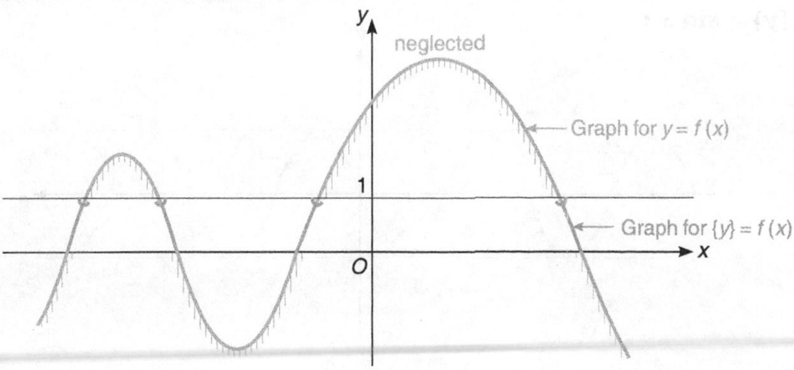

Fgi. 2.157

EXAMPLE 1 *Sketch the curve* $\{y\} = x^2$.

SOLUTION As we know the curve for $y = x^2$, is plotted as:

Graph for $y = x^2$:

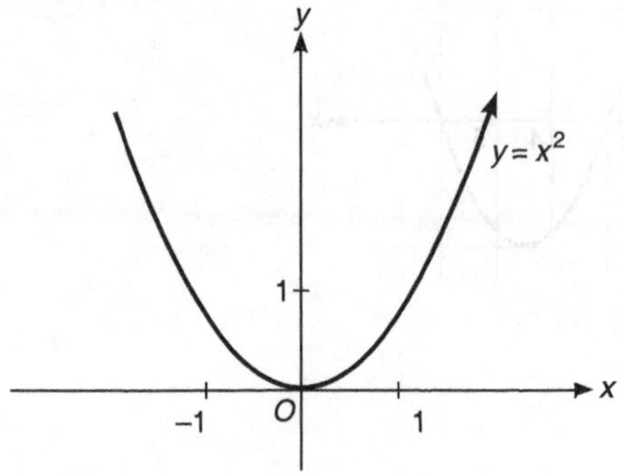

Fig. 2.158

Graph for $\{y\} = x^2$:

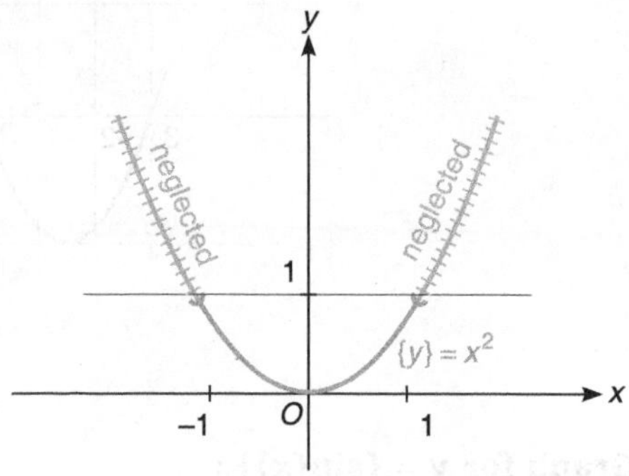

Fig. 2.159

EXAMPLE 2 *Sketch the curve* $\{y\} = \sin x$.

SOLUTION As we know the curve $y = \sin x$, is plotted as shown in Fig. 2.160 and 2.161.

Graph for $y = \sin x$:

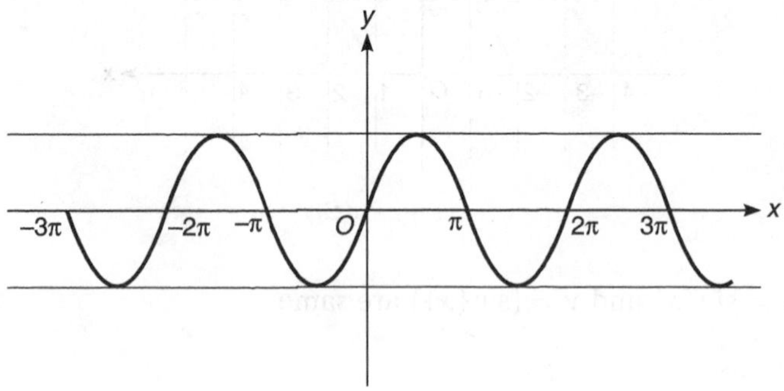

Fig. 2.160

Now to sketch $\{y\} = \sin x$: We retain the graph for $0 \le y < 1$ and neglect the graph for other values.

Graph for $\{y\} = \sin x$:

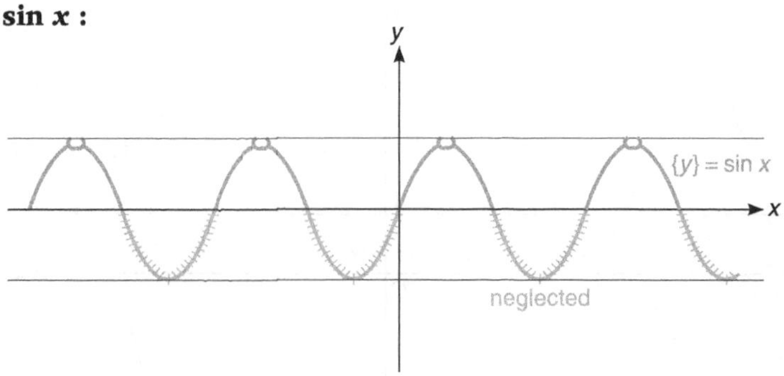

Fig. 2.161

(xxiii) $y = f(x)$ transforms to $\{y\} = \{f(x)\}$

As we have earlier discussed $y = \{f(x)\}$ (*i.e.*, transformation (xx), which shows $y = \{f(x)\}$ belongs to $[0, 1) \Rightarrow \{y\} = \{f(x)\}$. Thus, the graph of $y = \{f(x)\}$ and $\{y\} = \{f(x)\}$ are same.

EXAMPLE 1 *Sketch the curve* $\{y\} = \{x\}$.

SOLUTION As we know the curve for $y = x$.

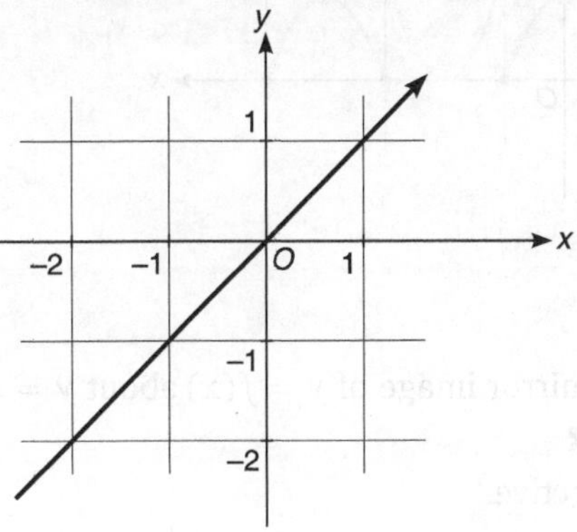

Fig. 2.162

Graph for $y = \{x\}$:

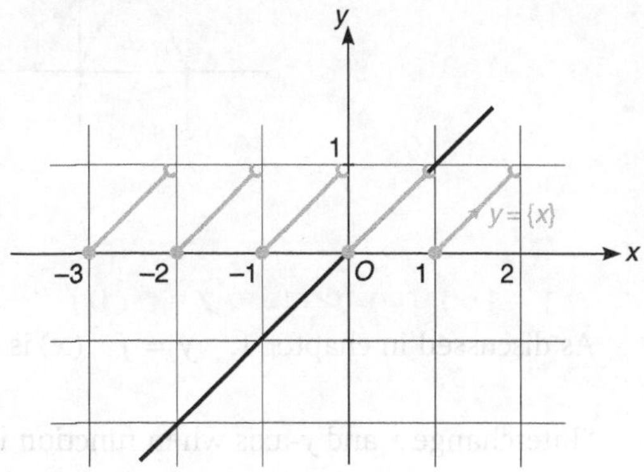

Fig. 2.163

Graph for $\{y\} = \{x\}$: From above figure we can see $y = \{x\}$ attains all values between [0, 1). Thus, graph remains same.

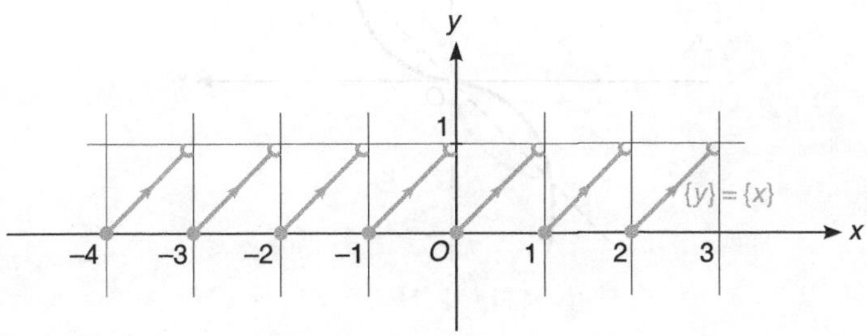

Fig. 2.164

EXAMPLE 2 *Sketch the curve* $\{y\} = \{\cos x\}$.

SOLUTION As we know the curve $y = \{\cos x\}$ is plotted as shown in Fig. 2.165.

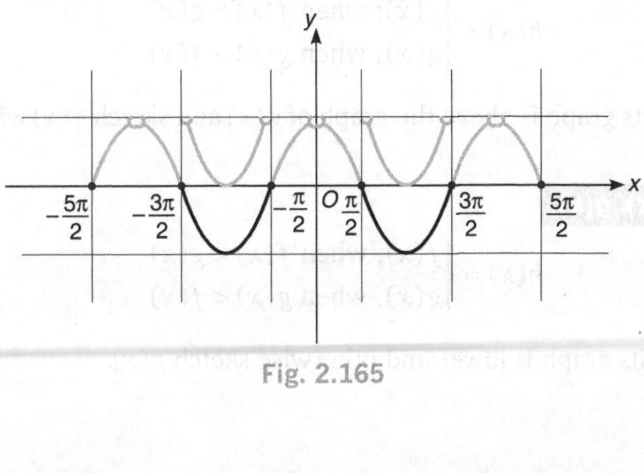

Fig. 2.165

From above figure; $y = \{\cos x\}$ lies between [0, 1), which shows $y = \{\cos x\}$ and $\{y\} = \{\cos x\}$ are same.

Thus, **graph for $\{y\} = \{\cos x\}$:**

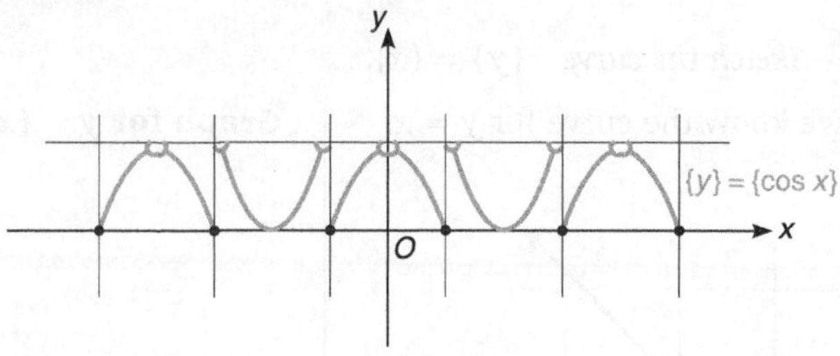

Fig. 2.166

(xxiv) $y = f(x)$ transforms to $y = f^{-1}(x)$

As discussed in chapter 1. $y = f^{-1}(x)$ is the mirror image of $y = f(x)$ about $y = x$.

OR

"Interchange x and y-axis when function is bijective."

Graphically it could be stated as :

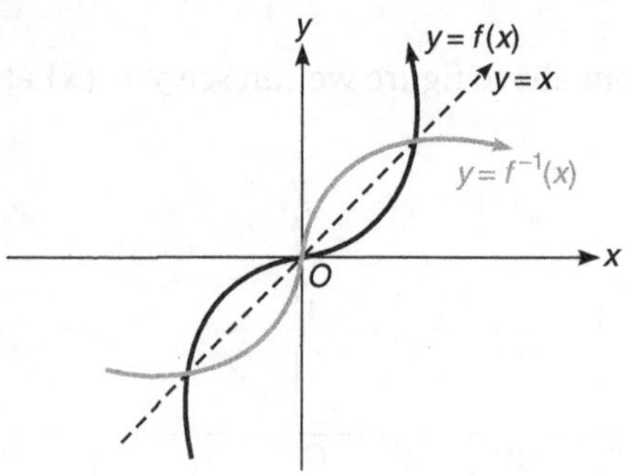

Fig. 2.167

2.4 SKETCHING $h(x)$ = MAXIMUM $\{f(x), g(x)\}$ AND $h(x)$ = MINIMUM $\{f(x), g(x)\}$

(i) $h(x)$ = maximum $\{f(x), g(x)\}$

$$\Rightarrow \qquad h(x) = \begin{cases} f(x); \text{ when } f(x) > g(x) \\ g(x); \text{ when } g(x) > f(x) \end{cases}$$

$\therefore$ Sketch $f(x)$ when its graph is above the graph of $g(x)$ and sketch $g(x)$ when its graph is above the graph of $f(x)$.

(ii) $h(x)$ = minimum $\{f(x), g(x)\}$

$$\Rightarrow \qquad h(x) = \begin{cases} f(x), \text{ when } f(x) < g(x) \\ g(x), \text{ when } g(x) < f(x) \end{cases}$$

$\therefore$ Sketch $f(x)$ when its graph is lower and otherwise sketch $g(x)$.

> **Note** One must remember the formula we can write;
>
> $$\max\{f(x), g(x)\} = \frac{f(x)+g(x)}{2} + \left|\frac{f(x)-g(x)}{2}\right|$$
>
> $$\min\{f(x), g(x)\} = \frac{f(x)+g(x)}{2} - \left|\frac{f(x)-g(x)}{2}\right|$$

OR

"To draw the graph of functions of the form $y = \max\{f(x), g(x)\}$ or $y = \min\{f(x), g(x)\}$."

We first draw the graphs of both the functions $f(x)$ and $g(x)$ and their points of intersections.

Then we find any two consecutive points of intersection. In between these points either $f(x) > g(x)$ or $f(x) < g(x)$, then, in order to **$\max\{f(x), g(x)\}$ we take those segments of $f(x)$ for which $f(x) > g(x)$,** between any two consecutive points of intersection of $f(x)$ and $g(x)$.

Similarly, in order to **$\min\{f(x), g(x)\}$, we take those segments of $f(x)$ for which $f(x) < g(x)$,** between any two consecutive points of intersection of $f(x)$ and $g(x)$.

EXAMPLE 1 *Sketch the graph of* $y = \max\{\sin x, \cos x\}$, $\forall\ x \in \left(-\pi, \frac{3\pi}{2}\right)$.

SOLUTION First plot both $y = \sin x$ and $y = \cos x$ by a dotted curve as can be seen from the graph in the interval $\left(-\pi, \frac{3\pi}{2}\right)$ and then darken those dotted lines for which $f(x) > g(x)$ or $g(x) > f(x)$.

From adjacent figure the point of intersections are A, B, C.

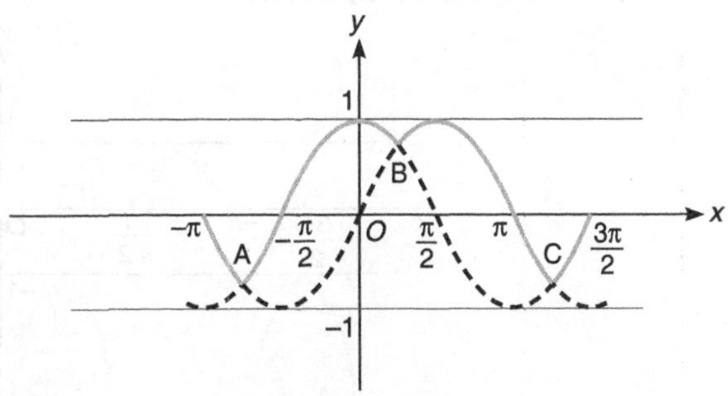

Fig. 2.168

$\therefore$ **Graph of max {sin x, cos x}**

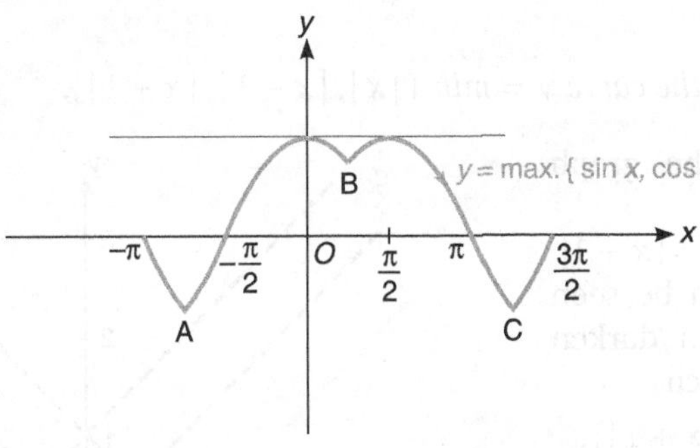

Fig. 2.169

EXAMPLE 2 *Sketch the graph for* $y = \min\{\tan x, \cot x\}$.

SOLUTION First plot both $f(x) = \tan x$ and $g(x) = \cot x$ by a dotted curves as can be seen from the graph and then darken those dotted lines for which $f(x) < g(x)$ and $g(x) < f(x)$.

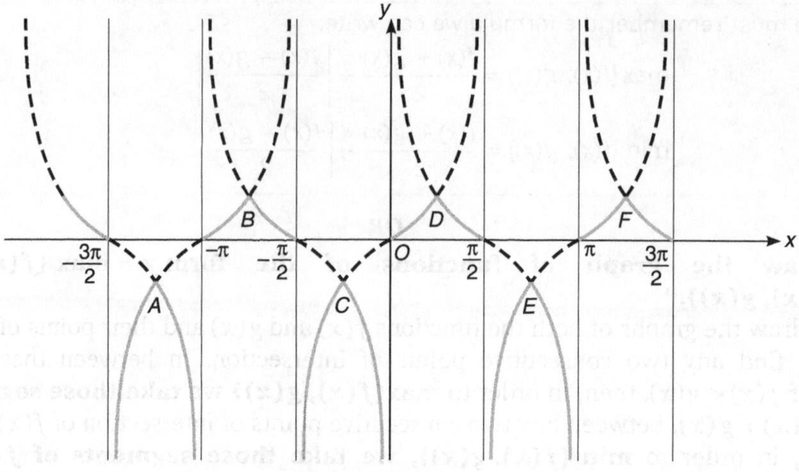

Fig. 2.170

From above figure we obtain the graph of min{tan x, cot x}.

Graph of min {tan x, cot x} :

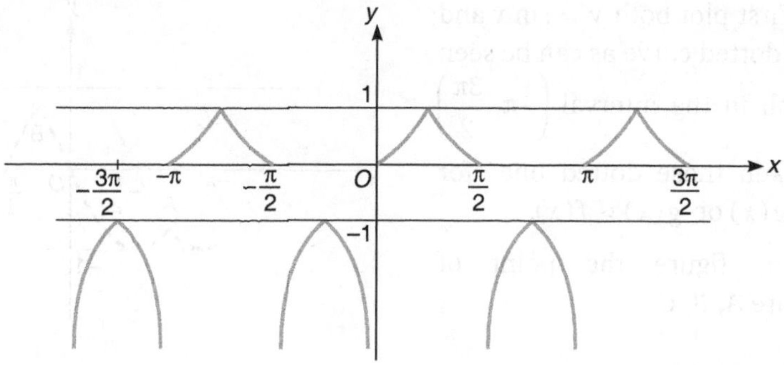

Fig. 2.171

EXAMPLE 3 *Sketch the curve* $y = \min\{|x|, |x-1|, |x+1|\}$.

SOLUTION First plot the graph for:

$y = |x|$, $y = |x-1|$, $y = |x+1|$ by a dotted curve as can be seen from the graph and then darken those dotted lines for which

$$|x| < \{|x-1|, |x+1|\};$$
$$|x-1| < \{|x|, |x+1|\}$$

and $$|x+1| < \{|x|, |x-1|\}.$$

Graph for

$\boldsymbol{y = |x|,\ y = |x-1|,\ y = |x+1|}$.

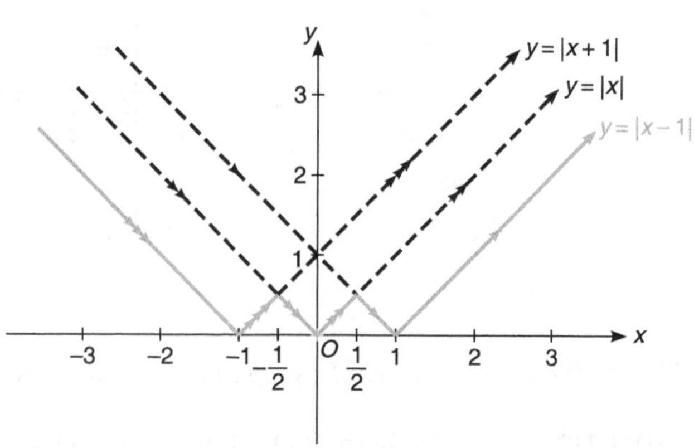

Fig. 2.172

As from the above curve graph for $y = \min\{|x-1|, |x|, |x+1|\}$ is plotted as;

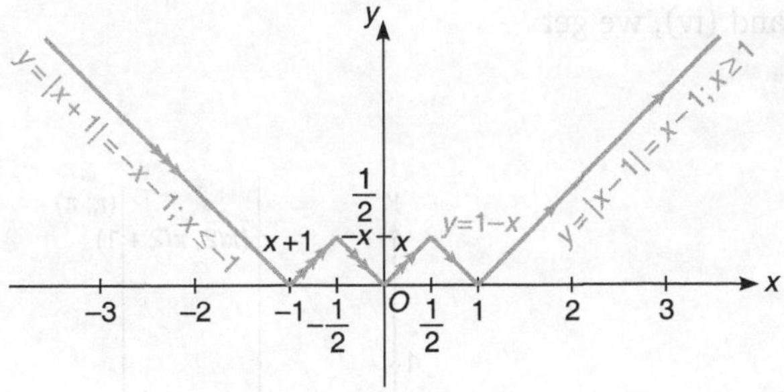

Fig. 2.173

From above figure;

$$\min\{|x-1|, |x|, |x+1|\} = \begin{cases} -(x+1); & x \le -1 \\ (x+1); & -1 \le x \le -\frac{1}{2} \\ -(x); & -\frac{1}{2} \le x \le 0 \\ (x); & 0 \le x \le \frac{1}{2} \\ (1-x); & \frac{1}{2} \le x \le 1 \\ (x-1); & x \ge 1 \end{cases}$$

2.5 WHEN $f(x), g(x) \longrightarrow f(x) + g(x) = h(x)$

There is no direct approach; but we can use following steps if minimum or maximum value of any one is known.

Step 1. Find maximum and minimum value of $g(x)$ say; $a \le g(x) \le b$.

Step 2. Plot the curve $h(x) = f(x) + g(x)$ between $f(x) + a$ to $f(x) + b$.

i.e.,
$$f(x) + a \le h(x) \le f(x) + b$$

Step 3. Check $g(x) = 0 \Rightarrow h(x) = f(x)$.

Step 4. When $g(x) > 0 \Rightarrow h(x) > f(x)$.

Step 5. When $g(x) < 0 \Rightarrow h(x) < f(x)$.

EXAMPLE 1 *Plot the curve* $y = x + \sin x$.

SOLUTION Here, $y = x + \sin x = f(x) + g(x)$

as we know;
$$g(x) = \sin x \in [-1, 1]$$

$\therefore$
$$x - 1 \le y \le x + 1 \qquad \ldots(i)$$

$\Rightarrow$ To sketch the curve between two parallel lines $y = x + 1$ and $y = x - 1$ (called bounded limits)

also;
$$g(x) = 0 \Rightarrow y = x \qquad \ldots(ii)$$
$$g(x) > 0 \Rightarrow y = x + \sin x > x \qquad \ldots(iii)$$

$$g(x) < 0 \Rightarrow y = x + \sin x < x \qquad \ldots(\text{iv})$$

Using Eqs. (i), (ii), (iii) and (iv), we get

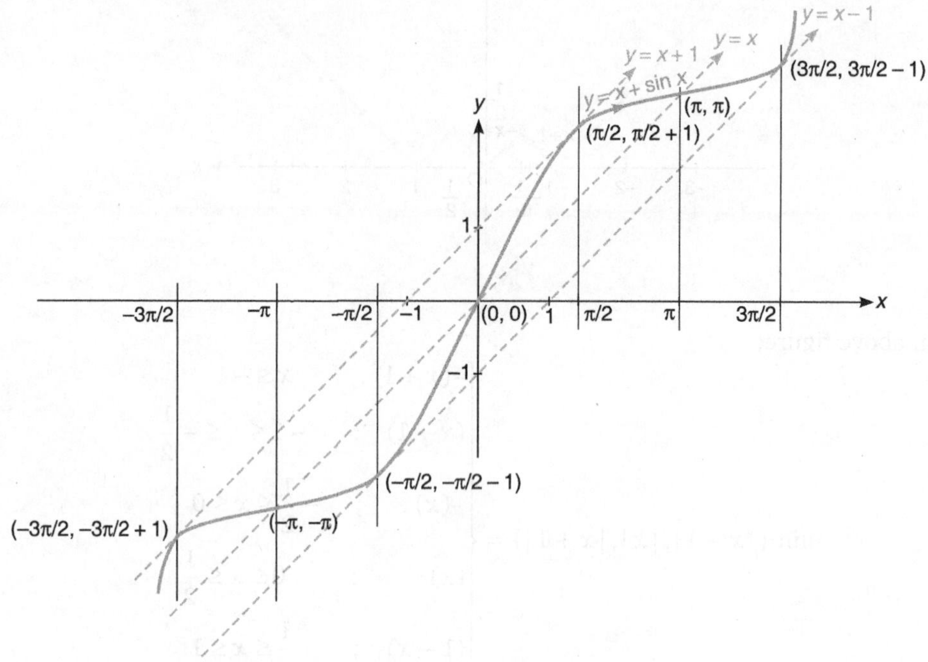

Fig. 2.174

2.6 WHEN $f(x), g(x) \longrightarrow f(x)\cdot g(x) = h(x)$

There is no direct approach but we can use the following steps if minimum and maximum of any one is known.

Step 1. Find the minimum and maximum of any one of them say $a \le g(x) \le b$.

Step 2. From step 1; $a\cdot f(x) \le h(x) \le b\cdot f(x)$

Step 3. Check $g(x) = 0 \Rightarrow h(x) = 0$.

EXAMPLE 1 *Sketch the curve;* $y = x\sin x$.

SOLUTION Here; $y = x\sin x$;

where, $-1 \le \sin x \le 1$

$\Rightarrow \quad -x \le y \le x \qquad \ldots(\text{i})$

also; $\sin x = 0$

$\Rightarrow \quad x = -2\pi, -\pi, 0, \pi, 2\pi, \ldots$

$\therefore\ y = 0$ when $x = -2\pi, -\pi, 0, \pi, \ldots$

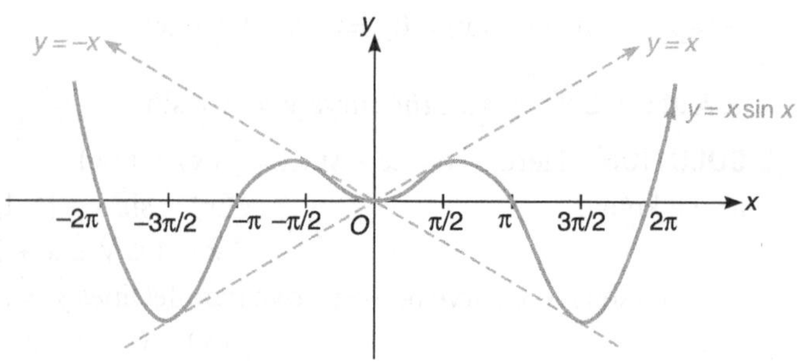

Fig. 2.175

and $\qquad y = x \qquad$ when $\quad x = \frac{\pi}{2}, \frac{5\pi}{2}, \ldots$

$\qquad y = -x \qquad$ when $\quad x = -\frac{\pi}{2}, \frac{3\pi}{2}, \ldots$

EXAMPLE 2 *Draw the graph of the function* $y = \frac{\sin x}{x}$.

SOLUTION As we know; $\quad -1 \le \sin x \le 1 \;\Rightarrow\; y = \frac{\sin x}{x}$ lies between $-\frac{1}{x}$ to $\frac{1}{x}$.

or $\qquad$ to sketch the curve where $\quad -\frac{1}{x} \le y \le \frac{1}{x} \qquad \ldots$(i)

Here at $x = 0$; y is not defined but as;

$$x \longrightarrow 0 \;\Rightarrow\; y = \frac{\sin x}{x} \longrightarrow 1 \qquad \ldots\text{(ii)}$$

also; $\qquad y = 0 \;$ at $\; x = n\pi\,; \;\; n \in z - \{0\} \qquad \ldots$(iii)

Using Eqs. (i), (ii) and (iii), the curve could be plotted as;

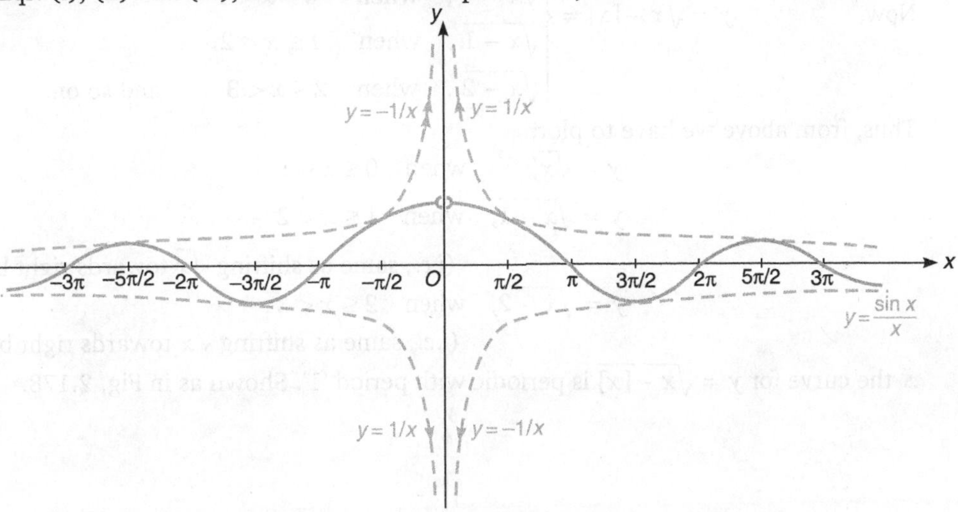

Fig. 2.176

EXAMPLE 3 *Draw the curve:* $\; y = e^{-x} \sin x$.

SOLUTION As we know;

$$-1 \le \sin x \le 1$$

$\therefore \qquad y = e^{x} \sin x$

$\Rightarrow \qquad -e^{-x} \le y \le e^{-x} \qquad \ldots$(i)

Thus, y is bounded between

$$y = -e^{-x} \quad \text{and} \quad y = e^{-x}.$$

Shown as in Fig. 2.177;

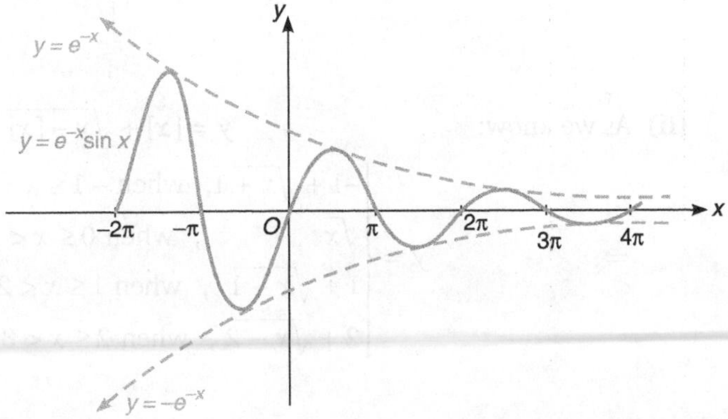

Fig. 2.177

SOME MORE SOLVED EXAMPLES

EXAMPLE 1 *Sketch the curves of the following :*

(i) $y = \sqrt{x - [x]}$ (ii) $y = [x] + \sqrt{x - [x]}$

(iii) $y = |[x] + \sqrt{x - [x]}\,|$; *where [•] represent greatest integer function*

SOLUTION **(i)** As we know that;

$$0 \le x - [x] < 1 \quad \text{for all} \quad x \in R.$$

Also, for any $x \in [0, 1)$, we have

$$x^2 \le x \le \sqrt{x}$$

$$\therefore \quad x - [x] \le \sqrt{x - [x]}$$

Now,
$$y = \sqrt{x - [x]} = \begin{cases} \sqrt{x+1}, & \text{when } -1 \le x < 0 \\ \sqrt{x}\,, & \text{when } \ \ 0 \le x < 1 \\ \sqrt{x-1}, & \text{when } \ \ 1 \le x < 2 \\ \sqrt{x-2}, & \text{when } \ \ 2 \le x < 3 \end{cases} \quad \text{... and so on.}$$

Thus, from above we have to plot;

$$y = \sqrt{x}, \quad \text{when} \quad 0 \le x < 1$$

$$y = \sqrt{x-1}, \quad \text{when} \quad 1 \le x < 2$$

(*i.e.*, same as shifting $\sqrt{x}$ towards right by 1 unit.)

$$y = \sqrt{x-2}, \quad \text{when} \quad 2 \le x < 3$$

(*i.e.*, same as shifting $\sqrt{x}$ towards right by 2 units)

$\therefore$ the curve for $y = \sqrt{x - [x]}$ is periodic with period '1'. Shown as in Fig. 2.178.

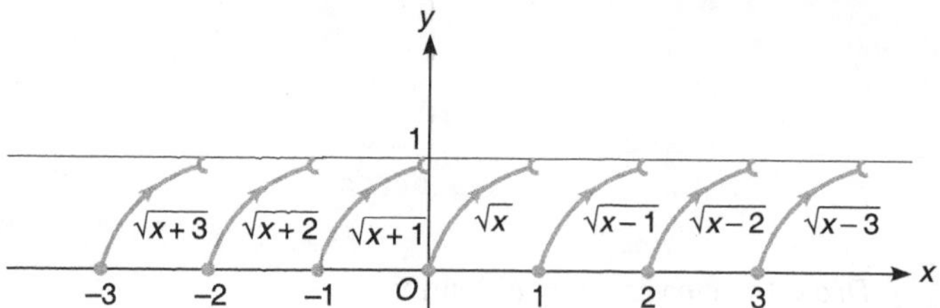

Fig. 2.178

(ii) As we know;
$$y = [x] + \sqrt{x - [x]}$$

$$\Rightarrow \quad y = \begin{cases} -1 + \sqrt{x+1}, & \text{when } -1 \le x < 0 \\ \sqrt{x}\,, & \text{when } 0 \le x < 1 \\ 1 + \sqrt{x-1}\,, & \text{when } 1 \le x < 2 \\ 2 + \sqrt{x-2}\,, & \text{when } 2 \le x < 3 \end{cases} \quad \text{... and so on.}$$

Thus, the graph for $y = [x] + \sqrt{x - [x]}$ is obtained by the graph of $y = \sqrt{x - [x]}$ by translating it by $[x]$ units in upward or downward direction according as $[x] > 0$ or $[x] < 0$.

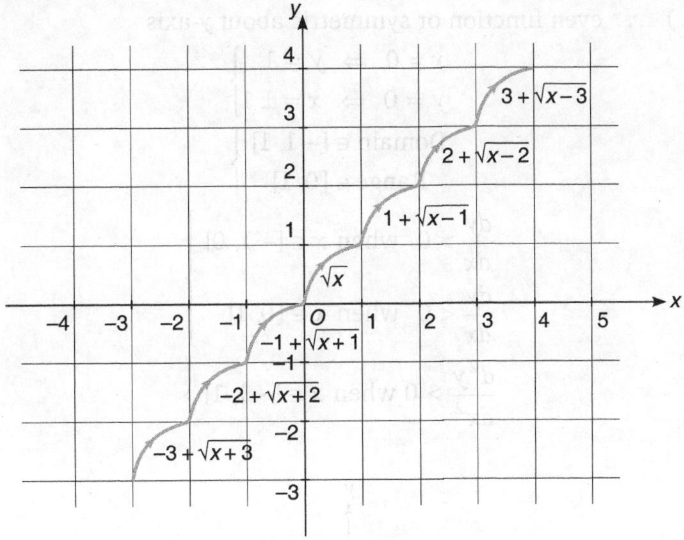

Fig. 2.179

Thus, the curve for $y = [x] + \sqrt{x - [x]}$

From above curve we could discuss that;

$y = [x] + \sqrt{x - [x]}$ is continuous and differentiable for all x.

(iii) The graph for $y = \left|[x] + \sqrt{x - [x]}\right|$ is obtained by reflecting the portion lying below x-axis of the graph of $y = [x] + \sqrt{x - [x]}$ about x-axis and keeping the portion lying above x-axis (as it is).

Thus, the graph for $y = \left|[x] + \sqrt{x - [x]}\right|$.

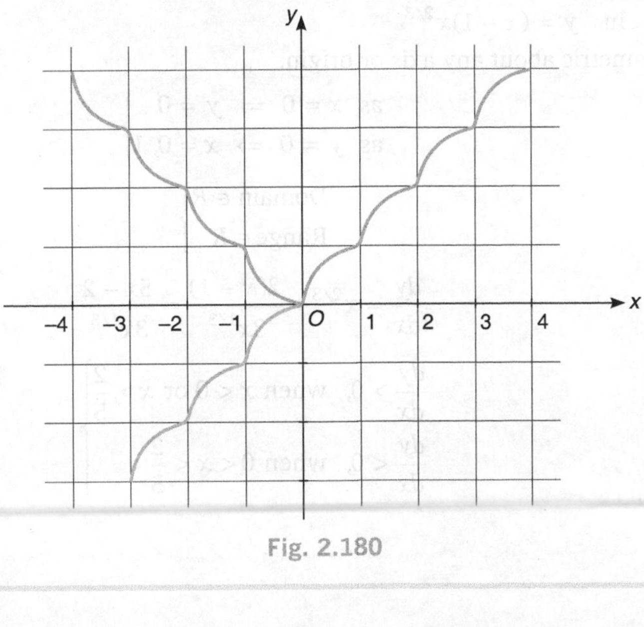

Fig. 2.180

EXAMPLE 2 *Sketch the curve* $y = (1 - x^{2/3})^{3/2}$.

SOLUTION Here;

(a) $f(-x) = f(x)$ $\therefore$ even function or symmetric about y-axis ...(i)

(b) When $\left.\begin{aligned} x = 0 &\Rightarrow y = 1 \\ y = 0 &\Rightarrow x = \pm 1 \end{aligned}\right\}$ ...(ii)

(c) $\left.\begin{aligned} \text{Domain} &\in [-1, 1] \\ \text{Range} &\in [0, 1] \end{aligned}\right\}$ ...(iii)

(d) $\left.\begin{aligned} \frac{dy}{dx} &> 0 \text{ when } x \in [-1, 0] \\ \frac{dy}{dx} &< 0 \text{ when } x \in [0, 1] \end{aligned}\right\}$ ...(iv)

(e) $\left.\frac{d^2y}{dx^2} > 0 \text{ when } x \in [-1, 1]\right\}$ ...(v)

From above;

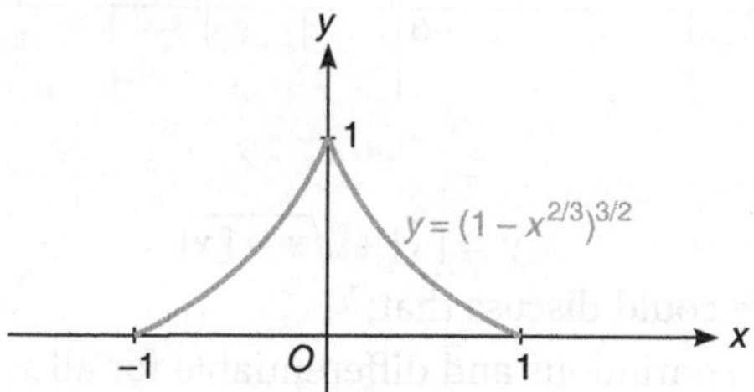

Fig. 2.181

EXAMPLE 3 *Sketch the graph for* $y = (x - 1)x^{2/3}$.

SOLUTION In $y = (x - 1)x^{2/3}$

(a) Not symmetric about any axis or origin. ...(i)

(b) $\left.\begin{aligned} \text{as } x = 0 &\Rightarrow y = 0 \\ \text{as } y = 0 &\Rightarrow x = 0, 1 \end{aligned}\right\}$ ...(ii)

(c) $\left.\begin{aligned} \text{Domain} &\in R \\ \text{Range} &\in R \end{aligned}\right\}$ ...(iii)

(d) $$\frac{dy}{dx} = x^{2/3} + \frac{2(x-1)}{3x^{1/3}} = \frac{5x - 2}{3x^{1/3}}$$

$\Rightarrow$ $\left.\begin{aligned} \frac{dy}{dx} &> 0, \text{ when } x < 0 \text{ or } x > \frac{2}{5} \\ \frac{dy}{dx} &< 0, \text{ when } 0 < x < \frac{2}{5} \end{aligned}\right\}$ ...(iv)

(e) $\dfrac{d^2y}{dx^2} = \dfrac{1}{3}\left\{\dfrac{x^{1/3}\cdot 5 - (5x-2)\dfrac{1}{3}(x)^{-2/3}}{x^{2/3}}\right\}$

$= \dfrac{10x+2}{9x^{4/3}}$

$$\Rightarrow \left.\begin{array}{l} \dfrac{d^2y}{dx^2} > 0, \text{ when } x < -\dfrac{2}{5} \text{ or } x > 0 \\ \dfrac{d^2y}{dx^2} < 0, \text{ when } -\dfrac{2}{5} < x < 0 \end{array}\right\} \quad \ldots(\text{v})$$

Thus, curve for $y = (x-1)\,x^{2/3}$ is shown in Fig. 2.182.

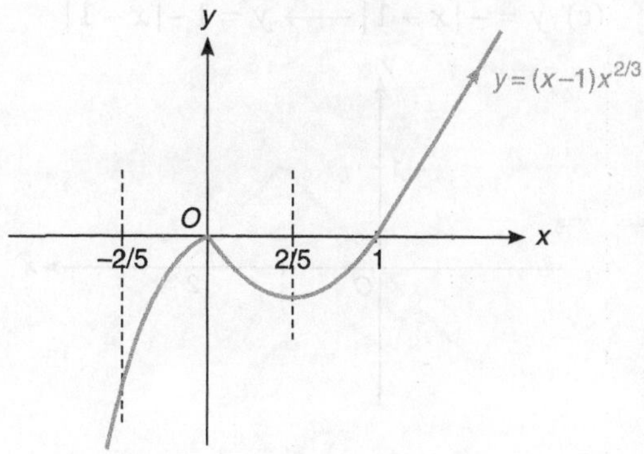

Fig. 2.182

EXAMPLE 4 *Draw the graph for:*

(i) $y = |1 - |x-1||$ *(ii)* $|y| = |1 - |x-1||$

SOLUTION As we know the graph for $y = x - 1$

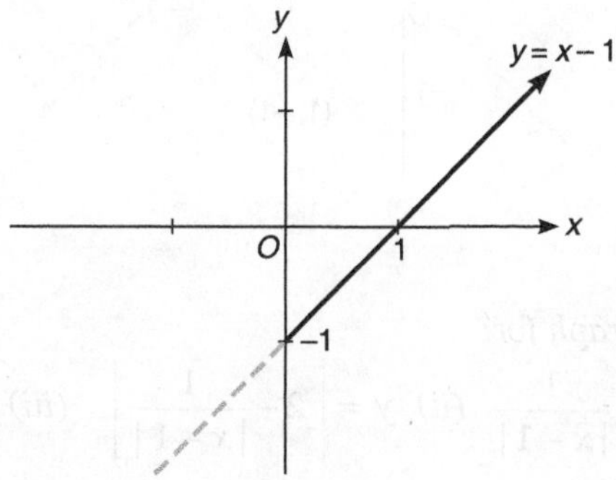

Fig. 2.183

(a) $y = (x-1) \longrightarrow y = |x-1|$

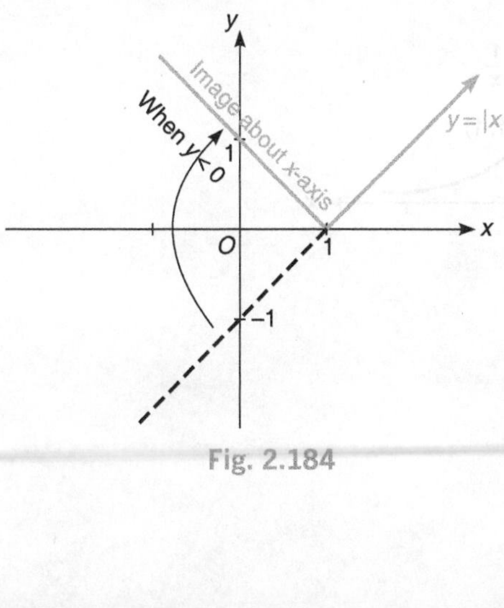

Fig. 2.184

(b) $y = |x-1| \longrightarrow y = -|x-1|$

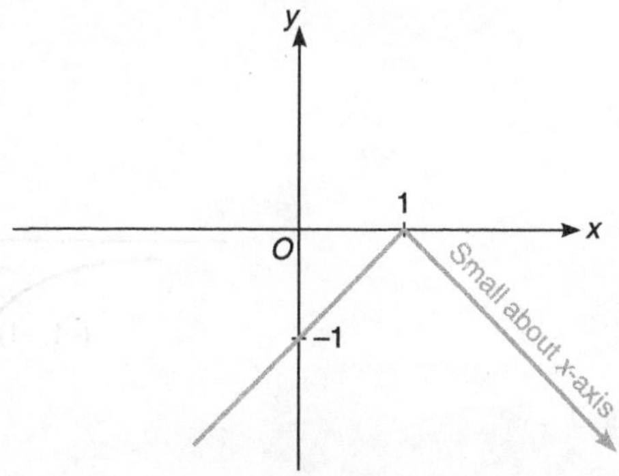

Fig. 2.185

(c) $y = -|x-1| \longrightarrow y = 1-|x-1|$

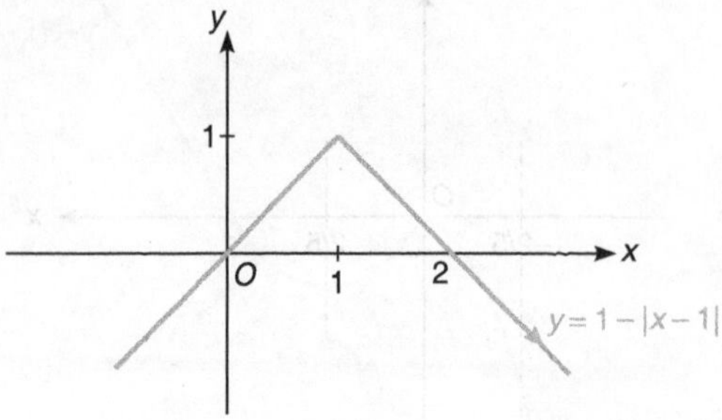

Fig. 2.186

(d) $y = 1-|x-1| \longrightarrow y = |1-|x-1||$

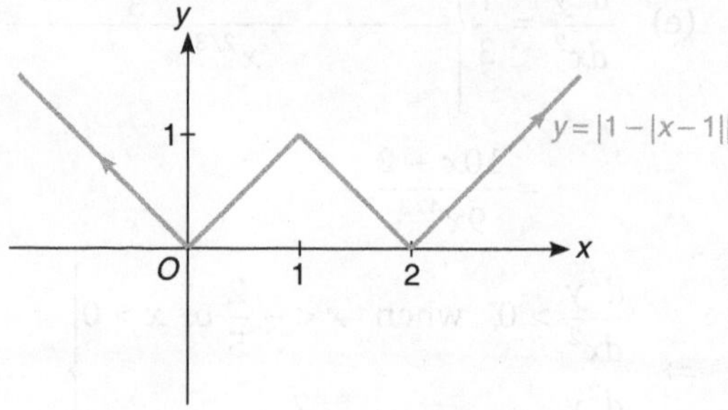

Fig. 2.187

(e) $y = |1-|x-1|| \longrightarrow |y| = |1-|x-1||$

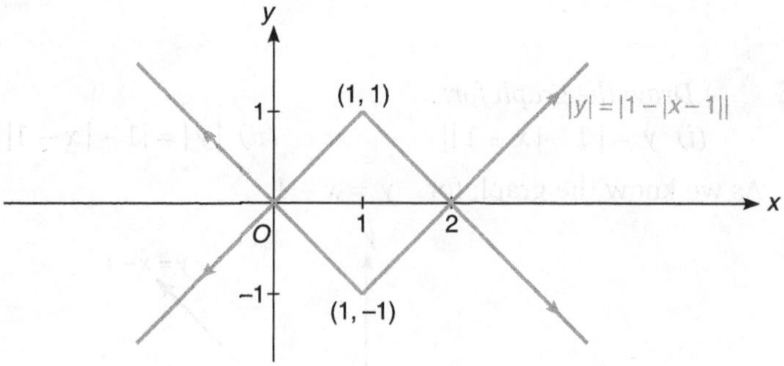

Fig. 2.188

EXAMPLE 5 *Draw the graph for:*

(i) $y = 2 - \dfrac{1}{|x-1|}$ (ii) $y = \left| 2 - \dfrac{1}{|x-1|} \right|$ (iii) $|y| = \left| 2 - \dfrac{1}{|x-1|} \right|$

SOLUTION We know the graph for $y = \dfrac{1}{x}$ is shown as;

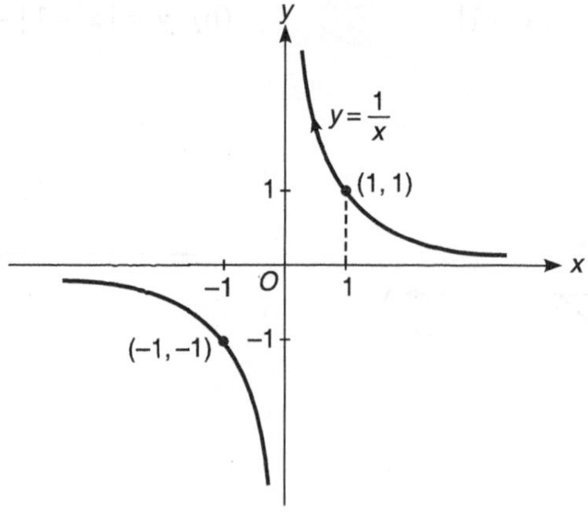

Fig. 2.189

(a) $y = \dfrac{1}{x} \longrightarrow y = \dfrac{1}{x-1}$

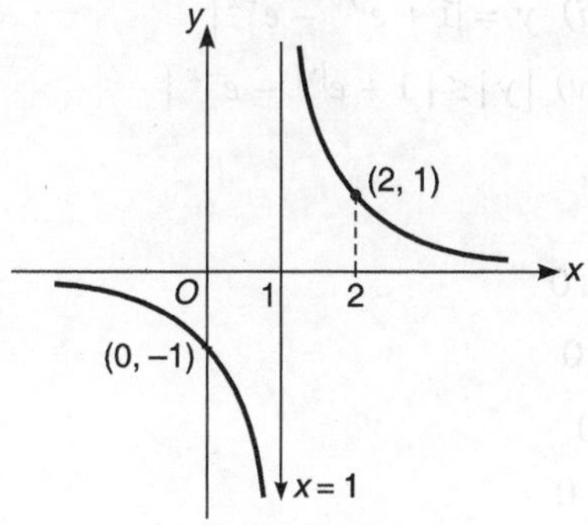

Fig. 2.190

(b) $y = \dfrac{1}{x-1} \longrightarrow y = \dfrac{1}{|x-1|}$

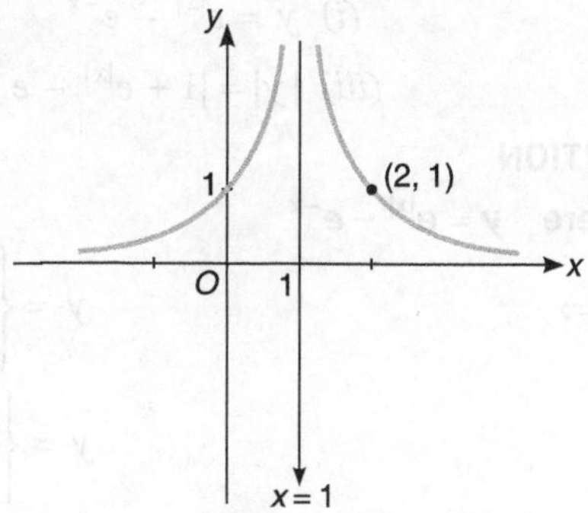

Fig. 2.191

(c) $y = \dfrac{1}{|x-1|} \longrightarrow y = -\dfrac{1}{|x-1|}$

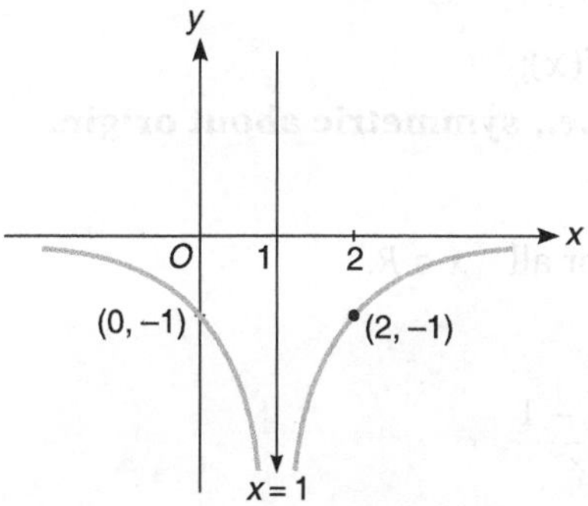

Fig. 2.192

(d) $y = -\dfrac{1}{|x-1|} \longrightarrow y = 2 - \dfrac{1}{|x-1|}$

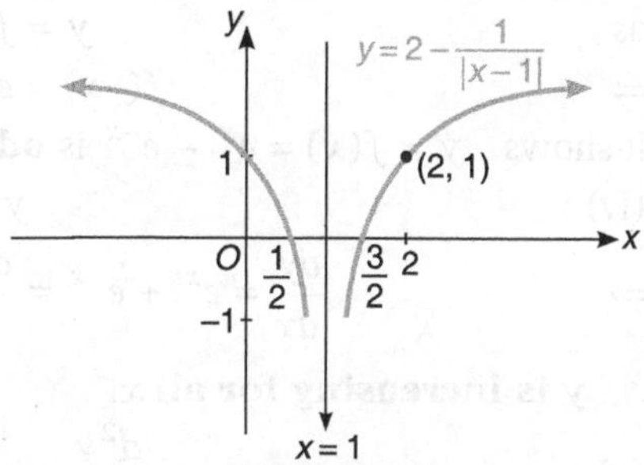

Fig. 2.193

(e) $y = 2 - \dfrac{1}{|x-1|} \longrightarrow y = \left| 2 - \dfrac{1}{|x-1|} \right|$

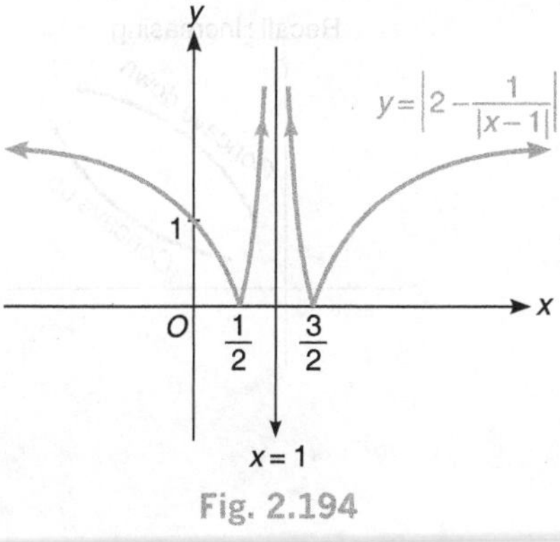

Fig. 2.194

(f) $y = \left| 2 - \dfrac{1}{|x-1|} \right| \longrightarrow |y| = \left| 2 - \dfrac{1}{|x-1|} \right|$

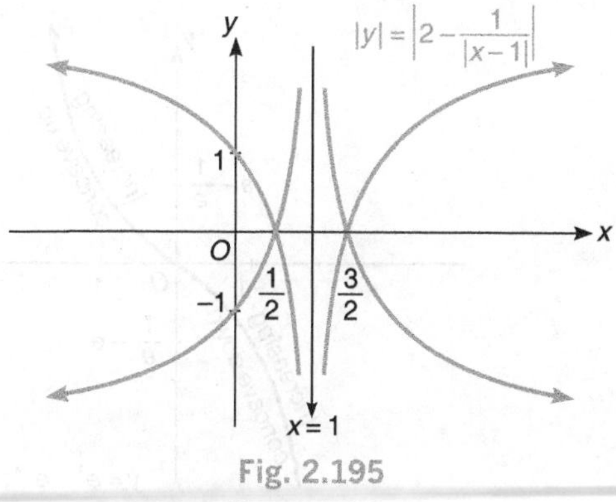

Fig. 2.195

EXAMPLE 6 *Sketch the graph for:*

(i) $y = e^{|x|} - e^{-x}$ (ii) $y = |1 + e^{|x|} - e^{-x}|$

(iii) $|y| = |1 + e^{|x|} - e^{-x}|$ (iv) $|y| \le |1 + e^{|x|} - e^{-x}|$

SOLUTION

(i) Here $y = e^{|x|} - e^{-x}$

$$\Rightarrow \quad y = \begin{cases} e^{x} - e^{-x} ; & x \ge 0 \\ e^{-x} - e^{-x} ; & x < 0 \end{cases}$$

$$y = \begin{cases} e^{x} - e^{-x} ; & x \ge 0 \\ 0 ; & x < 0 \end{cases}$$

To discuss; $\mathbf{y = e^{x} - e^{-x}}$

(i) when $x = 0 \Rightarrow y = 0$ **(it passes through origin)**

(ii) when $y = 0 \Rightarrow e^{2x} - 1 = 0 \Rightarrow x = 0$

(iii) $f(-x) = -f(x)$

as $y = f(x) = e^{x} - e^{-x}$

$\Rightarrow$ $f(-x) = e^{-x} - e^{x} = -f(x);$

it shows $y = f(x) = e^{x} - e^{-x}$ is **odd function, *i.e.*, symmetric about origin.**

(iv) $y = e^{x} - e^{-x}$

$$\Rightarrow \quad \frac{dy}{dx} = e^{x} + e^{-x} = \frac{e^{2x} + 1}{e^{x}} > 0 \quad \text{for all} \quad x \in R.$$

$\therefore$ ***y* is increasing for all *x*.**

(v) $$\frac{d^2y}{dx^2} = e^{x} - e^{-x} = \frac{e^{2x} - 1}{e^{x}}$$

$\Rightarrow \dfrac{d^2y}{dx^2} > 0$ **when** $\mathbf{x > 0}$; **concave up** and increasing .

also $\dfrac{d^2y}{dx^2} < 0$ **when** $\mathbf{x < 0}$; **concave down** and increasing from above discussion $\mathbf{y = e^{x} - e^{-x}}$; is plotted as shown in Fig. 2.196.

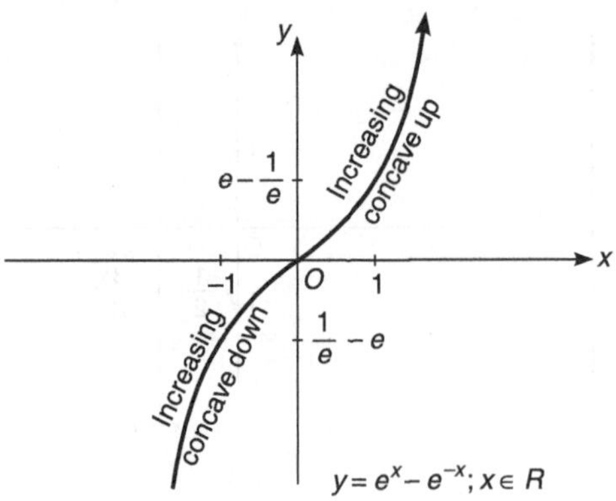

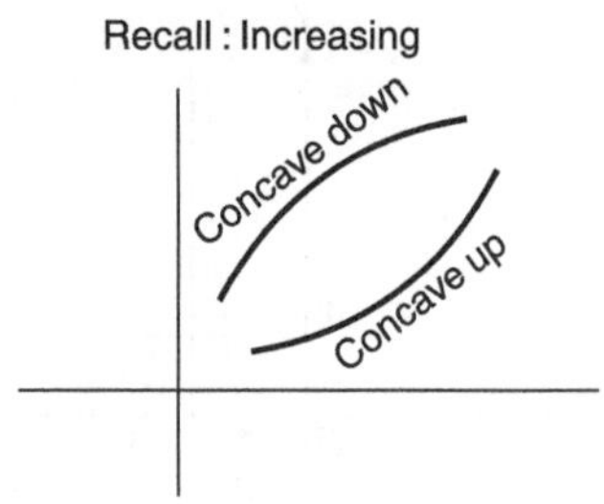

Fig. 2.196

Now;

$$y = e^{|x|} - e^{-x} = \begin{cases} e^x - e^{-x}; & x \geq 0 \\ 0 \quad ; & x < 0. \end{cases}$$

Thus from Fig. 2.197.

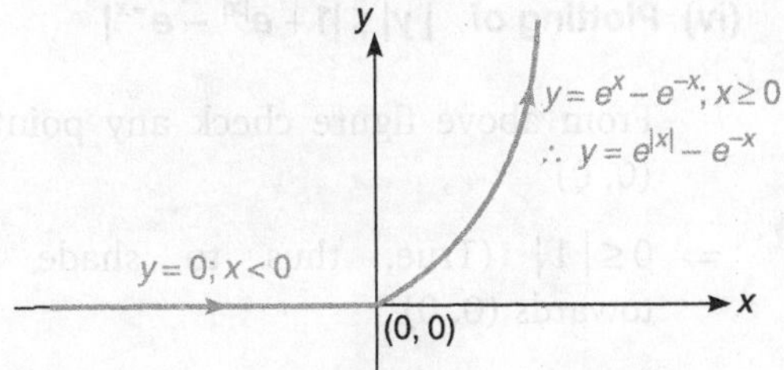

Fig. 2.197

(ii) Plotting of $y = |1 + e^{|x|} - e^{-x}|$

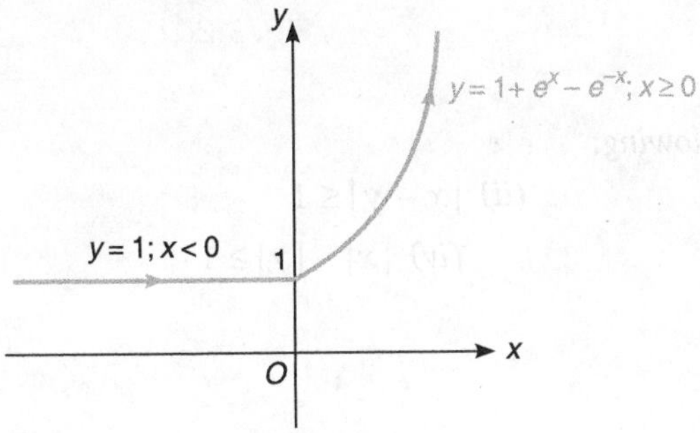

Fig. 2.198

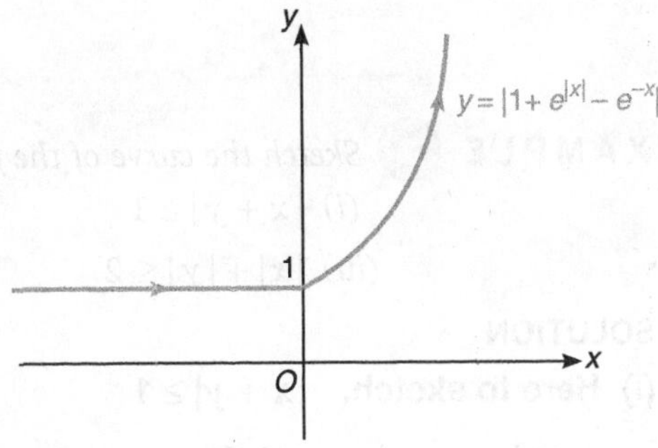

Fig. 2.199

From above Fig. 2.198.

$$y = e^{|x|} - e^{-x} \longrightarrow y = 1 + e^{|x|} - e^{-x}$$

$$\Rightarrow \qquad y = \begin{cases} 1 + e^x - e^{-x}; & x \geq 0 \\ 1 \quad ; & x < 0 \end{cases}$$

Thus, the graph for; $\quad y = |1 + e^{|x|} - e^{-x}|$

is same as; $\quad y = 1 + e^{|x|} - e^{-x}$ $\quad$ {as $y \geq 1$ for all $x \in R$}

$\therefore$ Graph for $y = |1 + e^{|x|} - e^{-x}|$ is shown in Fig. 2.199.

(iii) Plotting of $|y| = |1 + e^{|x|} - e^{-x}|$

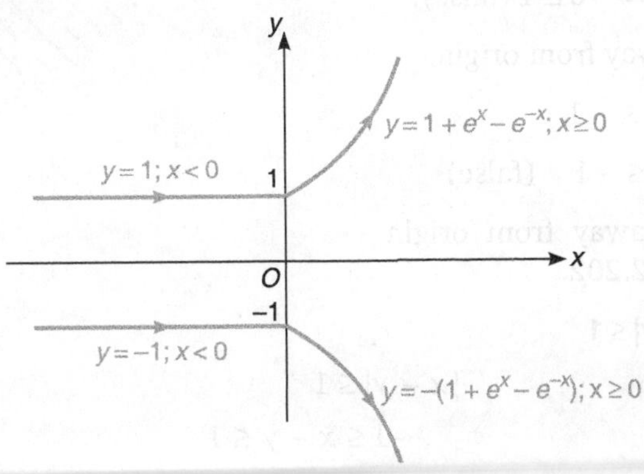

Fig. 2.200

(iv) Plotting of $|y| \le |1 + e^{|x|} - e^{-x}|$

From above figure check any point say (0, 0)

$\Rightarrow$ $0 \le |1|$ (True, thus to shade area towards (0, 0).

From given figure shade part represents the area bounded between two curves.

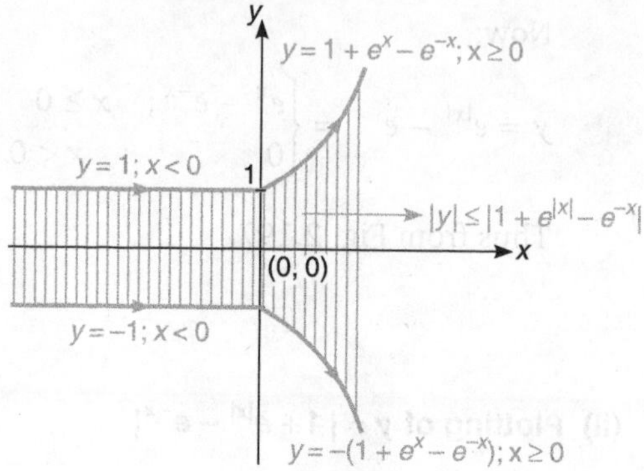

Fig. 2.201

EXAMPLE 7 *Sketch the curve of the following:*

(i) $|x + y| \ge 1$ (ii) $|x - y| \le 1$

(iii) $|x| + |y| \le 2$ (iv) $|x| - |y| \ge 1$

SOLUTION

(i) Here to sketch, $|x + y| \ge 1$

we know; $|x + y| \ge 1$

$$\Rightarrow \begin{cases} x + y \ge 1 \\ x + y \le -1 \end{cases}$$

or $x + y \ge 1$

and $x + y \le -1$

First we shall draw the graph for $x + y = 1$ and $x + y = -1$, now we shall consider any fixed point say (0, 0) and check $x + y \ge 1$ and $x + y \le -1$ holds or not.

As; $x + y \ge 1 \Rightarrow 0 \ge 1$ (false).

$\therefore$ shaded part is away from origin.

again as; $x + y \le -1$

$\Rightarrow$ $0 \le -1$ (false)

$\therefore$ Shaded part is away from origin shown as in Fig. 2.202.

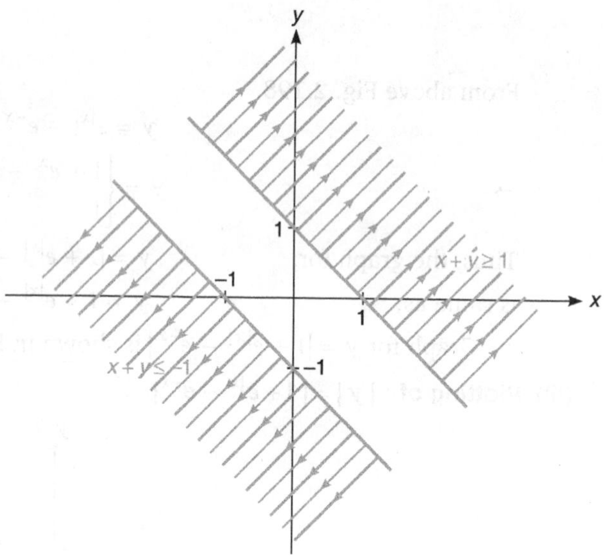

Fig. 2.202

(ii) To sketch $|x - y| \le 1$

we know $|x - y| \le 1$

$\Rightarrow$ $-1 \le x - y \le 1$

$\Rightarrow$ $-1 - x \le -y \le 1 - x$

or $\qquad (x-1) \le y \le (x+1)$

Thus, to plot y between $(x-1)$ to $(x+1)$.

In figure shaded parts is towards origin as; putting any fixed point say $(0, 0)$; in $|x-y| \le 1$.

$\Rightarrow \qquad |0-0| \le 1$

$\Rightarrow \qquad 0 \le 1 \qquad$ (true)

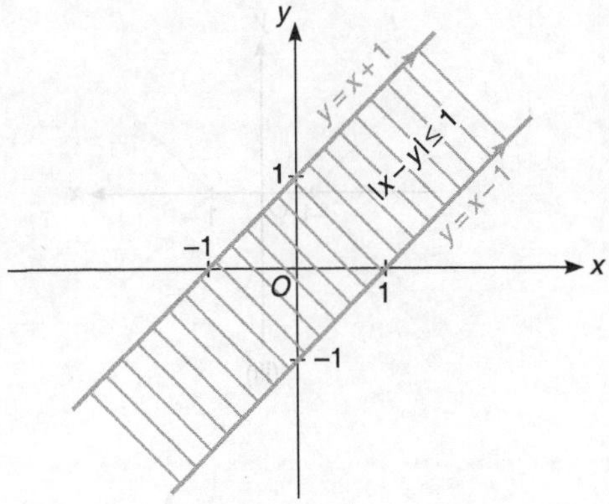

|x − y| ≤ 1

Fig. 2.203

(iii) To sketch $|x|+|y| \le 2$

Here; $|y| = 2-|x|$ is plotted as;

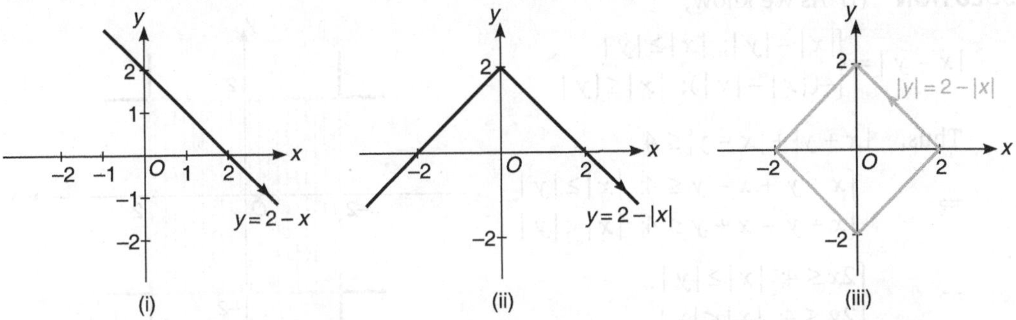

Fig. 2.204

From above graph of $|y| = 2-|x|$, we can check shading of

$$|y| \le 2-|x| \quad \text{or} \quad |x|+|y| \le 2.$$

as at $(0, 0) \Rightarrow 0 \le 2$ (true)

$\therefore$ shading towards origin; shown as in Fig. 2.205.

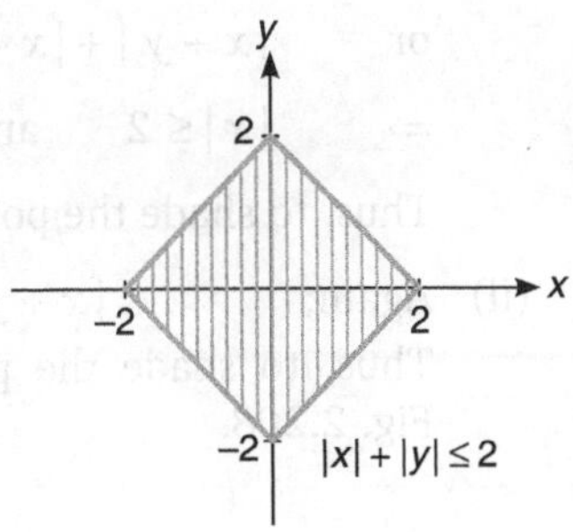

Fig. 2.205

(iv) Sketching of $|x|-|y| \ge 1$

To sketch $|x|-|y| \ge 1$; we proceed as:

$$y = (x-1) \to y = |x|-1 \to |y| = |x|-1 \to |x|-|y| \ge 1$$

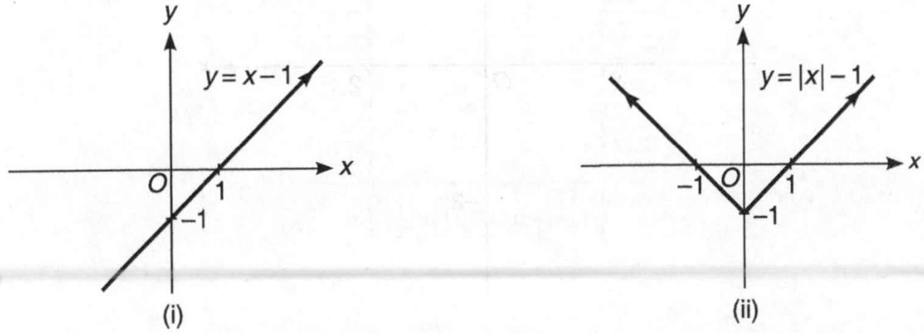

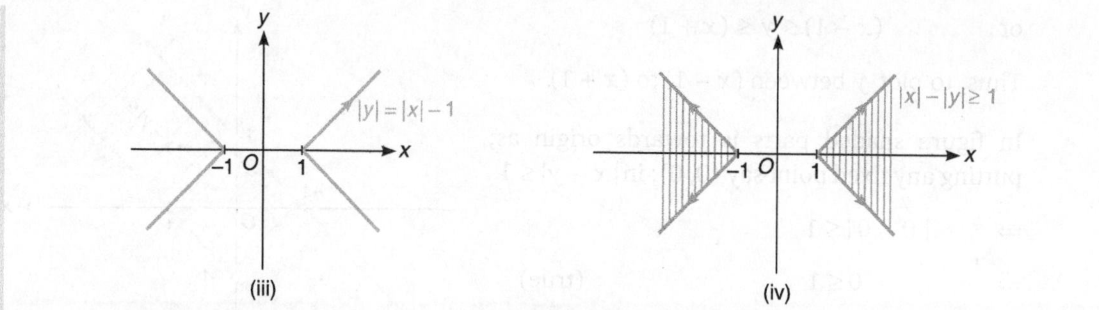

Fig. 2.206

EXAMPLE 8 *Sketch the curve;*

(i) $|x+y|+|x-y| \le 4$ (ii) $|x+y|+|x-y| \ge 4$

SOLUTION (i) As we know;

$$|x-y| = \begin{cases} |x|-|y|; & |x| \ge |y| \\ -(|x|-|y|); & |x| \le |y| \end{cases}$$

Thus; $|x+y|+|x-y| \le 4$

$$\Rightarrow \quad \begin{cases} x+y+x-y \le 4; & |x| \ge |y| \\ x+y-x+y \le 4; & |x| < |y| \end{cases}$$

$$\Rightarrow \quad \begin{cases} 2x \le 4; & |x| \ge |y| \\ 2y \le 4; & |x| < |y| \end{cases}$$

or $\quad |x+y|+|x-y| \le 4$

$\Rightarrow \quad |x| \le 2 \quad$ and $\quad |y| \le 2$

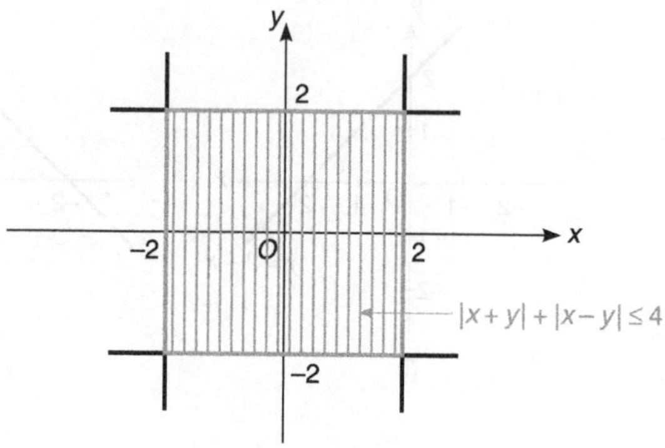

Fig. 2.207

Thus, to shade the portion when $-2 \le x \le 2$ and $-2 \le y \le 2$. Shown as in Fig. 2.207.

(ii) Again; $|x+y|+|x-y| \ge 4 \quad \Rightarrow \quad |x| \ge 2$ and $|y| \ge 2$

Thus, to shade the portion when $x \le -2$ or $x \ge 2$ and $y \le -2$ or $y \ge 2$. Shown as in Fig. 2.208.

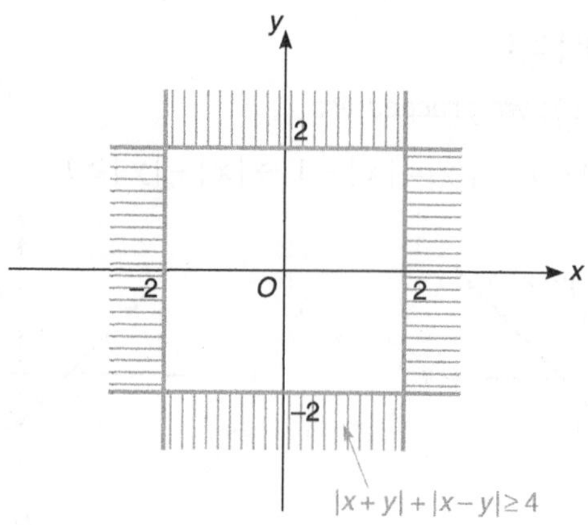

Fig. 2.208

EXAMPLE 9 *Sketch the curve for;*

$$\text{(i)}\ 2^{|x|}|y| + 2^{|x|-1} \le 1 \qquad \text{(ii)}\ 2^{|x|}|y| + 2^{|x|-1} \le 1;\quad |x| \le \frac{1}{2} \text{ and } |y| \le \frac{1}{2}.$$

SOLUTION **(i)** Here $2^{|x|}|y| + 2^{|x|-1} \le 1 \quad \Rightarrow \quad 2^{|x|}\left\{|y| + \frac{1}{2}\right\} \le 1$

$$\Rightarrow \qquad |y| + \frac{1}{2} \le 2^{-|x|}$$

Thus, to plot $|y| + \frac{1}{2} \le 2^{-|x|}$, we proceed as;

$$\underset{\text{(i)}}{y = \left(2^{-x} - \frac{1}{2}\right)} \longrightarrow \underset{\text{(ii)}}{y = 2^{-|x|} - \frac{1}{2}} \longrightarrow \underset{\text{(iii)}}{|y| = 2^{-|x|} - \frac{1}{2}} \longrightarrow \underset{\text{(iv)}}{|y| \le 2^{-|x|} - \frac{1}{2}}$$

Shown as;

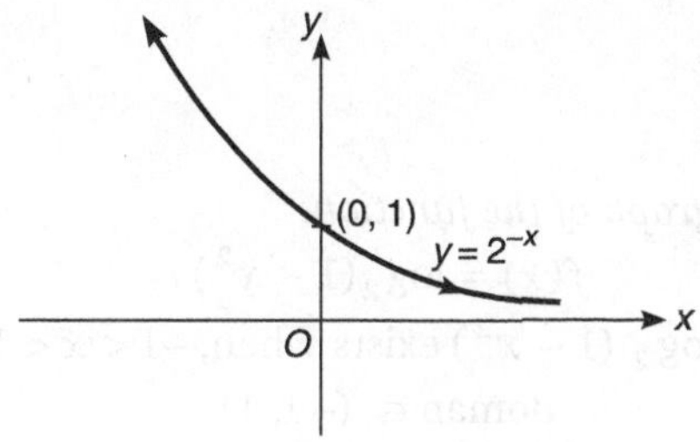

Fig. 2.209

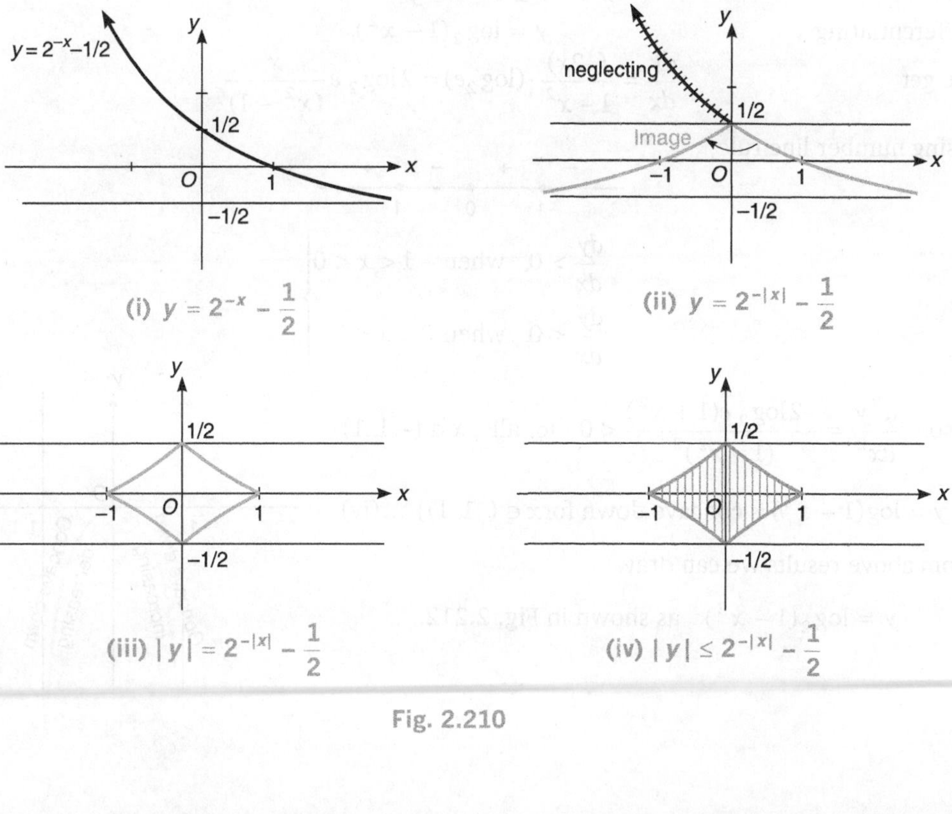

(i) $y = 2^{-x} - \frac{1}{2}$ (ii) $y = 2^{-|x|} - \frac{1}{2}$

(iii) $|y| = 2^{-|x|} - \frac{1}{2}$ (iv) $|y| \le 2^{-|x|} - \frac{1}{2}$

Fig. 2.210

Here, figure (iv) is shaded towards origin as putting $x = 0, y = 0$ in $|y| \le 2^{-|x|} - \frac{1}{2}$.

$\Rightarrow \; 0 \le 1 - \frac{1}{2}$ or $0 \le \frac{1}{2}$, which is true, therefore, shaded towards origin.

(ii) Plotting of $2^{|x|}|y| + 2^{|x|-1} \le 1$;

$$|x| \le \frac{1}{2} \text{ and } |y| \le \frac{1}{2}.$$

from above figure;

$$2^{|x|}|y| + 2^{|x|-1} \le 1$$

and $|x| \le \frac{1}{2}$, $|y| \le \frac{1}{2}$ shown as in Fig. 2.211.

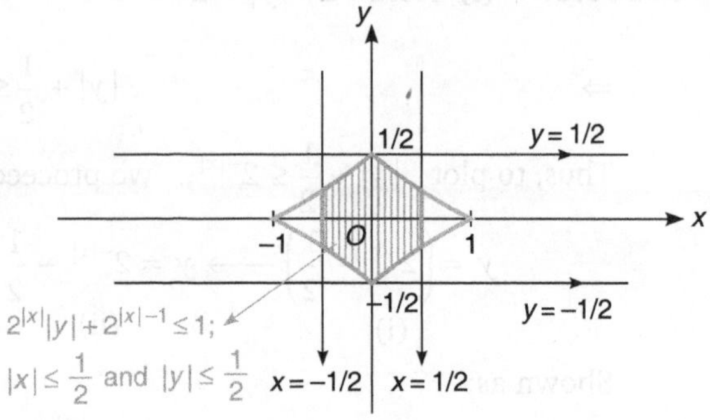

Fig. 2.211

EXAMPLE 10 *Sketch the graph of the function:*

$$f(x) = \log_2(1 - x^2)$$

SOLUTION Here, $\log_2 (1 - x^2)$ exists when, $-1 < x < 1$

i.e., doman $\in (-1, 1)$ ...(i)

and range $\in (-\infty, 0]$

as $x \longrightarrow \pm 1 \Rightarrow y \longrightarrow -\infty$ ...(ii)

differentiating $y = \log_2(1 - x^2)$,

we get $$\frac{dy}{dx} = \frac{(-2x)}{1 - x^2} \cdot (\log_2 e) = 2\log_2 e \frac{x}{(x^2 - 1)}$$

Using number line rule;

$$\underset{-1}{\overset{-\quad +}{\bullet}}\quad \underset{0}{\bullet} \quad \underset{1}{\overset{-\quad +}{\bullet}}$$

$$\left.\begin{aligned} \frac{dy}{dx} &> 0, \text{ when } -1 < x < 0 \\ \frac{dy}{dx} &< 0, \text{ when } 0 < x < 1 \end{aligned}\right\} \quad \text{...(iii)}$$

also, $\frac{d^2y}{dx^2} = \frac{-2\log_2 e(1 + x^2)}{(1 - x^2)^2} < 0$ for all $x \in (-1, 1)$

$\therefore \; y = \log(1 - x^2)$ is concave down for $x \in (-1, 1)\}$ ...(iv)

from above results we can draw

$y = \log_2(1 - x^2)$ as shown in Fig. 2.212.

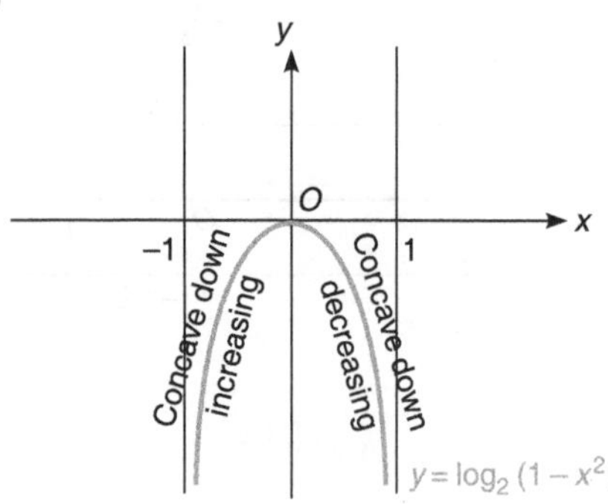

Fig. 2.212

EXAMPLE 11 *Sketch the graph of* $y = \log(\sin x)$.

SOLUTION Here;

(i) $y = \log(\sin x)$ is defined, when $\sin x > 0 \quad \Rightarrow \quad x \in (2n\pi, (2n+1)\pi)$.

(ii) Since, $\sin x$ is periodic with period 2π.

$\therefore$ to discuss $y = \log(\sin x)$, when $x \in (0, \pi)$ as exists when $x \in (0, \pi)$ and then plotted for entire number line.

(iii) as $x \longrightarrow 0 \quad \Rightarrow \quad y = \log(0) = -\infty$.

as $y \longrightarrow 0 \quad \Rightarrow \quad \sin x \longrightarrow 1 \quad \Rightarrow \quad x = \frac{\pi}{2}$.

(iv) $\frac{dy}{dx} = \frac{1}{\sin x} \cdot \cos x = \cot x$

$\Rightarrow \quad \frac{dy}{dx} > 0$ when $x \in \left(0, \frac{\pi}{2}\right)$

$\frac{dy}{dx} < 0$, when $x \in \left(\frac{\pi}{2}, \pi\right)$

(v) $\frac{d^2y}{dx^2} = -\operatorname{cosec}^2 x < 0$ for all $x \in (0, \pi)$

Thus, **increasing and concave down** $\boldsymbol{x \in (0, \pi/2)}$

decreasing and concave down $\boldsymbol{x \in (\pi/2, \pi)}$.

and to plot the curve only when $x \in (2n\pi, (2n+1)\pi)$. Shown as in Fig. 2.213.

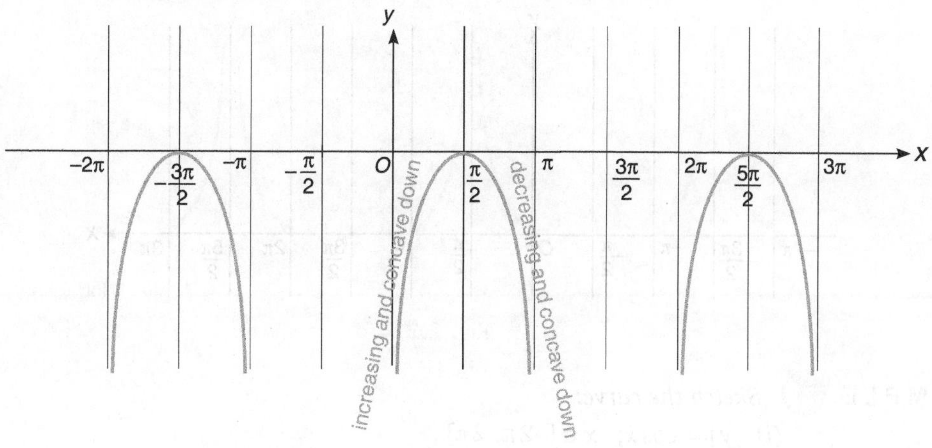

Fig. 2.213

EXAMPLE 12 *Sketch the graph for*

$$y = -\log_{\sin x} e.$$

SOLUTION (i) Here, $y = -\log_{\sin x} e = \frac{-1}{\log_e \sin x}$ is defined when,

$$\sin x > 0 \quad \text{and} \quad \sin x \neq 1.$$

i.e., $\boldsymbol{x \in (2n\pi, (2n+1)\pi) - \{2n\pi + \pi/2\};\ n \in N}$ ...(i)

Curvature and Transformations

(ii) $y = \dfrac{-1}{\log_e \sin x}$ is periodic with period 2π,

$\therefore$ to discuss the graph for $x \in (0, \pi) - \{\pi/2\}$ and sketch for the entire number line.

(iii) $\therefore$ $\dfrac{dy}{dx} = + \dfrac{1}{(\log \sin x)^2} \cdot \dfrac{1}{\sin x} \cdot \cos x \Rightarrow \dfrac{dy}{dx} = + \dfrac{\cot x}{(\log \sin x)^2}$

or $\dfrac{dy}{dx} > 0$ when $x \in \left(0, \dfrac{\pi}{2}\right)$

$\dfrac{dy}{dx} < 0$ when $x \in \left(\dfrac{\pi}{2}, \pi\right)$

(iv) $$\frac{d^2y}{dx^2} = + \left\{\frac{(\log \sin x)^2 \cdot (-\text{cosec}^2\, x) - \cot x \cdot 2\,(\log \sin x) \cdot \cot x}{(\log \sin x)^4}\right\}$$

$$= - \frac{\{(\log \sin x)^2 \text{cosec}^2 x + 2\,(\log \sin x) \cot^2 x\}}{(\log \sin x)^2}$$

$\Rightarrow$ $\dfrac{d^2y}{dx^2} > 0$ for all $x \in (0, \pi) - \left\{\dfrac{\pi}{2}\right\}$

Thus, $$y = \frac{1}{\log \sin x} \Rightarrow \begin{cases} \text{increasing and concave up } x \in \left(\dfrac{\pi}{2}, \pi\right) \\ \text{decreasing and concave up } x \in \left(0, \dfrac{\pi}{2}\right) \end{cases}$$

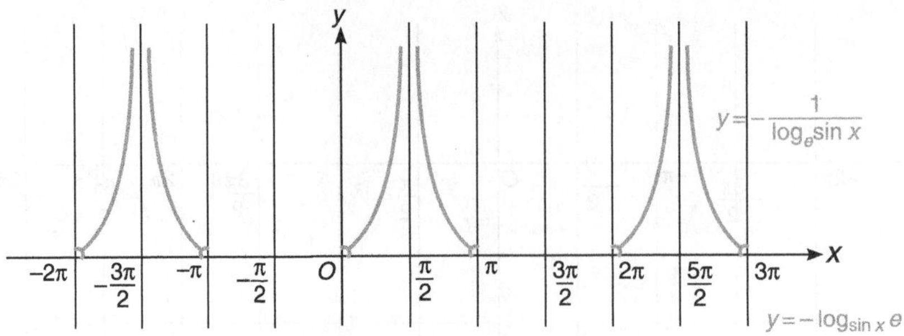

Fig. 2.214

EXAMPLE 13 *Sketch the curve:*

(i) $[y] = \cos x;\ x \in [-2\pi, 2\pi]$

(ii) $[y] = [\cos x];\ x \in [-2\pi, 2\pi]$; *where [·] denotes greatest integer function.*

SOLUTION As we know; $y = \cos x$ could be plotted as;

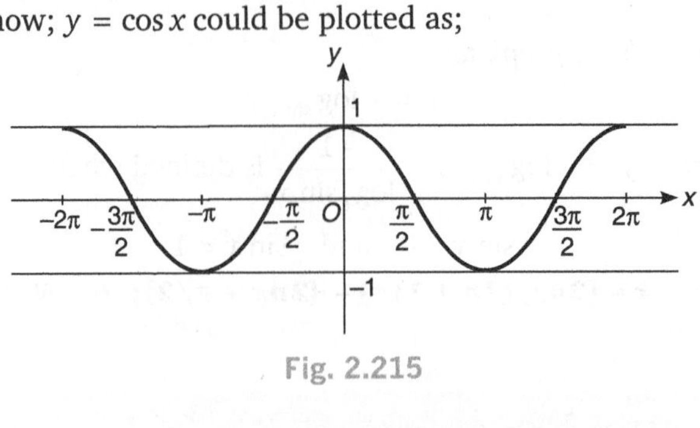

Fig. 2.215

(i) Sketching of $[y] = \cos x$

From above curve $\cos x = -1, 0, 1$ when $x = -2\pi, -\frac{3\pi}{2}, -\pi, -\frac{\pi}{2}, 0, \frac{\pi}{2}, \pi, \frac{3\pi}{2}, 2\pi$

$\Rightarrow$ $[y] = \cos x$ is possible only

when $x \in \left\{\pm 2\pi, \pm \frac{3\pi}{2}, \pm \pi, \pm \frac{\pi}{2}, 0\right\}$

Thus,

$\cos x = -1 \Rightarrow -1 \le y < 0$

$\cos x = 0 \Rightarrow 0 \le y < 0$

$\cos x = 1 \Rightarrow 1 \le y < 2$

Thus, **$[y] = \cos x$ is plotted as shown in Fig. 2.216.**

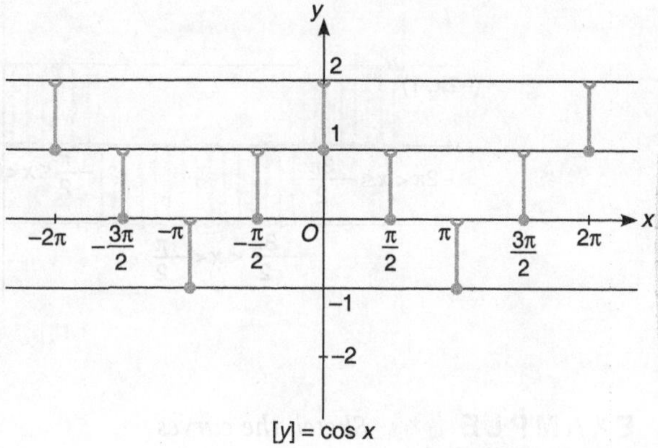

Fig. 2.216

(ii) Sketching of $[y] = [\cos x]$

First to plot $y = [\cos x]$. Shown as;

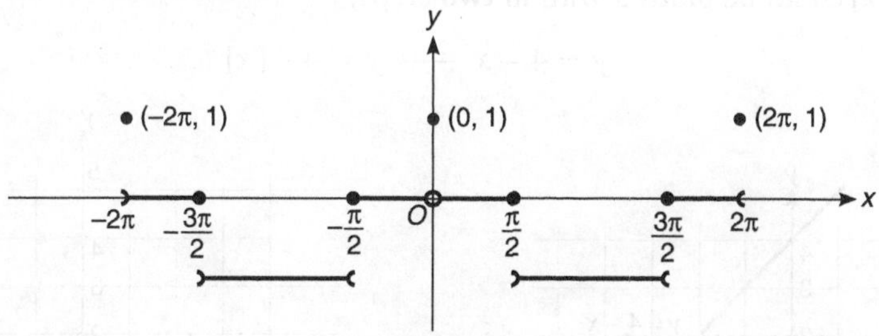

Fig. 2.217

From above figure $y = [\cos x]$

$$\Rightarrow \quad y = \begin{cases} -1 \; ; & \frac{\pi}{2} < |x| < \frac{3\pi}{2} \\ 0 \; ; & 0 < |x| \le \frac{\pi}{2} \text{ and } \frac{3\pi}{2} \le |x| < 2\pi \\ 1 \; ; & |x| = 0, 2\pi \end{cases}$$

Thus, when $y = 0 \Rightarrow [y] \in [0, 1)$

$y = 1 \Rightarrow [y] \in [1, 2)$

$y = -1 \Rightarrow [y] \in [-1, 0)$

Thus, **graph for $[y] = [\cos x]$**

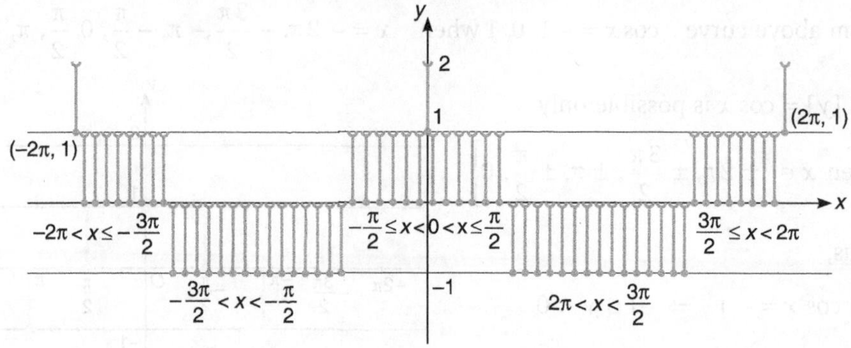

Fig. 2.218

EXAMPLE 14 *Sketch the curves*

(i) $y = 4 - [x]$

(ii) $[y] = 4 - [x]$

(iii) $[|y|] = 4 - [|x|]$

SOLUTION As we know;

(i) $y = 4 - [x]$ could be plotted with in two steps;

$$y = 4 - x \longrightarrow y = 4 - [x]$$

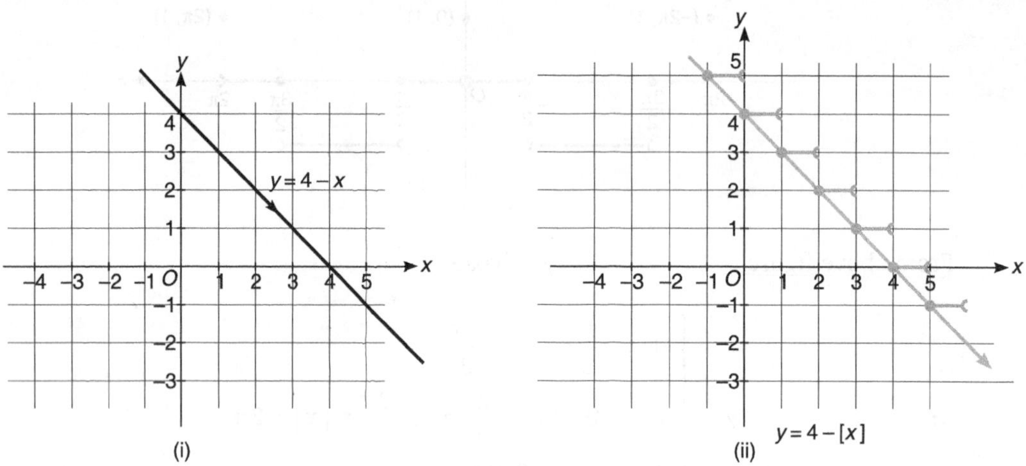

Fig. 2.219

(ii) Since, we know from above curve $y = 4 - [x]$

$\Rightarrow$ $y \in$ integer, thus, on taking integral part, say

$$[y] = I \Rightarrow I \leq y < (I + 1)$$

∴ **Graph for $[y] = 4 - [x]$**

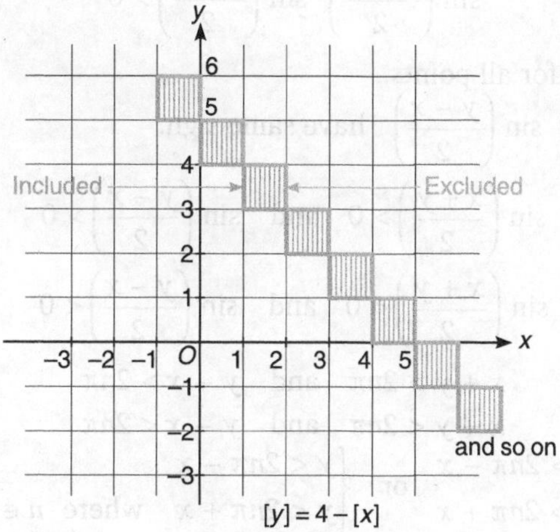

Fig. 2.220

In above curve lower boundary are included and upper are excluded.

(iii) To sketch $[|y|] = 4 - [|x|]$

From above figure we can say;

$$[y] = 4 - [x] \longrightarrow [y] = 4 - [|x|] \longrightarrow [|y|] = 4 - [|x|]$$

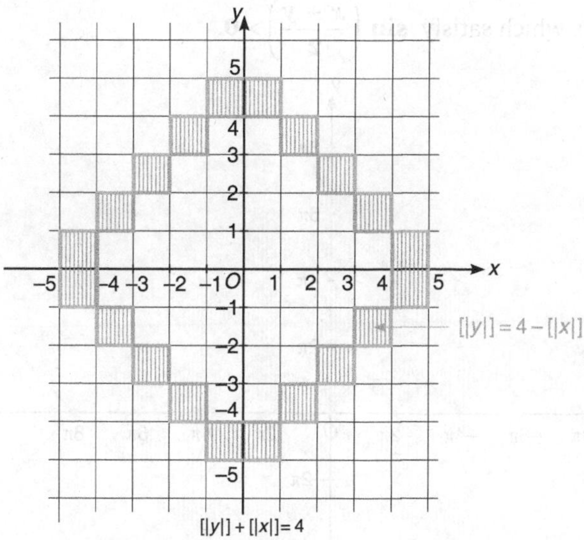

Fig. 2.221

EXAMPLE 15 *Shade the region whose co-ordinates x and y satisfy the equation.*

$$\cos x - \cos y > 0$$

SOLUTION Here, $\cos x - \cos y > 0$ can be written as, $2\sin\left(\frac{x+y}{2}\right)\sin\left(\frac{y-x}{2}\right) > 0$

$$\Rightarrow \qquad \sin\left(\frac{x+y}{2}\right)\sin\left(\frac{y-x}{2}\right) > 0$$

This inequality holds true for all points,

when $\sin\left(\frac{x+y}{2}\right)$ and $\sin\left(\frac{y-x}{2}\right)$ have same sign.

i.e., $\sin\left(\frac{x+y}{2}\right) > 0$ and $\sin\left(\frac{y-x}{2}\right) > 0$

or $\sin\left(\frac{x+y}{2}\right) < 0$ and $\sin\left(\frac{y-x}{2}\right) < 0$

$\Rightarrow$ $x + y > 2n\pi$ and $y - x > 2n\pi$

or $x + y < 2n\pi$ and $y - x < 2n\pi$

$$\Rightarrow \quad \begin{cases} y > 2n\pi - x \\ y > 2n\pi + x \end{cases} \text{ or } \begin{cases} y < 2n\pi - x \\ y < 2n\pi + x \end{cases} \text{ where } n \in z$$

Here, the equation $x + y = 2k\pi$ represents a system of parallel straight lines corresponding to different values of k.

Say $k = 0, 1 \Rightarrow y = -x$ and $y = -x + 2\pi$.

i.e., the set of points whose coordinates satisfy the inequality

$$0 < x + y < 2\pi$$

Similarly, for general points: $\mathbf{2k\pi < x + y < (2k + 1)\pi}$ ...(i)

Now, the set of points which satisfy $\mathbf{\sin\left(\frac{x+y}{2}\right) > 0}$.

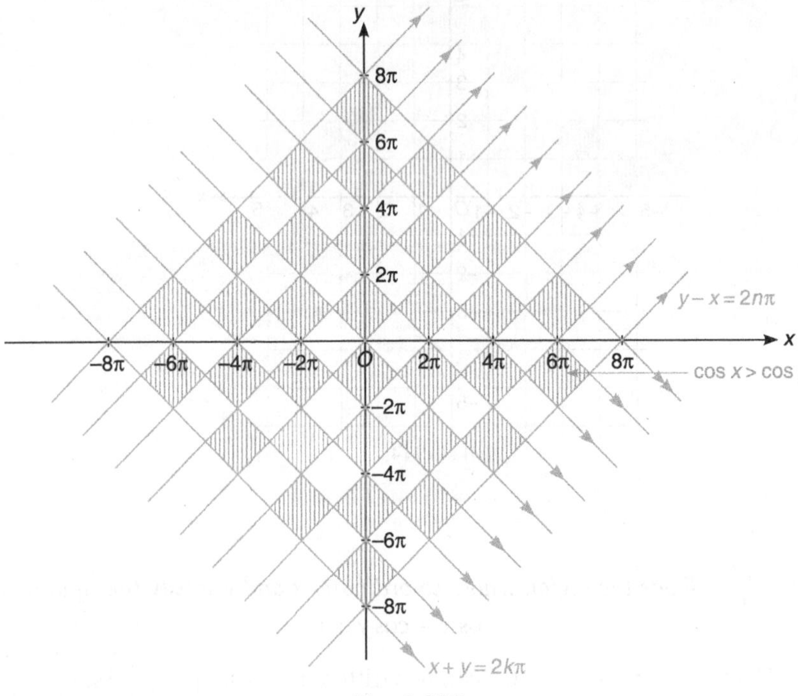

Fig. 2.222

$$\Rightarrow \quad 2n\pi < \frac{x+y}{2} < (2n+1)\pi$$

$$\Rightarrow \quad \mathbf{2(2n\pi) < x + y < 2(2n+1)\pi} \quad \text{...(ii)}$$

$\Rightarrow$ only those points of x, y which satisfy Eq. (ii).

Similarly, $\quad \mathbf{2(2p-1)\pi < y - x < 2(2p)\pi} \quad$ ...(iii)

From above results, graph for $\mathbf{\cos x > \cos y}$ is shown in Fig. 2.222.

EXAMPLE 16 *Sketch the curve* $x^{2/3} + y^{2/3} = a^{2/3}$.

SOLUTION As to sketch, $x^{2/3} + y^{2/3} = a^{2/3}$.

(i) Curve is symmetric about y-axis (as when x is replaced by $(-x)$ curve remains same)

(ii) Curve is symmetric about x-axis (as when y is replaced by $(-y)$ curve remains same).

(iii) Curve is symmetric about origin (as when x is replaced by y and y by x curve remains same).

(iv) When, $\quad x = 0 \Rightarrow y = \pm a$

when, $\quad y = 0 \Rightarrow x = \pm a$.

(v) $\quad y^{2/3} = (a^{2/3} - x^{2/3}),$

or $\quad y^2 = (a^{2/3} - x^{2/3})^3;$

differentiating both sides, $\quad 2y\frac{dy}{dx} = 3(a^{2/3} - x^{2/3})^2 \cdot \left(-\frac{2}{3}x^{-1/3}\right)$

$\Rightarrow \frac{dy}{dx} < 0$ when $x > 0$, $y > 0$ [to discuss $0 < x \le a$ and to take symmetry for rest of graph using Eqs. (i), (ii) and (iii)].

(vi) From above $\quad y\frac{dy}{dx} = -x^{-1/3}(a^{2/3} - x^{2/3})^2$

Differentiating again, we get;

$$y\frac{d^2y}{dx^2} + \left(\frac{dy}{dx}\right)^2 = -x^{-1/3} \cdot 2(a^{2/3} - x^{2/3}) \cdot \left(-\frac{2}{3}x^{-1/3}\right) + \frac{1}{3}x^{-4/3} \cdot (a^{2/3} - x^{2/3})^2$$

$$= \frac{4}{3}x^{-2/3}(a^{2/3} - x^{2/3}) + \frac{1}{3}x^{-4/3}(a^{2/3} - x^{2/3})^2$$

$$= \frac{1}{3}x^{-2/3}(a^{2/3} - x^{2/3})\{4 + x^{-2/3}(a^{2/3} - x^{2/3})\}$$

$\frac{d^2y}{dx^2} = +$ ve whenever; $\quad 0 < x \le a$ and $y > 0$.

Thus; when $\quad 0 < x \le a$ and $y > 0$

$\Rightarrow \quad \frac{dy}{dx} < 0$ and $\frac{d^2y}{dx^2} > 0$,

i.e., **decreasing and concave up.**

$\therefore$ **Graph for** $\mathbf{x^{2/3} + y^{2/3} = a^{2/3}}$

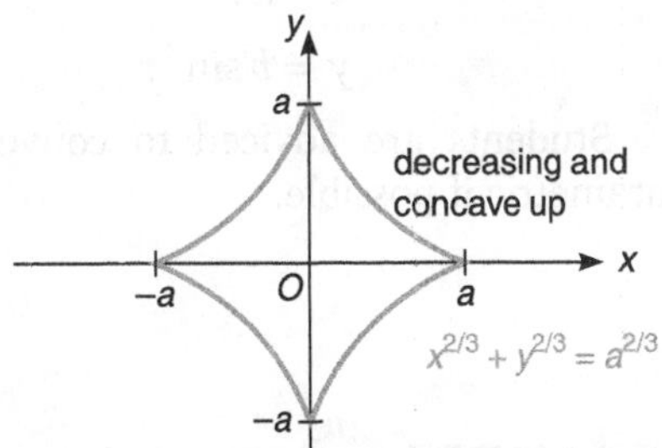

Fig. 2.223

Note Above curve is known as **Astroid** represented by the following parametric equations:

$$\left.\begin{array}{l} x = a\cos^3 t \\ y = a\sin^3 t \end{array}\right\} \quad 0 \le t \le 2\pi, \ a > 0$$

Eliminating $\cos t$ and $\sin t$ from above equations, we get $x^{2/3} + y^{2/3} = a^{2/3}$.
where the quantities x and y are defined for all values of t. But since, the functions $\cos^3 t$ and $\sin^3 t$ are periodic with period 2π, it is sufficient to discuss the curve for $t \in [0, 2\pi]$.

Thus, $\left.\begin{array}{l}\text{Domain of function} \in [0, 2\pi]\\ \text{Range of function} \in [-a, a]\end{array}\right\}$ ...(i)

Now, $\left.\begin{array}{l}\dfrac{dx}{dt} = -3a\cos^2 t \sin t\\ \dfrac{dy}{dt} = +3a\sin^2 t \cos t\end{array}\right\}$ ...(ii)

Here, $\dfrac{dx}{dt} = 0, \dfrac{dy}{dx} = 0 \Rightarrow t = 0, \dfrac{\pi}{2}, 8\pi, \dfrac{3\pi}{2}, 2\pi.$

$\therefore$ $\left.\dfrac{dy}{dx} = -\tan t\right\}$ ...(iii)

$\left.\dfrac{d^2y}{dx^2} = \dfrac{1}{3a\cos^4 t \sin t}\right\}$ ...(iv)

From Eqs. (i), (ii), (iii) and (iv), we construct a table;

Range of t	Domain (x)	Range (y)	Sign of $\dfrac{dy}{dx}$	Sign of $\dfrac{d^2y}{dx^2}$
$0 < t < \dfrac{\pi}{2}$	$0 < x < a$	$0 < y < a$	−	+
$\dfrac{\pi}{2} < t < \pi$	$-a < x < 0$	$0 < y < a$	+	+
$\pi < t < \dfrac{3\pi}{2}$	$-a < x < 0$	$-a < y < 0$	−	−
$\dfrac{3\pi}{2} < t < 2\pi$	$0 < x < a$	$-a < y < 0$	+	−

On the basis of above information we can sketch

$x = a\cos^3 t$

$y = b\sin^3 t$

Students are adviced to convert the cartesian into parametric if possible.

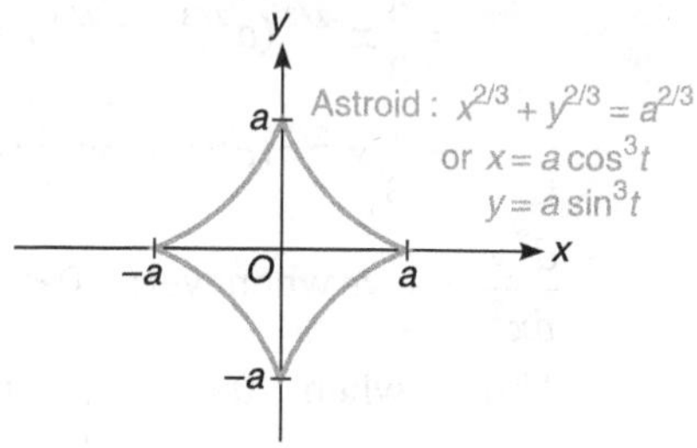

Fig. 2.224

EXAMPLE 17 *Sketch the curves:*

(i) $y = \dfrac{e^x + e^{-x}}{2}$ (ii) $y = \dfrac{e^x - e^{-x}}{2}$ (iii) $y = \dfrac{e^x - e^{-x}}{e^x + e^{-x}}$ (iv) $y = \dfrac{e^x + e^{-x}}{e^x - e^{-x}}$

SOLUTION (i) Sketching of $f(x) = y = \dfrac{e^x + e^{-x}}{2}$

(a) $f(-x) = f(x)$, $\therefore$ even or symmetric about y-axis ...(i)

Play with Graphs

(b) When $x = 0 \Rightarrow y = 1$ ...(ii)

Differentiating, we get; $\dfrac{dy}{dx} = \dfrac{e^x - e^{-x}}{2} = \dfrac{e^{2x} - 1}{2e^x}$

$\Rightarrow \dfrac{d^2y}{dx^2} = \dfrac{e^x + e^{-x}}{2} = \dfrac{e^{2x} + 1}{2e^x}$

From above; $\left.\begin{aligned} \frac{dy}{dx} > 0 \text{ whenever } x > 0 \\ \frac{dy}{dx} < 0 \text{ whenever } x < 0 \end{aligned}\right\}$ ...(iii)

$\left.\dfrac{d^2y}{dx^2} > 0 \text{ for all } x\right\}$ ...(iv)

Thus, from Eqs. (i), (ii), (iii) and (iv) graph of $y = \dfrac{e^x + e^{-x}}{2}$ is;

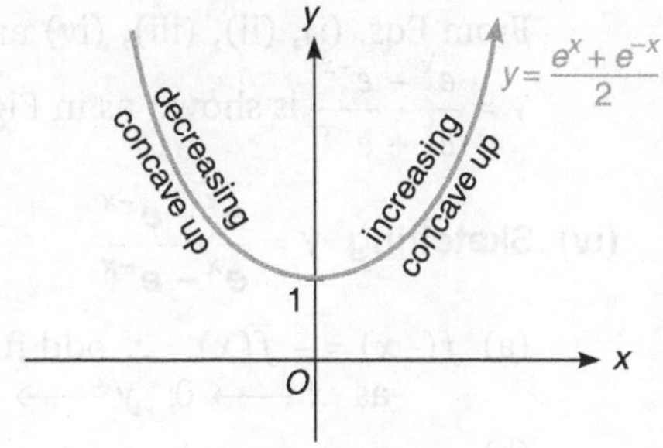

Fig. 2.225

(ii) Sketching $y = \dfrac{e^x - e^{-x}}{2}$

(a) $f(-x) = -f(x)$,

$\therefore$ odd function or symmetric about origin ...(i)

(b) $x = 0 \Rightarrow y = 0$ ...(ii)

(c) $\dfrac{dy}{dx} = \dfrac{e^x + e^{-x}}{2} > 0$ for all x

...(iii)

(d) $\dfrac{d^2y}{dx^2} = \dfrac{e^x - e^{-x}}{2}$

$\Rightarrow \left.\begin{aligned} \frac{d^2y}{dx^2} > 0 \text{ when } x > 0 \\ \frac{d^2y}{dx^2} < 0 \text{ when } x < 0 \end{aligned}\right\}$ ...(iv)

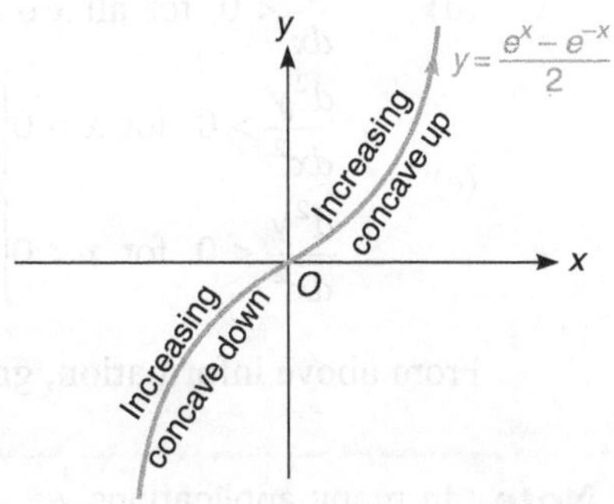

Fig. 2.226

From Eqs. (i), (ii), (iii) and (iv) graph of $y = \dfrac{e^x - e^{-x}}{2}$.

(iii) Sketching of $y = \dfrac{e^x - e^{-x}}{e^x + e^{-x}}$

(a) $f(-x) = -f(x)$, $\therefore$ odd function or symmetric about origin ...(i)

(b) as $x = 0 \Rightarrow y = 0$ ...(ii)

(c) $\left.\begin{aligned} \text{Domain} \in R \\ \text{Range} \in (-1, 1) \end{aligned}\right\}$ ...(iii)

(d) $\dfrac{dy}{dx} > 0$ for all $x \in R$ ...(iv)

(e) $\left.\begin{array}{l}\dfrac{d^2y}{dx^2} < 0 \text{ for } x > 0 \\ \dfrac{d^2y}{dx^2} > 0 \text{ for } x < 0\end{array}\right\}$ ...(v)

From Eqs. (i), (ii), (iii), (iv) and (v) graph of $y = \dfrac{e^x - e^{-x}}{e^x + e^{-x}}$ is shown as in Fig. 2.227.

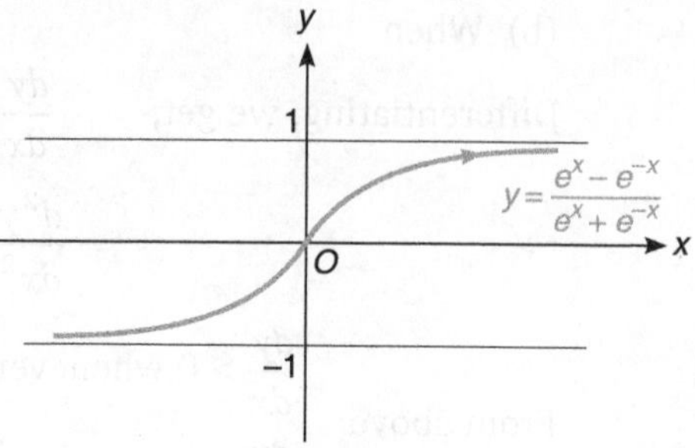

Fig. 2.227

(iv) Sketching $y = \dfrac{e^x + e^{-x}}{e^x - e^{-x}}$

(a) $f(-x) = -f(x)$, $\therefore$ odd function or symmetric about origin ...(i)

(b) $\left.\begin{array}{l}\text{as } x \longrightarrow 0;\ y \longrightarrow \infty \\ \text{as } y \longrightarrow 1;\ x \longrightarrow \infty \\ y \longrightarrow -1;\ x \longrightarrow -\infty\end{array}\right\}$ ...(ii)

(c) $\left.\begin{array}{l}\text{Domain} \in R - \{0\} \\ \text{Range} \in R - [-1, 1]\end{array}\right\}$ ...(iii)

(d) $\dfrac{dy}{dx} < 0$ for all $x \in R - \{0\}$ ...(iv)

(e) $\left.\begin{array}{l}\dfrac{d^2y}{dx^2} > 0 \text{ for } x > 0 \\ \dfrac{d^2y}{dx^2} < 0 \text{ for } x < 0\end{array}\right\}$ ...(v)

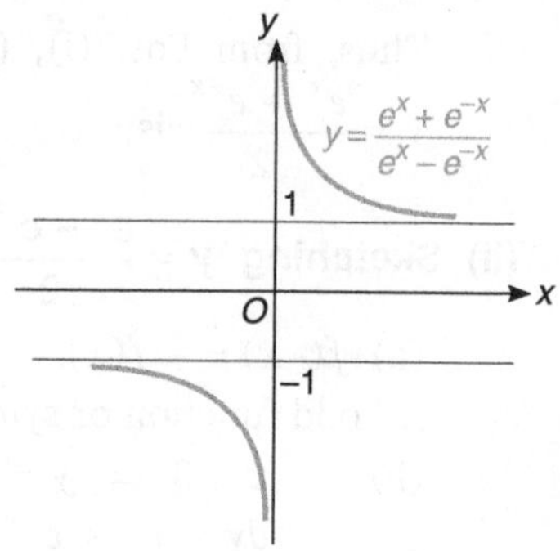

Fig. 2.228

From above information, graph for $y = \dfrac{e^x + e^{-x}}{e^x - e^{-x}}$ as shown in Fig. 2.228.

Note In many applications we come across exponential functions of the form $\frac{1}{2}(e^x - e^{-x})$ and $\frac{1}{2}(e^x + e^{-x})$ known as "hyperbolic functions" represented by;

$$\left.\begin{array}{l}\sin hx = \dfrac{e^x - e^{-x}}{2} \\ \cos hx = \dfrac{e^x + e^{-x}}{2}\end{array}\right\} \quad ...(i)$$

called hyperbolic sine and hyperbolic cosine.

also, $$\left.\begin{array}{l}\tan hx = \dfrac{e^x - e^{-x}}{e^x + e^{-x}} \\ \cot hx = \dfrac{e^x + e^{-x}}{e^x - e^{-x}}\end{array}\right\} \quad ...(ii)$$

called hyperbolic tangent and hyperbolic cotangent. Where if; $x = \cos ht$, $y = \sin ht$

$\Rightarrow$ $\mathbf{x^2 - y^2 = \cos^2 ht - \sin^2 ht = 1}$ {Using Eq. (i)}

which is equation of hyperbola.

Thus, $\sin hx = \dfrac{e^x - e^{-x}}{2}$ and $\cos hx = \dfrac{e^x + e^{-x}}{2}$ are hyperbolic functions.

EXAMPLE 18 *Sketch the curve* $y = 2\sin x + \cos 2x$.

SOLUTION Since, the function is periodic with period 2π, it is sufficient to investigate the function in the interval $[0, 2\pi]$.

(a) $$\frac{dy}{dx} = 2\cos x - 2\sin 2x = 2(\cos x - 2\sin x \cos x)$$

$\therefore$ $$\frac{dy}{dx} = 2\cos x(1 - 2\sin x)$$

Here, $\frac{dy}{dx} = 0 \Rightarrow x = \frac{\pi}{6}, \frac{\pi}{2}, \frac{5\pi}{6}, \frac{3\pi}{2}$

(b) $$\frac{d^2y}{dx^2} = -2\sin x - 4\cos 2x$$

$$\left(\frac{d^2y}{dx^2}\right)_{\text{at } x = \frac{\pi}{6}} = -3 < 0 \qquad \text{where } y = \frac{3}{2}$$

$\therefore$ at $x = \frac{\pi}{6}$ is maximum at $x = \frac{\pi}{6}$. $\left(\frac{d^2y}{dx^2}\right)_{\text{at } x = \frac{\pi}{2}} = 2 > 0;$

$\therefore$ at $x = \frac{\pi}{2}$ is minimum at $x = \frac{\pi}{2}$, where $y = 1$

at $x = \frac{5\pi}{6}$ we have, $\frac{d^2y}{dx^2} = -3 < 0$

and $y = \frac{3}{2}$ (maximum)

at $x = \frac{3\pi}{2}$, we have, $\frac{d^2y}{dx^2} = 6 > 0$

and $y = -3$ (minimum)

Thus, curve for $y = 2\sin x + \cos 2x$;

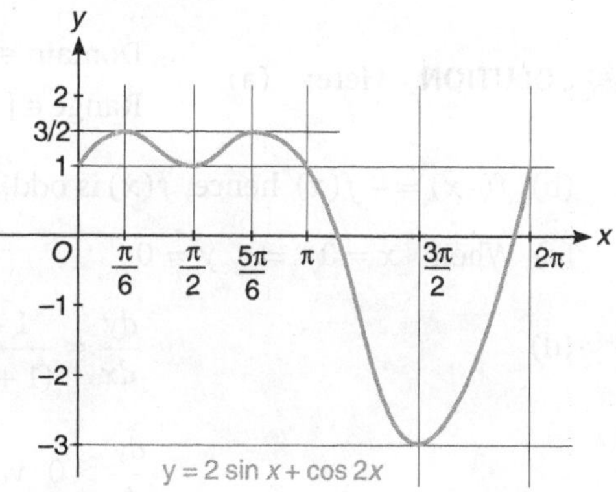

Fig. 2.229

EXAMPLE 19 *Sketch the curve* $y = e^{-x^2}$ (*Gaussian curve*)

SOLUTION As the curve $y = f(x) = e^{-x^2}$

(a) Symmetric about y-axis {as $f(-x) = f(x)$} ...(i)

(b) $$\left.\begin{array}{l} \text{as } x \longrightarrow 0 \Rightarrow y \longrightarrow 1 \\ \text{as } y \longrightarrow 0 \Rightarrow x \longrightarrow \pm\infty \end{array}\right\}$$ ...(ii)

(c) $$\frac{dy}{dx} = e^{-x^2}(-2x) \Rightarrow \left.\begin{array}{l} \frac{dy}{dx} < 0 \Rightarrow x > 0 \\ \frac{dy}{dx} > 0 \Rightarrow x < 0 \end{array}\right\}$$ ...(iii)

(d) $$\frac{d^2y}{dx^2} = -2e^{-x^2} - 2x(-2xe^{-x^2}) = 2e^{x^2}(2x^2 - 1)$$

$$\Rightarrow \left.\begin{aligned} &\frac{d^2y}{dx^2} > 0 \Rightarrow x < -\frac{1}{\sqrt{2}} \text{ or } x > \frac{1}{\sqrt{2}} \\ &\frac{d^2y}{dx^2} < 0 \Rightarrow -\frac{1}{\sqrt{2}} < x < \frac{1}{\sqrt{2}} \end{aligned}\right\} \quad \ldots\text{(iv)}$$

(e) $y = e^{-x^2}$ assumes maximum at

$x = 0 \Rightarrow y = 1$

and Domain $\in R$
Range $\in (0, 1]$ ...(v)

From above discussion;

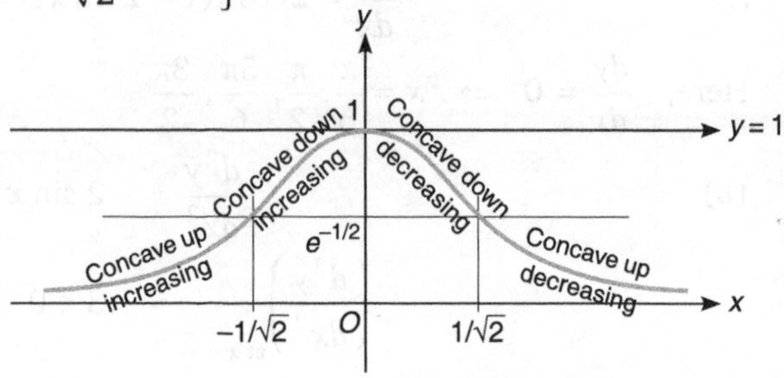

Fig. 2.230

EXAMPLE 20 *Sketch the curve* $y = \dfrac{x}{1+x^2}$.

SOLUTION Here; (a) $$\left.\begin{aligned}&\text{Domain} \in R \\ &\text{Range} \in [-0.5,\ 0.5]\end{aligned}\right\} \quad \ldots\text{(i)}$$

(b) $f(-x) = -f(x)$, hence, $f(x)$ is odd, symmetric about origin ...(ii)

(c) When $x = 0 \Rightarrow y = 0$ ...(iii)

(d) $$\frac{dy}{dx} = \frac{1 - x^2}{(1 + x^2)^2}$$

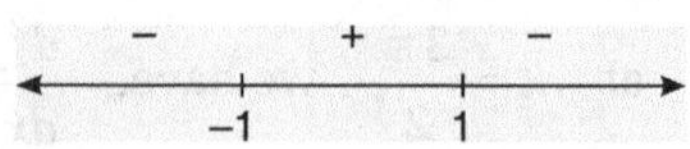

Fig. 2.231

$$\text{or} \quad \left.\begin{aligned} &\frac{dy}{dx} > 0 \text{ when } -1 < x < 1 \\ &\frac{dy}{dx} < 0 \text{ when } x < -1 \text{ or } x > 1 \end{aligned}\right\} \quad \ldots\text{(iv)}$$

From above maximum at $x = 1$ and minimum at $x = -1$} ...(v)

(e) $$\frac{d^2y}{dx^2} = \frac{2^x(x^2 - 3)}{(1 + x^2)^3}$$

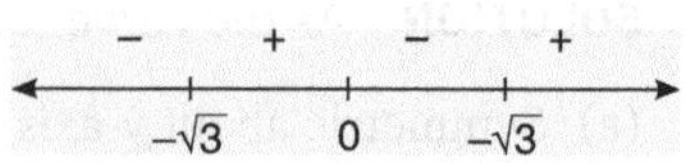

Fig. 2.232

$$\Rightarrow \left.\begin{aligned} &\frac{d^2y}{dx^2} > 0, \text{ when } x \in (-\sqrt{3}, 0) \text{ and } (\sqrt{3}, \infty) \\ &\frac{d^2y}{dx^2} < 0, \text{ when } x \in (-\infty, -\sqrt{3}) \text{ and } (0, \sqrt{3}) \end{aligned}\right\} \quad \ldots\text{(vi)}$$

From above conditions graph for $y = \dfrac{x}{1+x^2}$

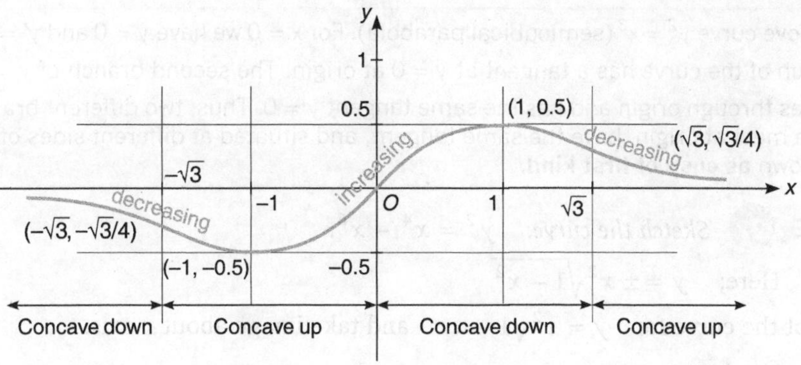

Fig. 2.233

In above figure :

x	dy/dx	d^2y/dx^2
$-\infty < x < -\sqrt{3}$	$-$	$-$
$-\sqrt{3} < x < -1$	$-$	$+$
$-1 < x < 0$	$+$	$+$
$0 < x < 1$	$+$	$-$
$1 < x < \sqrt{3}$	$-$	$-$
$\sqrt{3} < x < \infty$	$-$	$+$

EXAMPLE 21 *Sketch the curve* $y^2 = x^3$. *(Semicubical parabola).*

SOLUTION To plot the curve we discuss;

(i) Symmetric about x-axis (as $y \longrightarrow -y$ curve remains same).

(ii) Domain $\in [0, \infty)$

(iii) Range $\in R$.

Here, to plot $y^2 = x^3 \Rightarrow y = \sqrt{x^3}$ and $y = -\sqrt{x^3}$, we draw $y = \sqrt{x^3}$ and take image about x-axis for $y = -\sqrt{x^3}$. (*i.e.*, to discuss curve when $x, y \geq 0$).

(iv) $\dfrac{dy}{dx} = \dfrac{3}{2}x^{1/2} = \dfrac{3}{2}\sqrt{x} > 0$ for all $x, y > 0$.

(v) $\dfrac{d^2y}{dx^2} = \dfrac{3}{4\sqrt{x}} > 0$ for all $x, y > 0$.

$\Rightarrow$ Increasing and concave up when $x > 0$.

From Eqs. (i), (ii), (iii), (iv) and (v).

x	y	dy/dx	d^2y/dx^2
$+$	$+$	$+$	$+$
$+$	$-$	$-$	$-$

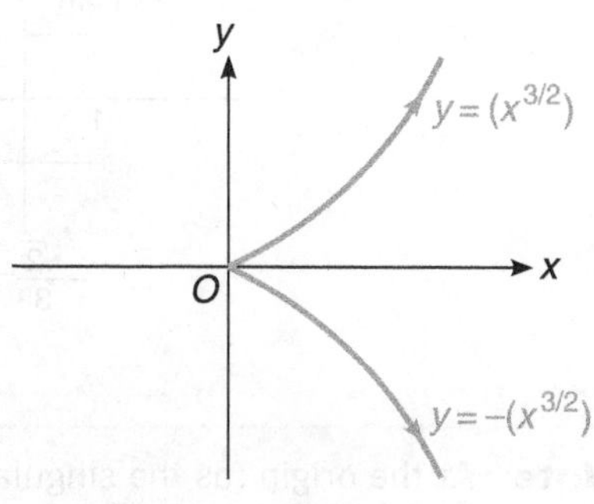

Fig. 2.234

> **Note** In above curve $y^2 = x^3$ (semicubical parabola). For $x = 0$ we have $y = 0$ and $y' = 0$, thus, the branch of the curve has a tangent at $y = 0$ at origin. The second branch of $y = -\sqrt{x^3}$ also passes through origin and has the same tangent $y = 0$. Thus, two different branches of the curve meet at origin, have the same tangent, and situated at different sides of origin. This is known **as cusp of first kind.**

EXAMPLE 22 *Sketch the curve:* $y^2 = x^4 - x^6$.

SOLUTION Here; $y = \pm x^2\sqrt{1 - x^2}$

Thus, to plot the curve for $y = x^2\sqrt{1 - x^2}$ and take image about x-axis.

(i) Symmetric about x and y-axis.

(ii) Domain $\in [-1, 1]$

(iii) Range $\in \left[-\frac{2}{3\sqrt{3}}, \frac{2}{3\sqrt{3}}\right]$.

(iv) When $x = 0 \Rightarrow y = 0$

$y = 0 \Rightarrow x = 0, \pm 1.$

(v) $2y\frac{dy}{dx} = 4x^3 - 6x^5 = 2x^3(2 - 3x^2) = 2x^3(\sqrt{2} - \sqrt{3}x)(\sqrt{2} + \sqrt{3}x)$

$\Rightarrow \quad \frac{dy}{dx} > 0$ when; $0 < x < \sqrt{\frac{2}{3}}$ and $y > 0.$

$\frac{dy}{dx} < 0$ when $\sqrt{\frac{2}{3}} < x < 1$ and $y > 0.$

Here, we are sketching the curve only when $x, y \geq 0$ and then take image about x and y-axis.

(vi) $\frac{d^2y}{dx^2} > 0$ when $0 < x < \sqrt{\frac{2}{5}}$ and $y > 0$.

$\frac{d^2y}{dx^2} < 0$ when $\sqrt{\frac{2}{5}} < x < 1$ and $y > 0.$

From above discussion.

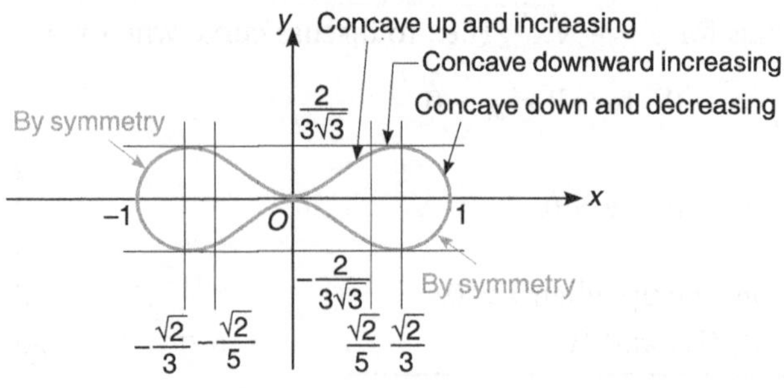

Fig. 2.235

> **Note** At the origin (as the singular point) the two branches of the curve corresponding to plus and minus in front of the radical sign are mutually tangent. **Known as point of osculation or tacnode or double cusp.**

EXERCISE

1. Sketch the curves; (where $[\cdot]$ denotes the greatest integer function).
   (i) $y = |2 - |x - 1||$
   (ii) $y = 2 - \dfrac{4}{|x - 1|}$
   (iii) $|y| = 2 - \dfrac{4}{|x - 1|}$
   (iv) $y = \left| 2 - \dfrac{4}{|x - 1|} \right|$
   (v) $|y| = \left| 2 - \dfrac{4}{|x - 1|} \right|$
   (vi) $y = |e^{|x|} - 2|$
   (vii) $|y| = |e^{|x|} - 2|$
   (viii) $y = x - [x]$
   (ix) $y = \sqrt{x - [x]}$
   (x) $y = (x - [x])^2$
   (xi) $|y| = \sqrt{x - [x]}$
   (xii) $|y| = (x - [x])^2$
   (xiii) $y = |x - 1| + |x + 1|$
   (xiv) $|y| = |x - 1| + |x + 1|$
   (xv) $y = [|x - 1|]$
   (xvi) $|y| = [|x - 1|]$
   (xvii) $y = x + [x]$
   (xviii) $y = |x| + [|x|]$
   (xix) $|y| = x + [x]$
   (xx) $|y| = |x| + [|x|]$
2. Sketch the curves;
   (i) $y = \sqrt{\sin x}$
   (ii) $|y| = \sqrt{\sin x}$
   (iii) $y = |\sin x| + |\cos x|$
   (iv) $|y| = \cos x + |\cos x|$
   (v) $y = \sin^2 x - 2\sin x$
   (vi) $y = 2^{\sin x}$
   (vii) $y = \log_2(|\sin x|)$
   (viii) $|y| = \log_2|\sin x|$
   (ix) $y = \log_{\sin x}\left(\dfrac{1}{2}\right)$
   (x) $|y| = \log_{\sin x}\left(\dfrac{1}{2}\right)$
3. Sketch the curve $y = \sin^{-1}\left(\dfrac{1 - x^2}{1 + x^2}\right)$
4. Sketch the curve $y = \dfrac{2x}{x^2 + 1}$
5. Sketch the curves;
   (i) $y = x^2 - 2|x|$ (ii) $y = e^{-|x|}$
   (iii) $y = e^{|x|}$ (iv) $|y| = x$
   (v) $y = x^3 - x$ (vi) $y^2 = x - 1$
6. Construct the graph of the function
   $$y = f(x - 1) + f(x + 1)$$
   where $f(x) = \begin{cases} 1 - |x|, & \text{when } |x| \le 1 \\ 0, & \text{when } |x| > 1 \end{cases}$
7. Sketch the curves;
   (i) $y = \dfrac{1}{x - 2}$ (ii) $y = \dfrac{1}{|x| - 2}$
   (iii) $y = \left| \dfrac{1}{|x| - 2} \right|$ (iv) $|y| = \left| \dfrac{1}{|x| - 2} \right|$
8. Find number of solutions of $2\cos x = |\sin x|$. When $x \in [0, 4\pi]$.
9. Find the number of solutions of;
   $$\sin \pi x = |\log|x||.$$
10. Sketch the curve
   $$y = \frac{\sin 2x}{2} + \cos x.$$

ANSWERS

8. 2 solutions, **9.** 6 solutions.

Graph Tells Us

Remark 1

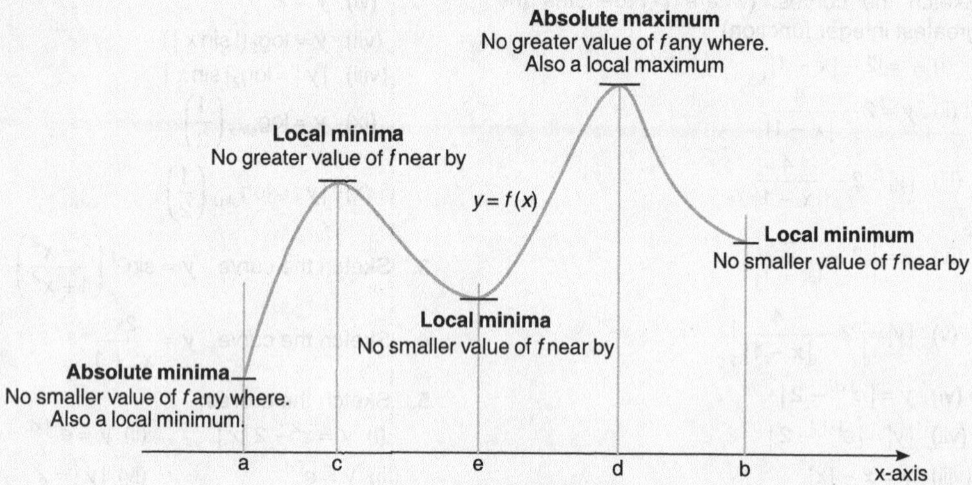

Local vs absolute (Global) Extrema

Fig. 1

Remark 2

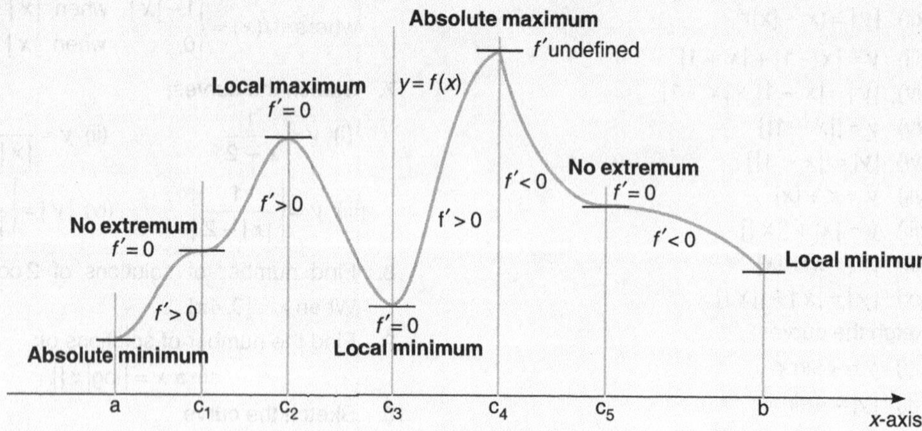

Local vs absolute (Global) Extrema

Fig. 2

Remark 3

The First derivative Test for Local Extreme Value

(i) If f' changes from positive to negative at C ($f' > 0$ for $x < c$ and $f' < 0$ for $x > c$), then f has **a local maximum value at c**.

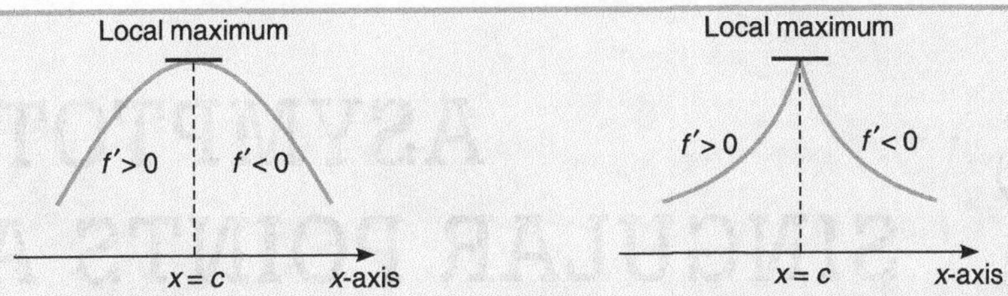

(a) $f'(c) = 0$ (b) $f'(c)$ undefined

Fig. 3

(ii) If f' changes from negative to positive at c ($f' < 0$ for $x < c$ and $f' > 0$ for $x > c$), then f has **a local minimum at $x = c$.**

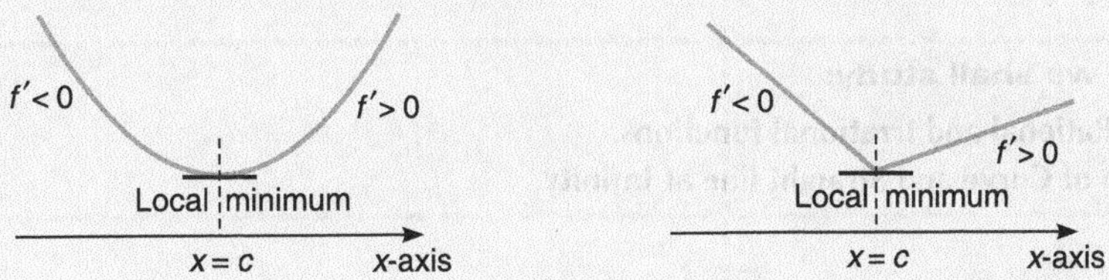

(a) $f'(c) = 0$ (b) $f'(c)$ undefined

Fig. 4

(iii) If f' does not change sign at c (f' has the same sign on both sides of c), then f has **no local extreme value at c.**

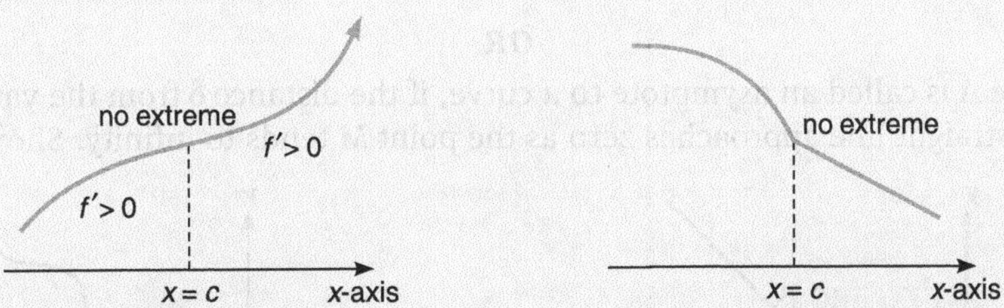

(a) $f'(c) = 0$ (b) $f'(c)$ undefined

Fig. 5

(iv) **At a left end point 'a':** If $f' < 0$ ($f' > 0$) for $x > a$, then f has **local maximum (minimum) value at $x = a$.**

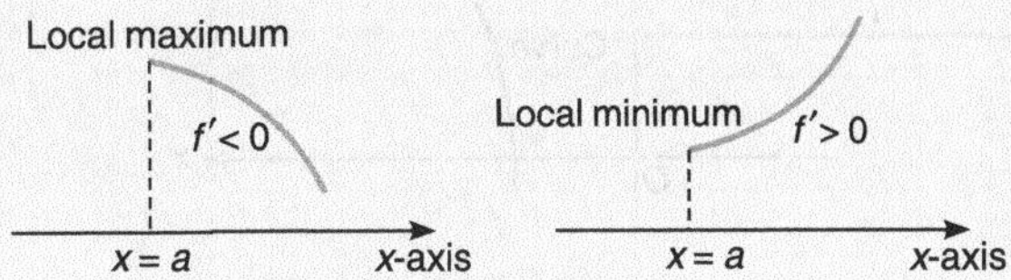

Fig. 6

(v) **At a right end point 'b':** If $f' < 0$ ($f' > 0$) for $x < b$, then f has **local minimum (maximum) at $x = b$.**

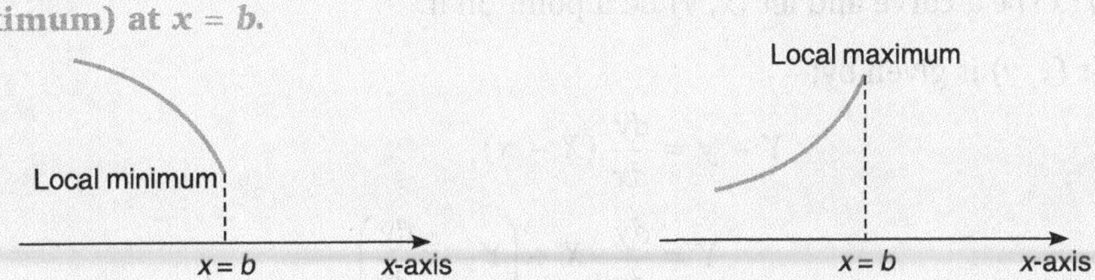

Fig. 7

Chapter 3

ASYMPTOTES, SINGULAR POINTS AND CURVE TRACING

In this chapter we shall study:

- Plotting of Rational and Irrational functions.
- Intersection of Curve and Straight line at Infinity.

3.1 ASYMPTOTES

A straight line, at a finite distance from origin, is said to be an asymptote of the curve $y = f(x)$, if the perpendicular distance of the point P on the curve from the line tends to zero when x or y both tends to infinity.

OR

A straight line A is called an asymptote to a curve, if the distance δ from the variable point M of the curve to this straight line approaches zero as the point M tends to infinity. Shown as:

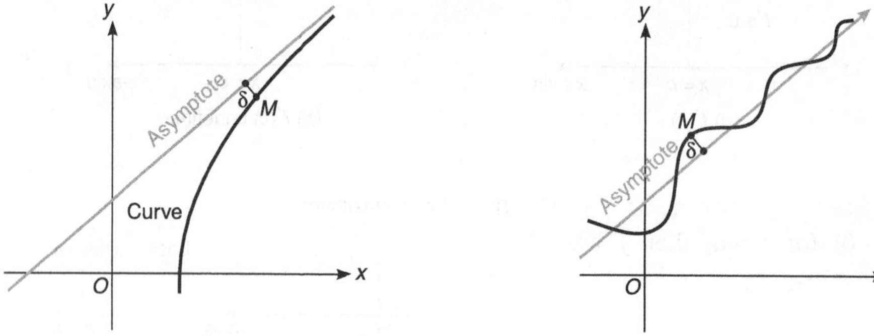

Fig. 3.1

Mathematically

Let $y = f(x)$ be a curve and let (x, y) be a point on it.

Tanget at (x, y) is given by;

$$Y - y = \frac{dy}{dx}(X - x)$$

$$Y = \frac{dy}{dx} \cdot X + \left(Y - x\frac{dy}{dx}\right) \qquad \ldots \text{(i)}$$

Now, if asymptote exists, then $x \to \infty$

$$\Rightarrow \quad \frac{dy}{dx} \text{ and } \left(y - x\frac{dy}{dx}\right) \longrightarrow \text{finite limit say } m \text{ and } c$$

Say $$\frac{dy}{dx} \to m \quad \text{and} \quad y - \frac{x\,dy}{dx} \to c$$

$\therefore$ Eq. (i) reduces to, $\mathbf{Y = mX + c}$ is asymptote of equation.

Now we shall discuss the following cases

(i) Asymptote parallel to x-axis.
(ii) Asymptote parallel to y-axis.
(iii) Asymptote of algebraic curves or oblique asymptotes.
(iv) Asymptote by inspection.
(v) Intersection of curve and its Asymptotes.
(vi) Asymptote by Expansion.
(vii) The position of the curve with respect to asymptote.

3.1 (i) Asymptote parallel to x-axis

Let the equation of curve be,

$$(a_0x^n + a_1x^{n-1}y + a_2x^{n-2}y^2 + \ldots + a_ny^n) + (b_1x^{n-1} + b_2x^{n-2}y + \ldots + b_ny^{n-1})$$
$$+ (c_2x^{n-2} + c_3x^{n-2}y + \ldots + c_ny^{n-2}) + \ldots + \ldots = 0 \qquad \ldots\text{(i)}$$

then it can be arranged in descending powers of x as follows:

$$a_0x^n + (a_1y + b_1)x^{n-1} + (a_2y^2 + b_2y + c_2)x^{n-2} + \ldots = 0 \qquad \ldots\text{(ii)}$$

Now, if $a_0 = 0$, *i.e.*, the term consisting x^n is absent, then $a_1y + b_1 = 0$, *i.e.*, coefficient of $x^{n-1} = 0$ will make two roots of Eq. (i) infinite as coefficients of both x^n and x^{n-1} are zero.

Hence, $\mathbf{a_1y + b_1 = 0}$ **is an asymptote parallel to x-axis.**

Again if; both x^n and x^{n-1} are absent, then $a_2y^2 + b_2y + c_2 = 0$, *i.e.*, coefficient of x^{n-2} being zero will make three roots of Eq. (ii) infinite hence, $\mathbf{a_2y^2 + b_2y + c_2 = 0}$ **will give two asymptote parallel to x-axis.**

Method to find asymptote parallel to x-axis

To find the asymptote parallel to x-axis equate the coefficient of highest power of x to zero.

If the coefficient is constant, then there is no asymptote parallel to x-axis (horizontal).

3.1 (ii) Asymptote parallel to y-axis

From above article, if we need an asymptote parallel to y-axis, equate the coefficient of highest power of y to zero.

If this coefficient is constant, then there is no asymptote parallel to y-axis (vertical).

EXAMPLE 1 *Sketch the curve* $y = \dfrac{1}{x-5}$

SOLUTION Here; $y(x-5) = 1$

$\therefore$ **Asymptote parallel to x-axis.**

$\Rightarrow$ $y = 0$ (equating highest power of $x = 0$)

Asymptote parallel to y-axis.

$\Rightarrow$ $x = 5$ (equating highest power of $y = 0$)

Thus, $x = 5$ and y-axis are asymptotes shown as in figure.

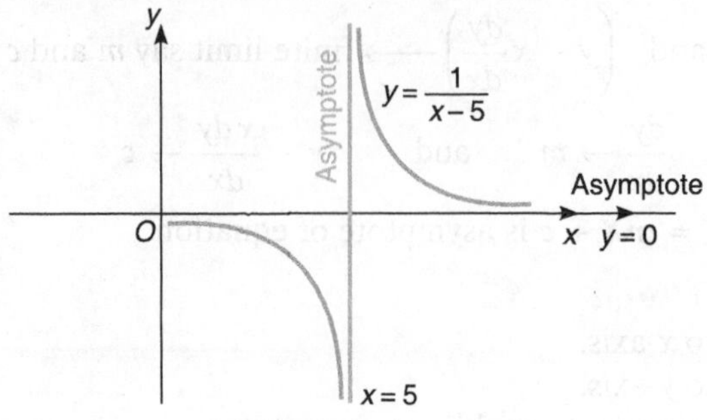

Fig. 3.2

EXAMPLE 2 *Show the curve $y = \tan x$ has an infinite number of vertical asymptote.*

SOLUTION $$y = \tan x$$

here $$y \to \pm\infty \quad \text{as} \quad x \to \pm\frac{\pi}{2}, \pm\frac{3\pi}{2}, \ldots$$

or $$\tan x \to \infty \quad \text{as} \quad x \to \pm\frac{\pi}{2}, \pm\frac{3\pi}{2}, \ldots$$

i.e., equating highest power of $y = 0$.

(as $y = \tan x \Rightarrow y \cot x = 1$, where $\cot x \to 0$).

Shown as:

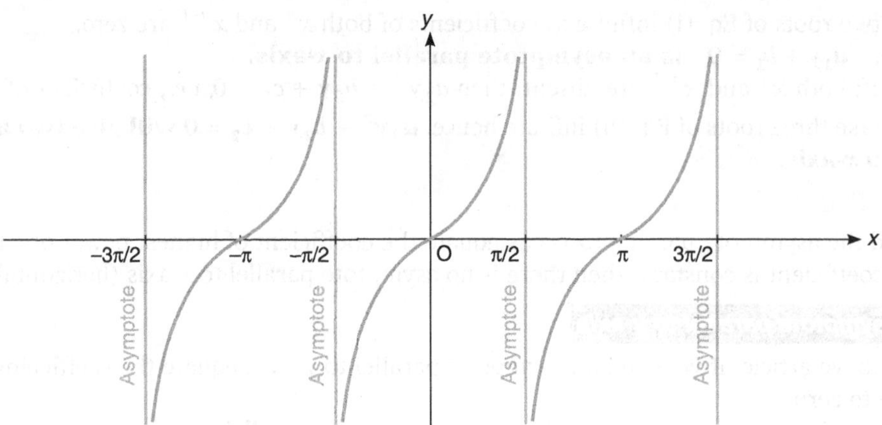

Fig. 3.3

EXAMPLE 3 *Show the curve $y = e^{1/x}$ has a vertical and horizontal asymptote.*

SOLUTION Here $$y = e^{1/x}$$

$$\Rightarrow \quad y \cdot e^{-1/x} = 0$$

or $$e^{-1/x} \to 0 \quad \text{as} \quad x \to 0 \qquad \left(\text{Since, } \lim_{x\to 0} e^{-1/x} \to 0\right)$$

From adjoining figure

$$y = e^{1/x}$$

$$\Rightarrow \qquad \frac{1}{x} = \log y$$

$$\Rightarrow \qquad x = \frac{1}{\log y}$$

which shows $x(\log y) = 1$ has an asymptote parallel to x-axis as

$$\log y = 0 \Rightarrow y = 1.$$

Thus, $y = e^{1/x}$ has two asymptote

$$x = 0 \text{ and } y = 1.$$

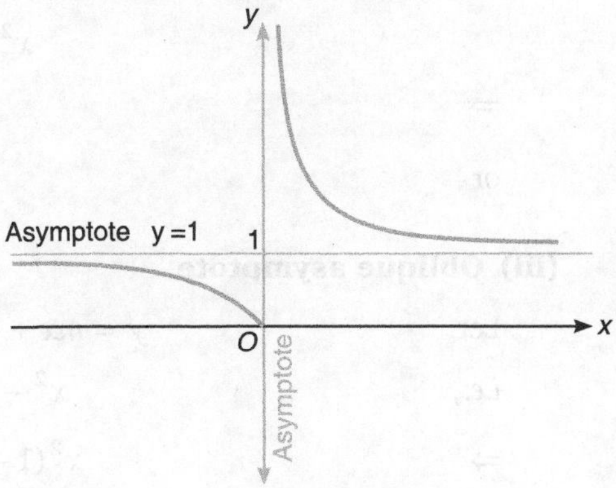

Fig. 3.4

3.1 (iii) Asymptote of algebraic curves or oblique asymptote

An asymptote which is not parallel to y-axis is called an oblique asymptote. Let $y = mx + c$ be an asymptote of $y = f(x)$, then

$$m = \lim_{\substack{x \to \infty \\ \text{or } x \to -\infty}} \frac{y}{x} \quad \text{and} \quad c = \lim_{\substack{x \to \infty \\ \text{or } x \to -\infty}} (y - mx)$$

Method to find oblique asymptote

Suppose $y = mx + c$ is an asymptote of the curve. Put $y = mx + c$ in the equation of the curve and arrange it in descending powers of x. Equate to zero the coefficients of two highest degree terms. Solve these two equations, find m and c. Put them in $y = mx + c$ to get asymptotes.

Note 1. Here, we will find non-parallel or non repeated asymptote only.
2. Neglect all imaginary values of m.

EXAMPLE 1 *Find the asymptotes to the curve* $y = x + \dfrac{1}{x}$ *and then sketch.*

SOLUTION Here, the given curve $y = x + \dfrac{1}{x}$

$$\Rightarrow \qquad xy = x^2 + 1$$

or $$x^2 - xy + 1 = 0$$

(i) Asymptote parallel to x-axis

Equating highest power coefficient of x to zero in $x^2 - xy + 1 = 0$

$$\Rightarrow \qquad 1 = 0 \qquad \text{(which is not true)}$$

$\therefore$ no asymptote parallel to x-axis.

(ii) Asymptote parallel to y-axis

Equating highest power coefficient of y to zero in

$$x^2 - xy + 1 = 0$$

$\Rightarrow$ $$-x = 0$$

or $x = 0$ (*i.e.*, y-axis) is asymptote for $y = x + \dfrac{1}{x}$

(iii) Oblique asymptote

Let $y = mx + c$ in $x^2 - xy + 1 = 0$

i.e., $$x^2 - mx^2 - xc + 1 = 0$$

$\Rightarrow$ $$x^2(1 - m) - (c)x + 1 = 0$$

Equating highest and second highest power of x to zero

i.e., $$1 - m = 0 \quad \text{and} \quad c = 0$$

$\therefore$ $$m = 1 \quad \text{and} \quad c = 0$$

or $$y = x$$

is oblique asymptote to $y = x + \dfrac{1}{x}$.

Now to trace the curve;

(iv) Symmetric about origin (as odd function)

(v) Domain $\in R - \{0\}$.

(vi) Range $\in (-\infty, -2] \cup [2, \infty)$

(vii) $\dfrac{dy}{dx} = 1 - \dfrac{1}{x^2} = \dfrac{x^2 - 1}{x^2}$ {using number line rule, $+ \quad -1 \quad - \quad 1 \quad +$ }

$$\frac{dy}{dx} > 0, \quad \text{when} \quad x < -1 \quad \text{or} \quad x > 1$$

$$\frac{dy}{dx} < 0, \quad \text{when} \quad -1 < x < 1 - \{0\}$$

which shows $\mathbf{y_{max}}$ at $\mathbf{x = -1}$

$\mathbf{y_{min}}$ at $\mathbf{x = 1}$

(viii) Also, $$\frac{d^2y}{dx^2} = \frac{2}{x^3}$$

$\Rightarrow$ $\dfrac{d^2y}{dx^2} > 0$, when $x > 0$ (concave up)

$\dfrac{d^2y}{dx^2} < 0$, when $x < 0$ (concave down)

Using above information we can trace $y = x + \frac{1}{x}$ as:

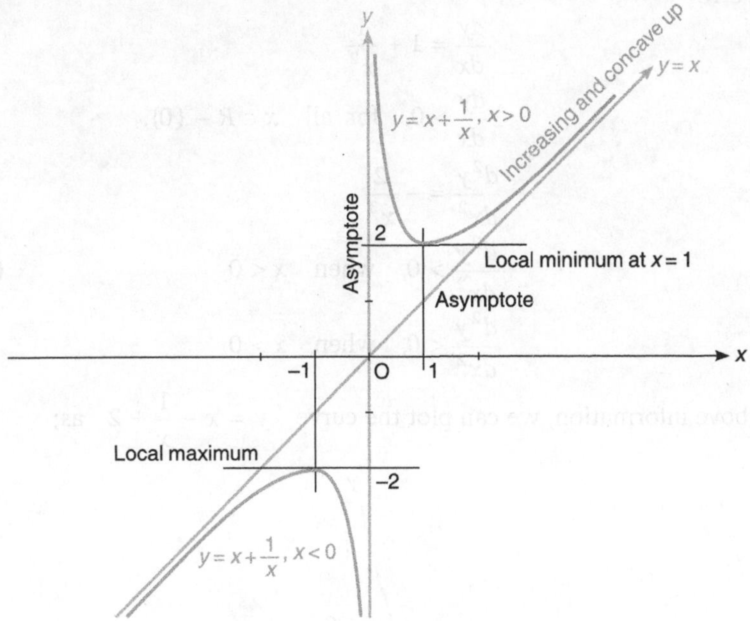

Fig. 3.5

EXAMPLE 2 *Find the asymptotes of the curve* $y = \dfrac{x^2 + 2x - 1}{x}$ *and hence, sketch.*

SOLUTION Here, the curve $y = \dfrac{x^2 + 2x - 1}{x}$ could be written as;

$$x^2 + 2x - yx - 1 = 0 \qquad \ldots\text{(i)}$$

(i) No asymptote parallel to x-axis.

(ii) Asymptote parallel to y-axis. $\Rightarrow$ $x = 0$.

(iii) Oblique asymptote

Let $y = mx + c$ be oblique asymptote

$$\therefore \quad x^2 + 2x - x(mx + c) - 1 = 0$$

$$x^2 - mx^2 + 2x - cx - 1 = 0$$

$$\Rightarrow \quad x^2(1 - m) + x(2 - c) - 1 = 0$$

For oblique asymptote equate highest power and second highest power of x to zero.

i.e., Coefficient of $x^2 = 0 \Rightarrow m = 1$

Coefficient of $x = 0 \Rightarrow c = 2$

$\therefore$ $y = x + 2$ is oblique asymptote to $y = x - \frac{1}{x} + 2$.

(iv) Neither symmetric about axis nor about origin.

(v) Domain $\in R - \{0\}$.

(vi) Range $\in R$.

(vii)
$$\frac{dy}{dx} = 1 + \frac{1}{x^2}$$

$$\Rightarrow \quad \frac{dy}{dx} > 0, \quad \text{for all} \quad x \in R - \{0\}.$$

(viii)
$$\frac{d^2y}{dx^2} = -\frac{2}{x^3}$$

$$\Rightarrow \quad \frac{d^2y}{dx^2} > 0, \quad \text{when} \quad x < 0$$ **(concave down)**

$$\frac{d^2y}{dx^2} < 0, \quad \text{when} \quad x > 0$$ **(concave up)**

Using above information, we can plot the curve $y = x - \frac{1}{x} + 2$ as;

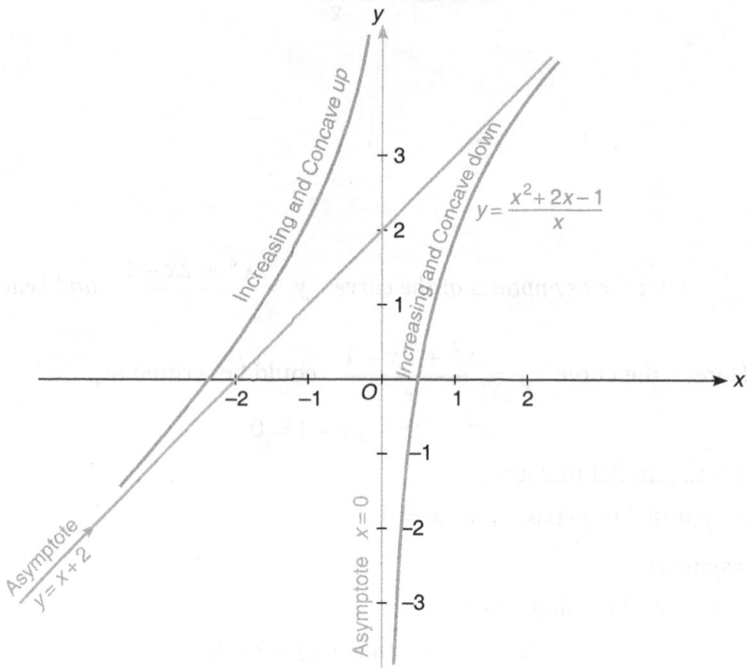

Fig. 3.6

3.1 (iv) Asymptote by inspection

If the equation of the curve be of the form $F_n + F_{n-2} = 0$, where F_n and F_{n-2} are expressions in x and y such that degree of $F_n = n$ and degree of $F_{n-2} \leq n - 2$, then every linear factor equated to zero will give an asymptote if no two straight lines represented by any other factor of F_n is parallel or coincident with it.

EXAMPLE 1 *Find the asymptote of the curve $x^2y + xy^2 = a^3$.*

SOLUTION Here, the given curve is.

$$x^2y + xy^2 = a^3 \quad \text{or} \quad x^2y + xy^2 - a^3 = 0$$

This equation is of the form $F_n + F_{n-2} = 0$

Here, $F_3 = x^2y + xy^2$ and $F_0 = -a^3$

$\therefore$ By inspection the asymptotes are given by

$$x^2y + xy^2 = 0 \quad \text{or} \quad xy(x+y) = 0$$

$\therefore$ The asymptotes are $\mathbf{x = 0,\ \ y = 0,\ \ x + y = 0.}$

EXAMPLE 2 *Find the asymptote of the curve $y = x + \frac{1}{x}$ (by Inspection).*

SOLUTION Here, the given curve is $x^2 - xy + 1 = 0$

This equation is of the form $F_n + F_{n-2} = 0$

Here $F_2 = x^2 - xy$

$F_0 = 1$

$\therefore$ By inspection the asymptotes are given by

$$x^2 - xy = 0 \quad \text{or} \quad x(x-y) = 0$$

$\therefore$ The asymptotes are $\mathbf{x = 0}$ and $\mathbf{x - y = 0}$.

3.1 (v) Intersection of curve and its asymptote

An asymptote of curve of nth degree cut the curve in $(n - 2)$ points provided the asymptote is not parallel to any asymptote.

Hence, if there be N asymptotes of the curve, then they cut the curve in $N(n - 2)$ points.

Note The number of asymptotes of an algebraic curve of nth degree can not be more than n.

EXAMPLE 1 *Show the asymptote of the curve $xy(x^2 - y^2) + x^2 + y^2 - 1 = 0$ cut at 8 points.*

SOLUTION The equation of the curve is,

$$xy(x^2 - y^2) + x^2 + y^2 - 1 = 0 \qquad \ldots\text{(i)}$$

Here $n = 4$

This equation is of the type $F_n + F_{n-2} = 0$

Hence, $F_n = xy(x^2 - y^2) = xy(x-y)(x+y)$

and $F_{n-2} = x^2 + y^2 - 1$

$\therefore$ $F_n = 0$

$\Rightarrow$ $x = 0,\ \ y = 0,\ \ x - y = 0$ and $x + y = 0$ are the equations of asymptotes.

The combined equation of the asymptotes is,

$$xy(x-y)(x+y) = 0 \qquad \ldots\text{(ii)}$$

Subtracting Eq. (ii) from (i), we get

$$x^2 + y^2 - 1 = 0$$

Thus, intersection of curve and asymptotes lie on this curve since, there are 4 asymptotes, *i.e.*, $N = 4$.

$\therefore$ **Point of intersection of curve and asymptotes = 4(4 – 2) = 8.**

3.1 (vi) Asymptote by expansion

If the equation of the curve is of the form

$$y = mx + c + \frac{A}{x} + \frac{B}{x^2} + \frac{C}{x^3} + \ldots$$

Then $y = mx + c$ will be an asymptote of the given curve.

EXAMPLE 1 *Find the asymptote of the curve* $y^3 = x^2(x - a)$.

SOLUTION The curve is, $y^3 = x^2(x-a) = x^3\left(1 - \frac{a}{x}\right)$

$\Rightarrow$ $$y = x\left(1 - \frac{a}{x}\right)^{1/3} \quad \text{or} \quad y = x\left(1 - \frac{1}{3}\frac{a}{x} - \frac{1}{9}\frac{a^2}{x^2}\ldots\right)$$

or $$y = x - \frac{a}{3} - \frac{1}{9}\frac{a^2}{x}\ldots$$ which is of the form

$$y = mx + c + \frac{A}{x} + \frac{B}{x^2} + \ldots$$

$\Rightarrow$ $$y = mx + c$$ is asymptote

Hence, $\boldsymbol{y = x - \frac{a}{3}}$ **is asymptote of the given curve.**

EXAMPLE 2 *Find the asymptote for* $y = x + \frac{1}{x}$.

SOLUTION Here; $y = x + \frac{1}{x}$ is of the form,

$$y = mx + c + \frac{A}{x} + \frac{B}{x^2} + \ldots$$

$\Rightarrow$ $y = x$ is asymptote of the curve

$$y = x + \frac{1}{x}.$$

Note Above method is useful to find oblique asymptote. Thus, students are adviced to find vertical and horizontal asymptote (*i.e.*, asymptote parallel to *x*-axis and *y*-axis).

3.1 (vii) The position of the curve with respect to an asymptote

Let the equation of the curve is of the form;

$$y = mx + c + \frac{A}{x} + \frac{B}{x^2} + \frac{C}{x^3} + \ldots, \text{ then}$$

(a) **The curve lies above the asymptote if**

(i) $A \neq 0$ and, A and x have same signs

or (ii) $A = 0,\ B > 0$

or (iii) $A = 0,\ B = 0,\ C \neq 0$ and C and x have same signs and

(b) **The curve lies below the asymptote if**

(i) $A \neq 0$ and, A and x have opposite signs.

or (ii) $A = 0,\ B < 0$

or (iii) $A = 0,\ B = 0,\ C \neq 0$ and C and x have opposite signs.

EXAMPLE 1 *For the curve* $y^5 = x^5 + 2x^4$; *show;*

(i) The curve lies above the asymptote $y = x + \frac{2}{5}$, *if* $x < 0$.

(ii) The curve lies below the asymptote $y = x + \frac{2}{5}$, *if* $x > 0$.

SOLUTION The given curve is, $y^5 = x^5 + 2x^4$

or $$y^5 = x^5\left(1 + \frac{2}{x}\right)$$

$$y = x\left(1 + \frac{2}{x}\right)^{1/5}$$

$$y = x\left(1 + \frac{2}{5}\cdot\frac{1}{x} - \frac{8}{25}\cdot\frac{1}{x^2}\cdots\right) = x + \frac{2}{5} - \frac{8}{25x} + \ldots$$

$\therefore$ The asymptote is $y = x + \frac{2}{5}$;

(i) Now if $A = -\frac{8}{25}$ and x have same sign $\Rightarrow$ $x < 0$. **Then the curve lie above the asymptote.**

(ii) Now if $A = -\frac{8}{25}$ and x have opposite sign $\Rightarrow x > 0$. **Then the curve lie below the asymptote.**

EXAMPLE 2 *For the curve* $y = x + \frac{1}{x}$ *show,*

(i) The curve lies above the asymptote $y = x$, *if* $x > 0$

(ii) The curve lies below the asymptote $y = x$, *if* $x < 0$

SOLUTION The given curve is,. $y = x + \frac{1}{x}$, is of the form

$$y = mx + c + \frac{A}{x} + \frac{B}{x^2} + \frac{C}{x^3}\ldots$$

Thus, $y = x$ is the asymptote to $y = x + \frac{1}{x}$.

(i) Now if $A = 1$ and x have same sign $\Rightarrow x > 0$, **then the curve lies above the asymptote.**

(ii) Now if $A = 1$ and x have opposite sign $\Rightarrow x < 0$, **then the curve lies below the asymptote.**

3.2 SINGULAR POINTS

Here, we shall discuss the following

(i) Multiple points
(ii) Double points
Types of double points :
(a) Node (b) Cusp (c) Isolated point
(iii) Tangent at the origin.
(iv) Necessary conditions for existence of double points.
(v) Types of cusps.

3.2 (i) Multiple points

A point on a curve is said to be a multiple point of order r, if r branches of the curve pass through this point.

If P is the multiple point of order r, then there will be r tangents at P, one of each of the r branches. These r tangents may be real, imaginary, distinct, coincident.

3.2 (ii) Double points

A point on a curve is said to be a double point of the curve, if two branches of the curve pass through this point.

Double points have two tangents, they may be real, imaginary, distinct or coincident.

Types of Double points

(a) Node

If the two branches of a curve pass through the double point and the tangents to them at the point are real and distinct, then the double point is called **a node** as shown in Fig. 3.7.

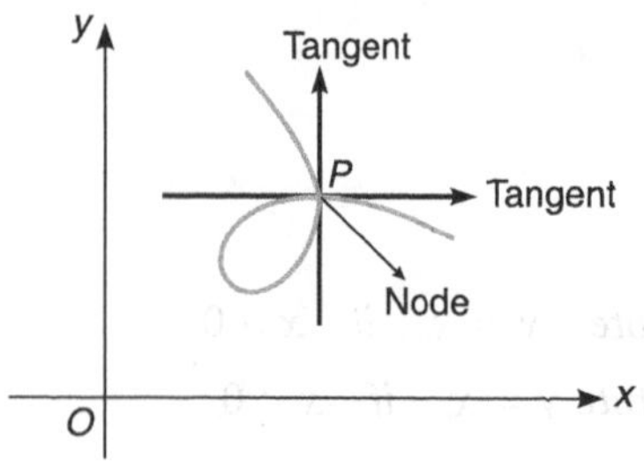

Fig. 3.7

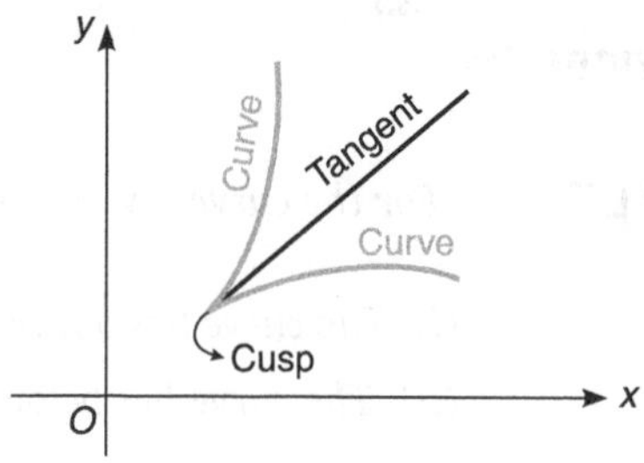

Fig. 3.8

(b) Cusp

If the two branches of the curve pass through the double point and the tangent to them are the point is real and coincident, then the double point is called **cusp** as shown in Fig. 3.8.

Cusps :

The graph of a continuous function $y = f(x)$ has a cusp at a point $x = c$ if the **concavity is same** on the both side of c and either.

1. $\lim_{x \to c^-} f'(x) = \infty$ and $\lim_{x \to c^+} f'(x) = -\infty$

OR

2. $\lim_{x \to c^-} f'(x) = -\infty$ and $\lim_{x \to c^+} f'(x) = \infty$ shown as:

1. $\lim_{x \to c^-} f'(x) = \infty$ and $\lim_{x \to c^+} f'(x) = -\infty$

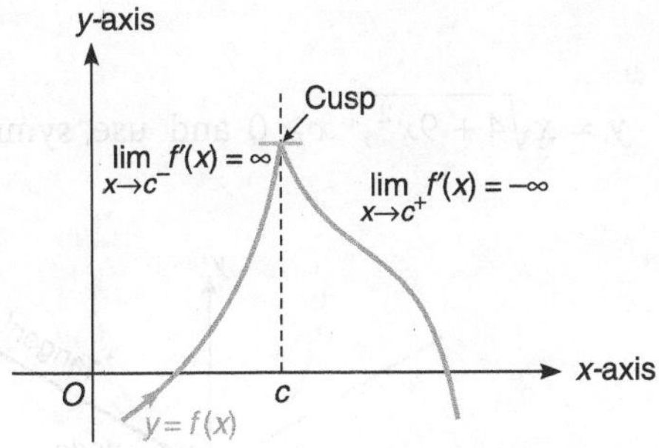

Fig. 3.9

2. $\lim_{x \to c^-} f'(x) = -\infty$ and $\lim_{x \to c^+} f'(x) = \infty$

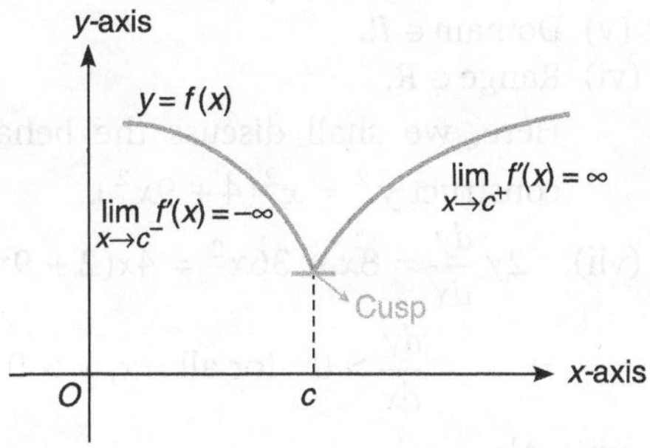

Fig. 3.10

Note A cusp can either be a local maximum (1) or a local minima as in (2).

(c) Isolated point

If there are no real point on the curve in the neighbourhood of a point P is called an **isolated or a conjugate point**.

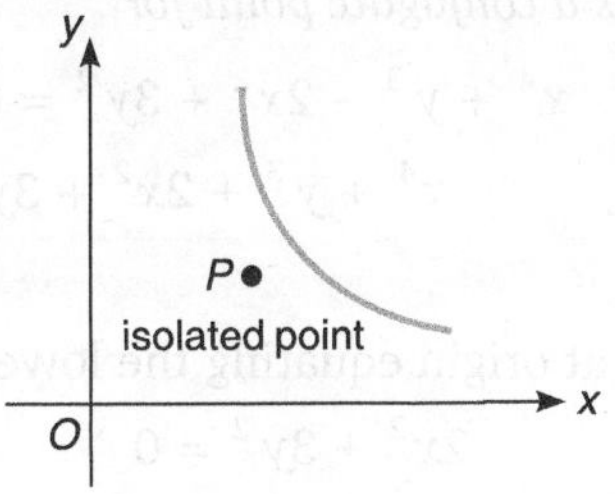

Fig. 3.11

3.2 (iii) Tangent at the origin

If an algebraic curve passes through the origin, the equation of tangent or tangents at the origin is obtained by equating to zero the lowest degree terms in the equation of the curve.

EXAMPLE 1 *Show that the curve* $y^2 = 4x^2 + 9x^4$ *has a node at origin and hence, sketch.*

SOLUTION The equation of the curve is,

$$y^2 = 4x^2 + 9x^4 \qquad \ldots(i)$$

It passes through the origin.

Now, equating to zero the lowest degree terms of the given curve, *i.e.*,

$$y^2 - 4x^2 = 0$$

$\Rightarrow$ $$y = 2x \quad \text{and} \quad y = -2x$$

There are two real and distinct tangents $y = 2x$ and $y = -2x$. Thus, two branches of curve passes through origin (0, 0).

$\therefore$ origin is node. ...(ii)

Now to sketch;

(iii) Symmetric about x-axis, y-axis and origin.

(iv) As $x \to 0 \Rightarrow y \to 0$

(v) Domain $\in R$.

(vi) Range $\in R$.

Here, we shall discuss the behaviour of $y = x\sqrt{4 + 9x^2}$, $x \geq 0$ and use symmetry to construct $y^2 = x^2(4 + 9x^2)$.

(vii) $2y\dfrac{dy}{dx} = 8x + 36x^3 = 4x(2 + 9x^2)$

$\Rightarrow$ $\dfrac{dy}{dx} > 0$ for all $x, y > 0$

(viii) Also,

$$\left(\frac{dy}{dx}\right)^2 + y\frac{d^2y}{dx^2} = 4 + 54x^2$$

$\Rightarrow$ $\dfrac{d^2y}{dx^2} > 0$ for all x

Thus, the graph for $y^2 = x^2(4 + 9x^2)$

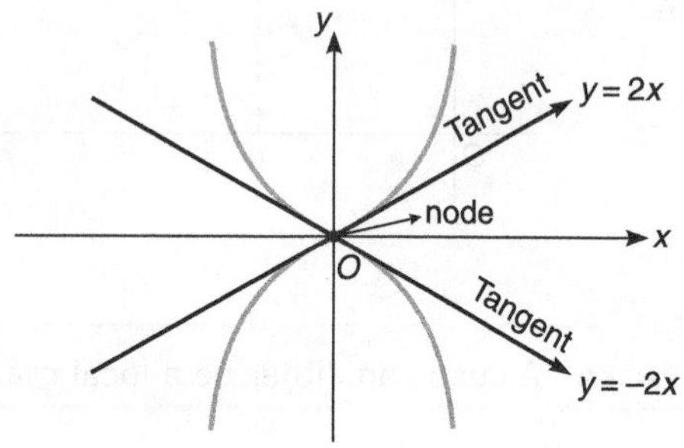

Fig. 3.12

EXAMPLE 2 *Show origin is a conjugate point for*

$$x^4 + y^3 + 2x^2 + 3y^2 = 0$$

SOLUTION The given curve is, $x^4 + y^3 + 2x^2 + 3y^2 = 0$

It passes through origin.

$\therefore$ To find equation of tangent at origin equating the lowest degree term to zero.

i.e., $$2x^2 + 3y^2 = 0$$

$\Rightarrow$ $$y = \pm i\sqrt{\frac{2}{3}}\,x$$

which are imaginary tangents.

Hence, **origin is a conjugate point of the curve.**

3.2 (iv) Necessary conditions for the existence of double points

Let (x, y) be a point on the given curve $f(x, y) = 0$.

The necessary and sufficient conditions for (x, y) to be a double points are:

$$f = 0, \quad \frac{\partial f}{\partial x} = 0, \quad \frac{\partial f}{\partial y} = 0 \quad \text{at} \quad (x, y)$$

Now, if $\dfrac{\partial^2 f}{\partial x\,\partial y}, \dfrac{\partial^2 f}{\partial x^2}$ and $\dfrac{\partial^2 f}{\partial y^2}$ are not all zero, then,

(i) Double point will be a node if

$$\left(\frac{\partial^2 f}{\partial x\,\partial y}\right)^2 - \left(\frac{\partial^2 f}{\partial x^2}\right)\left(\frac{\partial^2 f}{\partial y^2}\right) > 0$$

or $$f_{xy}^2 - f_{xx} f_{yy} > 0$$

(ii) The double point will be an isolated point, if

$$f_{xy}^2 - f_{xx} f_{yy} < 0$$

(iii) The double point will be a cusp if

$$f_{xy}^2 - f_{xx} f_{yy} = 0.$$

Here, if $f_{xx} = f_{xy} = f_{yy} = 0$ at (x, y), then it will be a multiple point of order greater than 2.

EXAMPLE 1 *For the curve* $x^3 + x^2 + y^2 - x - 4y + 3 = 0$, *find the double point and hence, whether the point is node or isolated point.*

SOLUTION Let $f(x, y) = x^3 + x^2 + y^2 - x - 4y + 3 = 0$

$\therefore$ $f_x = 3x^2 + 2x - 1$

$f_y = 2y - 4$ for a double point $f_x = 0, \quad f_y = 0$

$\therefore$ $f_x = 0 \Rightarrow 3x^2 + 2x - 1 = 0$

or $x = \dfrac{1}{3}, \; -1$

$f_y = 0 \Rightarrow 2y - 4 = 0 \Rightarrow y = 2.$

$\therefore$ Possible double points are $\left(\dfrac{1}{3}, 2\right), \; (-1, 2)$

$\because$ $f\left(\dfrac{1}{3}, 2\right) \neq 0$ and $f(-1, 2) = 0$

$\therefore$ **$f(-1, 2)$ is a double point.**

$f_{xx} = 6x + 2 \Rightarrow f_{xx}$ at $(-1, 2) = -4$

$f_{xy} = 0 \Rightarrow f_{xy}$ at $(-1, 2) = 0$

$f_{yy} = 2 \Rightarrow f_{yy}$ at $(-1, 2) = 2$

$\therefore$ $f_{xy} - f_{xx} f_{yy} = 0 - (-4)(2) = 8 > 0$

$\therefore$ **(– 1, 2) may be node.**

For shifting origin to (– 1, 2), substitute $x = X - 1, \quad y = Y + 2$ in the given equation,

we get, $$X^3 - 2X^2 + Y^2 = 0$$

or $$Y = \pm X\sqrt{2 - X}$$

For numerically small values of X, Y is real.

$\therefore$ **(–1, 2) is a node on the given curve.**

Asymptotes, Singular Points and Curve Tracing

EXAMPLE 2 *For the curve* $x^3 + 2x^2 + 2xy - y^2 + 5x - 2y = 0$, *find the double point and hence, check whether node, cusp or isolated point.*

SOLUTION Let $f(x, y) = x^3 + 2x^2 + 2xy - y^2 + 5x - 2y = 0$ ...(i)

$$f_x = \frac{\partial f}{\partial x} = 3x^2 + 4x + 2y + 5$$

$$f_y = \frac{\partial f}{\partial y} = 2x - 2y - 2$$

$$f_{xx} = \frac{\partial^2 f}{\partial x^2} = 6x + 4$$

$$f_{xy} = \frac{\partial^2 f}{\partial x\, \partial y} = 2$$

$$f_{yy} = \frac{\partial^2 f}{\partial y\, \partial x} = 2$$

For double points $f_x = f_y = f = 0$

$f_x = 0 \Rightarrow 3x^2 + 4x + 2y + 5 = 0$ ...(ii)

$f_y = 0 \Rightarrow 2x - 2y - 2 = 0$

or $2y = 2x - 2$ ...(iii)

Solving Eqs. (ii) and (iii), we get $3x^2 + 4x + 2x - 2 + 5 = 0$

$\Rightarrow$ $x = -1$

also $x = -1,\ y = -2$ satisfies the given equation.

$\therefore$ **(– 1, – 2) is a double point.**

At $(-1, -2)$, $f_{xx} = 6(-1) + 4 = -2$

$f_{xy} = 2,\ f_{yy} = -2$

$\therefore$ $f_{xy}^2 - f_{xx} f_{yy}$ at $(-1, -2) = 0$

$\therefore$ $(-1, -2)$ may be a cusp.

For shifting the origin to $(-1, -2)$ substitute $x = X - 1,\ y = Y - 2$ in the given equation.

$$(X - 1)^3 + 2(X - 1)^2 + 2(X - 1)(Y - 2) - (Y - 2)^2 + 5(X - 1) - 2(Y - 2) = 0$$

or $X^3 - X^2 + 2XY - Y^2 = 0$ ...(iv)

$\therefore$ $Y = X \pm X\sqrt{X}$

Y is real for all positive value of X.

$\therefore$ Two branches of (iv) pass through origin.

$\therefore$ Two branches of (i) pass through $(-1, -2)$.

$\Rightarrow$ **(– 1, – 2) is a cusp.**

EXAMPLE 3 *Find for the curve* $y^2 = x \sin x$ *origin is node, cusp or isolated point.*

SOLUTION Let $f(x, y) = y^2 - x \sin x$

$$f_x = -\sin x - x\cos x$$

$$f_y = 2y$$

$$f_{xx} = -\cos x + x\sin x - \cos x$$

$$f_{xy} = 0$$
$$f_{yy} = 2$$

$\therefore$ **at** $x = 0$ **:** $f_{xx} = -2,\ f_{xy} = 0,\ f_{yy} = 2$

$\therefore$ $f_{xy}^2 - f_{xx} f_{yy}$ at $(0, 0)$

$\Rightarrow$ $0 + 2(2) = 4 > 0$

$f_{xy}^2 - f_{xx} f_{yy} > 0$ $\therefore$ **origin is node.**

3.2 (v) Types of cusps

When two branches of a curve pass through a cusp and the tangents at cusp are coincident. Therefore, normal to the branches at a cusp would also be coincident.

Cusp can be of five kinds

(a) Single cusp

If the branches of the curve lie on the same side of the common normal, then the cusp is called a single cusp.

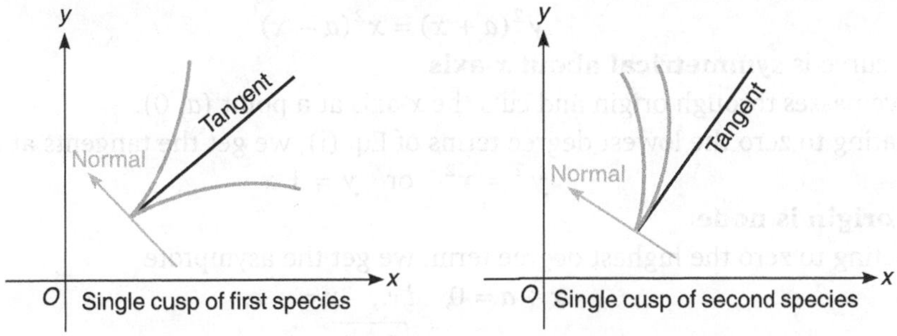

Fig. 3.13

(b) Double cusp

If the branches of the curve lie on the both sides of the common normal, then the cusp is called double cusp.

Here, both the branches of the curve lie on the both sides of common tangent, then the cusp is of first kind.

Also if, the branches of the curve lie on the same side of the common tangent, then the cusp is called **cusp of second species or Ramphoid cusp.**

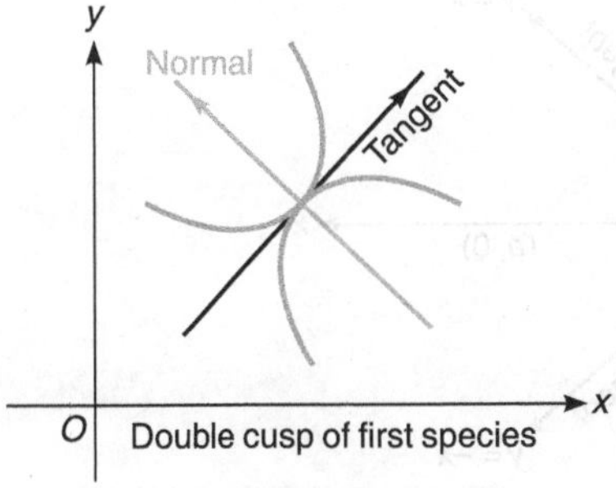

Fig. 3.14

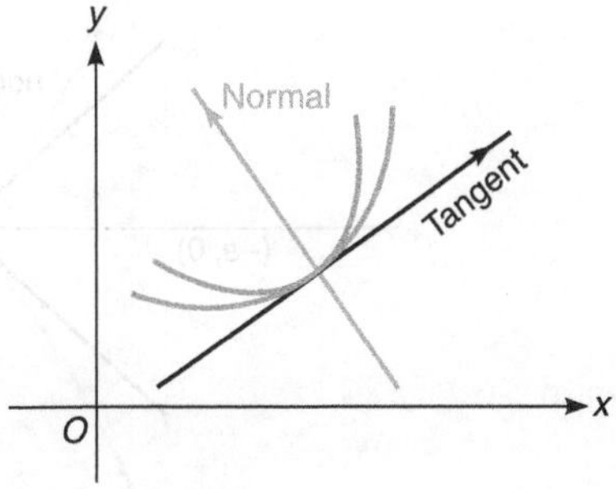

Fig. 3.15

(c) Point of oscu-inflexion

A double cusp of both the species is called a point of oscu-inflexion.

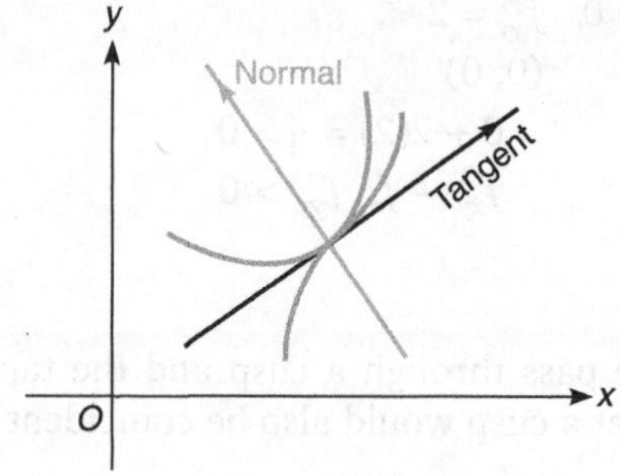

Fig. 3.16

EXAMPLE 1 *Sketch the curve* $y^2(a + x) = x^2(a - x)$.

SOLUTION Here, the curve is

$$y^2(a + x) = x^2(a - x) \qquad \ldots(i)$$

1. The curve is **symmetrical about x-axis.**
2. Curve passes through origin and cuts the x-axis at a point $(a, 0)$.
3. Equating to zero the lowest degree terms of Eq. (i), we get the tangents at origin.

$$\therefore \quad y^2 = x^2 \quad \text{or} \quad y = \pm x$$

$\therefore$ **origin is node.**

4. Equating to zero the highest degree term, we get the asymptote

$$x + a = 0, \quad i.e., \quad x = -a.$$

5. From Eq. (i)

$$y = \pm x\sqrt{\frac{a - x}{a + x}}$$

$\therefore$ y exists when $-a < x \le a \Rightarrow$ Domain $\in (-a, a]$.

6. As x increases from 0 to $a \Rightarrow y$ increases upto a point then decreases to zero.

$\Rightarrow$ y increases when $x \in [0, a]$
y decreases when $x \in (-a, 0]$

Thus, $y^2(a + x) = x^2(a - x)$ could be plotted as;

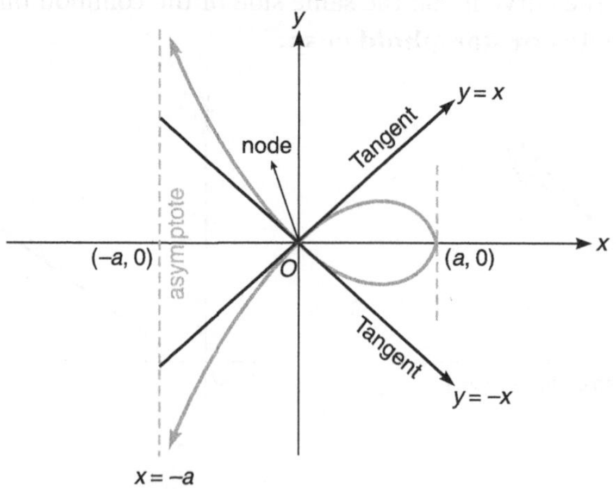

Fig. 3.17

3.3 REMEMBER FOR TRACING CARTESIAN EQUATION

1. Check symmetry

(a) A curve is symmetrical about x-axis, *i.e.*, y is replaced by $-y$ and curve remains same.

(b) A curve is symmetrical about y-axis, *i.e.*, $f(-x) = f(x)$.

(c) A curve is symmetrical about $y = x$, *i.e.*, on interchanging x and y curve remains same.

(d) A curve is symmetrical about $y = -x$. *i.e.*, on interchanging x by $-y$ and y by $-x$ curve remains same.

(e) The curve is symmetrical in opposite quadrants, *i.e.*, $f(-x) = -f(x)$.

2. Check for origin

Find whether origin lies on the curve or not.
If yes, check for multiple points (See Art. 3.2).

3. Point of intersection with x-axis and y-axis

Put $x = 0$ and find y, put $y = 0$ and find x. Also obtain the tangents at such points.

4. Asymptotes

Find the point at which asymptote meets the curve and equation of asymptote (see Art. 3.1)

5. Domain and range

To check in which part the curve lies.

6. Monotonicity and maxima minima

Find $\frac{dy}{dx}$ and check the interval in which y increases or decreases and the point at which it attains maximum or minimum.

7. Concavity and convexity

The interval in which,

$$\frac{d^2y}{dx^2} > 0$$

and

$$\frac{d^2y}{dx^2} < 0$$

Using all the above results we can sketch the curve

$$y = f(x).$$

SOME MORE SOLVED EXAMPLES

EXAMPLE 1 *Sketch the curve* $y^2(a^2 + x^2) = x^2(a^2 - x^2)$

SOLUTION Here, the curve is $y^2 = \dfrac{x^2(a^2 - x^2)}{(a^2 + x^2)}$

1. The curve is symmetric about x-axis and y-axis {as on replacing y by $-y$ curve remains same and on replacing x by $-x$ curve remains same thus, symmetric about x and y-axis respectively).

2. It passes through origin and $y = \pm x$ are two tangents at origin. **Thus, the origin is node.**

3. It meets x-axis at $(a, 0)$, $(0, 0)$ and $(-a, 0)$ and meets y-axis at $(0, 0)$ only. The tangents at $(a, 0)$ and $(-a, 0)$ are $x = a$ and $x = -a$ respectively.

4. The curve has no asymptote.

5. Here, $$y = \pm x\sqrt{\frac{a^2 - x^2}{a^2 + x^2}}$$

$\therefore$ Domain $\in [-a, a]$

6. $$\frac{dy}{dx} = \frac{a^4 - 2a^2x^2 - x^4}{(a^2 + x^2)^{3/2}(a^2 - x^2)^{1/2}}$$

$$\frac{dy}{dx} \to \infty \quad \text{as} \quad x \to \pm a$$

Also $\dfrac{dy}{dx} = 0$ when $a^4 - 2a^2x^2 - x^4 = 0$

i.e.,
$$\frac{dy}{dx} = \frac{a^4 - 2a^2x^2 - x^4}{(a^2 + x^2)^{3/2}(a^2 - x^2)^{1/2}}$$
$$= \frac{-\{x^4 + 2a^2x^2 + a^4 - 2a^4\}}{(a^2 + x^2)^{3/2}(a^2 - x^2)^{1/2}}$$
$$= \frac{-\{(x^2 + a^2)^2 - (\sqrt{2}a^2)^2\}}{(a^2 + x^2)^{3/2}(a^2 - x^2)^{1/2}}$$
$$= \frac{-\{x - \sqrt{(-1 + \sqrt{2})a}\}\{x + \sqrt{(-1 + \sqrt{2})a}\}\{x^2 + (1 + \sqrt{2})a^2\}}{(a^2 + x^2)^{3/2}(a^2 - x^2)^{1/2}}$$

$$\Rightarrow \quad \frac{dy}{dx} = \begin{cases} 0; & x = \pm\sqrt{(-1 + \sqrt{2})}\, a \\ +\text{ve}; & x \in (-\sqrt{(-1 + \sqrt{2})}\, a, \sqrt{(-1 + \sqrt{2})}\, a) \\ -\text{ve}; & x \in (-a, -\sqrt{(-1 + \sqrt{2})}\, a) \text{ or } (\sqrt{(-1 + \sqrt{2})}\, a, a) \end{cases}$$

i.e., y increasing when $x \in (-\sqrt{(-1 + \sqrt{2})}\, a, \sqrt{(-1 + \sqrt{2})}\, a)$

and y decreases when $x \in (-a, -\sqrt{(-1 + \sqrt{2})}\, a)$ or $(\sqrt{(-1 + \sqrt{2})}\, a, a)$

OR

$$\frac{dy}{dx} > 0, \quad \text{when} \quad x \in (-(0.6)\,a, (0.6)\,a)$$

$$\frac{dy}{dx} < 0, \quad \text{when} \quad x \in (-a, -(0.6)\,a) \quad \text{or} \quad ((0.6)\,a, a)$$

where $\sqrt{-1+\sqrt{2}} = (0.6)_{\text{approx}}$

Thus, the curve for $y^2(a^2 + x^2) = x^2(a^2 - x^2)$

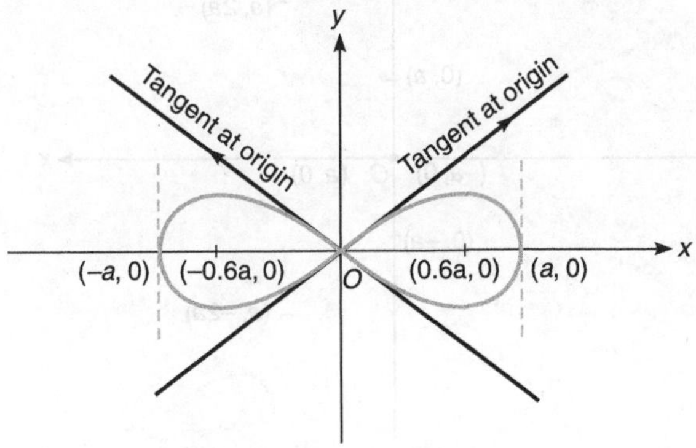

Fig. 3.18

EXAMPLE 2 *Sketch the curve* $y^2(x - a) = x^2(a + x)$.

SOLUTION Here, the curve is given by $y^2 = \dfrac{x^2(a + x)}{(x - a)}$

1. Symmetrical about x-axis only.
2. It passes through origin and $y^2 + x^2 = 0$, *i.e.*, $y = \pm ix$ are two imaginary tangents at origin. Thus, origin is **isolated point**.
3. It meets x-axis at $(-a, 0)$, $(0, 0)$ and y-axis at $(0, 0)$.
   The tangent at $(-a, 0)$ is $x = -a$.
4. $y = \pm(x - a)$ and $x = a$ are three asymptote.
5. $$y^2 = \frac{x^2(x + a)}{(x - a)} \Rightarrow y = \pm x\sqrt{\frac{x + a}{x - a}}$$

   Thus, for domain; $\dfrac{x + a}{x - a} \geq 0$ and $x \neq a$

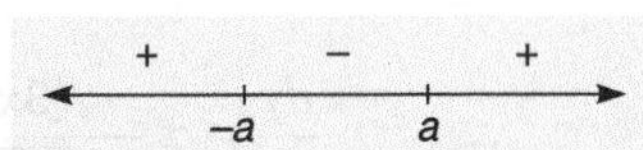

   i.e., $x \leq -a$ and $x > a$

   or Domain $\in (-\infty, -a] \cup (a, \infty) \cup \{0\}$

6. $$\frac{dy}{dx} = \pm \frac{x^2 - ax - a^2}{(x - a)^{3/2}(x + a)^{1/2}} = \pm \frac{\left\{x - \frac{1}{2}(1 + \sqrt{5})a\right\}\left\{x - \frac{1}{2}(1 - \sqrt{5})a\right\}}{(x - a)^{3/2}(x + a)^{1/2}}$$

   $$\Rightarrow \quad \frac{dy}{dx} > 0, \quad \text{when} \quad x \in (-\infty, -a] \cup \left[\frac{1}{2}(1 + \sqrt{5})a, \infty\right)$$

$$\frac{dy}{dx} < 0, \quad \text{when} \quad x \in \left(a, \frac{1}{2}(1+\sqrt{5})\, a\right]$$

Thus, the curve;

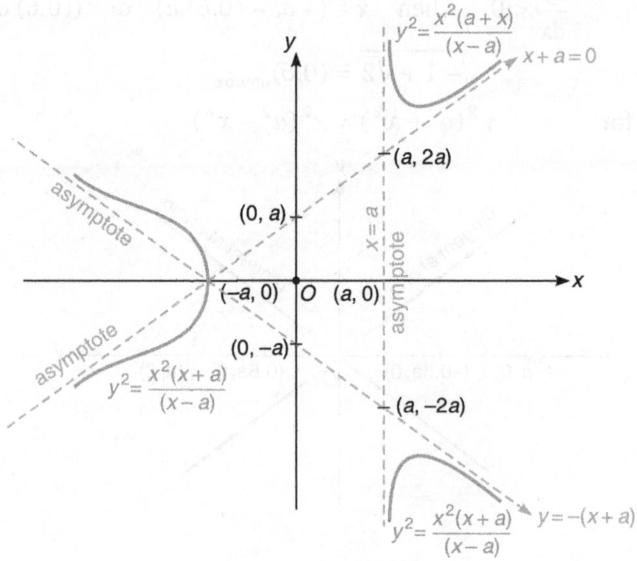

Fig. 3.19

EXAMPLE 3 *Sketch the curve* $y^2 = (x-1)(x-2)(x-3)$.

SOLUTION Here, $y^2 = (x-1)(x-2)(x-3)$

1. Symmetrical about x-axis.
2. It does not pass through origin.
3. It meets x-axis at (1, 0) (2, 0) and (3, 0) but it does not meet y-axis.
4. No asymptote.
5. For domain: $(x-1)(x-2)(x-3) \geq 0$

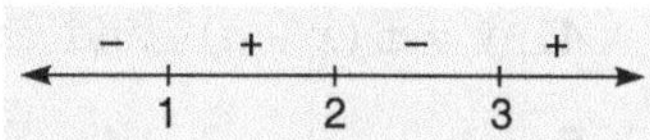

$\Rightarrow$ **Domain** $\in$ **[1, 2]** $\cup$ **[3, ∞)**

6. $y = \pm\sqrt{(x-1)(x-2)(x-3)}$

$$\therefore \quad \frac{dy}{dx} = \pm\frac{(3x^2 - 12x + 11)}{2\sqrt{(x-1)(x-2)(x-3)}} = \pm\frac{3\left(x - \frac{6-\sqrt{3}}{3}\right)\left(x - \frac{6+\sqrt{3}}{3}\right)}{2\sqrt{(x-1)(x-2)(x-3)}}$$

$$= \pm\frac{3(x-1.42)(x-2.5)}{2\sqrt{(x-1)(x-2)(x-3)}} \quad \left\{\text{as } \frac{6-\sqrt{3}}{3} = 1.42 \text{ and } \frac{6+\sqrt{3}}{3} = 2.5/\text{approx}\right\}$$

$$\Rightarrow \quad \frac{dy}{dx} > 0, \quad \text{when} \quad x \in (1, 1.42) \cup (3, \infty)$$

$$\frac{dy}{dx} < 0, \quad \text{when} \quad x \in (1.42, 2)$$

Thus, the curve $y^2 = (x-1)(x-2)(x-3)$

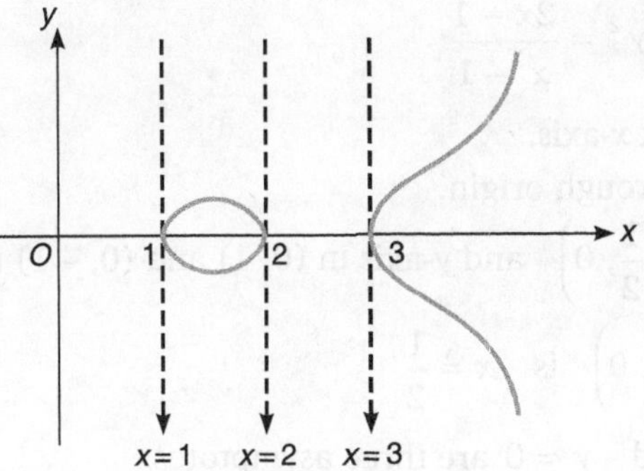

Fig. 3.20

EXAMPLE 4 *Sketch the curve* $y^2x^2 = x^2 - a^2$.

SOLUTION Here, $y^2 = \dfrac{x^2 - a^2}{x^2}$

1. Symmetrical about both the axis.
2. It does not pass through origin.
3. x-intercepts are $(a, 0)$ and $(-a, 0)$
   The tangent at $(a, 0)$ is $x = a$ and the tangent at $(-a, 0)$ is $x = -a$.
4. $y = \pm 1$ are the two asymptotes.
5. $$y = \pm \frac{\sqrt{x^2 - a^2}}{x} \Rightarrow \text{Domain} \in (-\infty, -a] \cup [a, \infty)$$
   $$\frac{dy}{dx} = \pm \frac{a^2}{x^2\sqrt{x^2 - a^2}} \Rightarrow \frac{dy}{dx} > 0, \text{ when } x \in (-\infty, -a) \cup (a, \infty)$$

Thus, the curve for $y^2 = \dfrac{x^2 - a^2}{x^2}$ is,

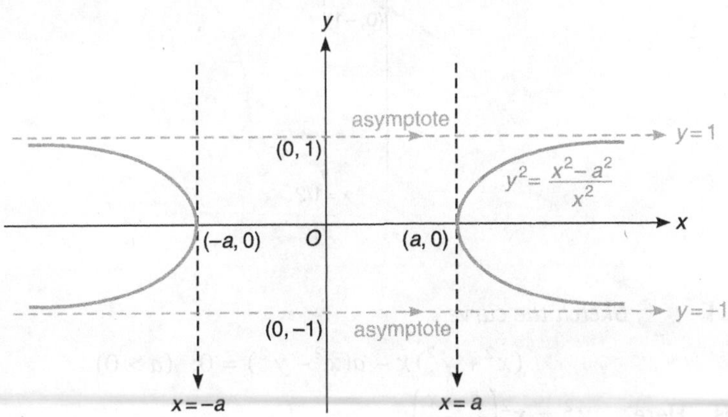

Fig. 3.21

EXAMPLE 5 *Sketch the curve* $y^2(x^2-1) = 2x - 1.$

SOLUTION Here, $y^2 = \dfrac{2x-1}{x^2-1}$

1. Symmetrical about x-axis.
2. It does not pass through origin.
3. It meets x-axis in $\left(\frac{1}{2}, 0\right)$ and y-axis in (0, 1) and (0, – 1) respectively.

   The tangent at $\left(\frac{1}{2}, 0\right)$ is $x = \frac{1}{2}$.
4. $x = 1$, $x = -1$ and $y = 0$ are three asymptotes.
5. $y^2 = \dfrac{2x-1}{x^2-1} \Rightarrow \text{Domain} \in \left(-1, \frac{1}{2}\right] \cup (1, \infty)$
6. $y = \pm\sqrt{\dfrac{2x-1}{x^2-1}} \Rightarrow \dfrac{dy}{dx} = \pm\left(\dfrac{-x^2+x+1}{(2x-1)^{1/2}(x^2-1)^{3/2}}\right)$

   $\Rightarrow \dfrac{dy}{dx} < 0$ when $x \in \left(-1, \frac{1}{2}\right] \cup (1, \infty)$

   $\therefore$ **y is decreasing in its domain.**

   Thus, the graph for $y^2 = \dfrac{2x-1}{x^2-1}$ is,

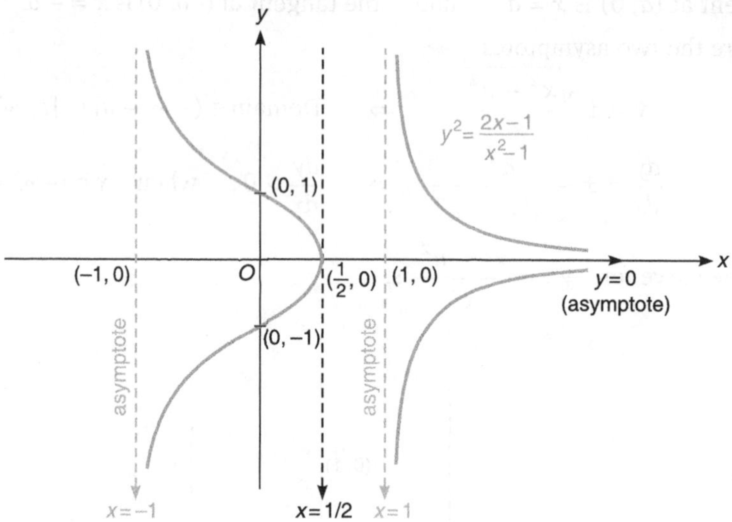

Fig. 3.22

EXAMPLE 6 *Sketch the curve :*

$$(x^2+y^2)x - a(x^2-y^2) = 0; \; (a > 0)$$

SOLUTION Here, $y^2 = x^2\left(\dfrac{a-x}{a+x}\right)$

1. Symmetric about x-axis.
2. Origin lies on the curve and $y = \pm x$ are two tangents at origin. So, origin is node.
3. x-intercept are (0, 0) and (a, 0). The tangent at (a, 0) is $x = a$.
4. $x = -a$ is the only asymptote.

5. $$y = \pm x\sqrt{\frac{a-x}{a+x}}$$

$\therefore$ Domain $\in (-a, a]$

6. $$\frac{dy}{dx} = \pm \frac{a^2 - ax - x^2}{(a+x)\sqrt{a^2 - x^2}}$$

$\Rightarrow \quad \frac{dy}{dx} > 0, \quad$ when $\quad x \in \left(-a, \frac{-1+\sqrt{5}}{2}a\right)$

$\Rightarrow \quad \frac{dy}{dx} < 0, \quad$ when $\quad x \in \left(\frac{-1+\sqrt{5}}{2}a, a\right)$.

Thus, the graph for

$y^2 = x^2\left(\frac{a-x}{a+x}\right)$ as shown in Fig. 3.23.

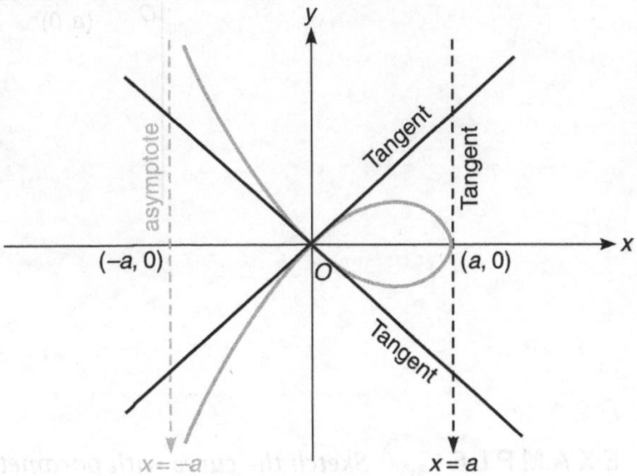

Fig. 3.23

EXAMPLE 7 *Sketch the curve* $x^3 + y^3 = 3ax^2 \quad (a > 0)$.

SOLUTION Here, $x^3 + y^3 = 3ax^2$
1. No line of symmetry.
2. Origin is cusp and $x = 0$ is tangent.
3. x-intercept, (0, 0) (3a, 0)
   The tangent at (3a, 0) is $x = 3a$.
4. $y = a - x$ is asymptote and the curve meets asymptote at $\left(\frac{a}{3}, \frac{2a}{3}\right)$.
5. Here; $$x^3 + y^3 = 3ax^2$$
$\Rightarrow \quad 3ax^2 > 0 \quad$ (as, $a > 0$)
$\therefore \quad x^3 + y^3 > 0$
i.e., x and y both cannot be negative (thus, curve would not lie in third quadrant).

6. $$y^2\frac{dy}{dx} = x(2a - x)$$

$\Rightarrow \quad \frac{dy}{dx} > 0, \quad$ when $\quad x \in (0, 2a)$

$\frac{dy}{dx} < 0, \quad$ when $\quad x \in (-\infty, 0) \cup (2a, \infty)$

Thus, the curve $y^3 + x^3 = 3ax^2$ is,

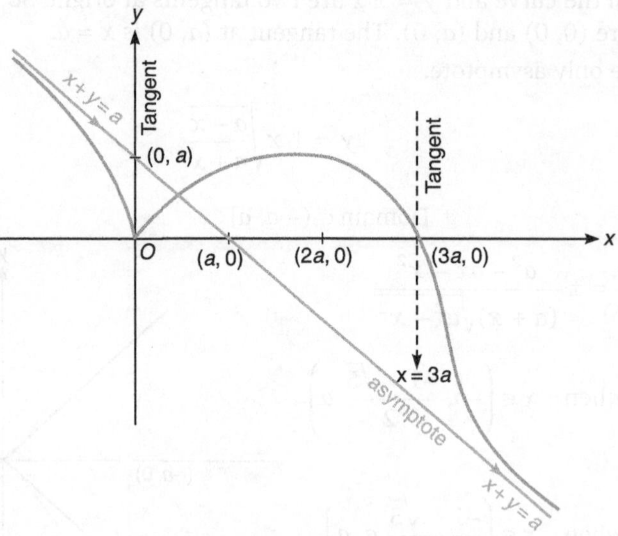

Fig. 3.24

EXAMPLE 8 *Sketch the curve with parametric equation θ.*

$$x = a(\theta + \sin\theta), \quad y = a(1 + \cos\theta); \quad x \in (-\pi, \pi).$$

SOLUTION Here, $x = a(\theta + \sin\theta)$ and $y = a(1 + \cos\theta)$ gives the following table for x and y with θ.

θ	$-\pi$	0	π
x	$-a\pi$	0	$a\pi$
y	0	$2a$	0

So, that we have,

$$-\pi \le \theta \le 0$$

⇒ (x, y) starting from $(-a\pi, 0)$ moves to the right and upwards to $(0, 2a)$.

$$0 \le \theta \le \pi$$

⇒ the point (x, y) starting from $(0, 2a)$ moves to the right and downward to $(a\pi, 0)$.

Also $$\frac{dx}{d\theta} = a(1 + \cos\theta)$$

and $$\frac{dy}{d\theta} = -a\sin\theta$$

Now, $\frac{dx}{d\theta} = 0$ if $\theta = \pi, -\pi$

$$\frac{dy}{dx} = -\frac{\tan\theta}{2},$$

except for the values $\mp\pi$ of θ for which $\frac{dx}{d\theta} = 0$.

Also, tangent at $\theta = \pi$ and $\theta = -\pi$ are $x = \pi$ and $x = -\pi$.

Thus, the curve for $x = a(\theta + \sin\theta)$ and $y = a(1 + \cos\theta)$.

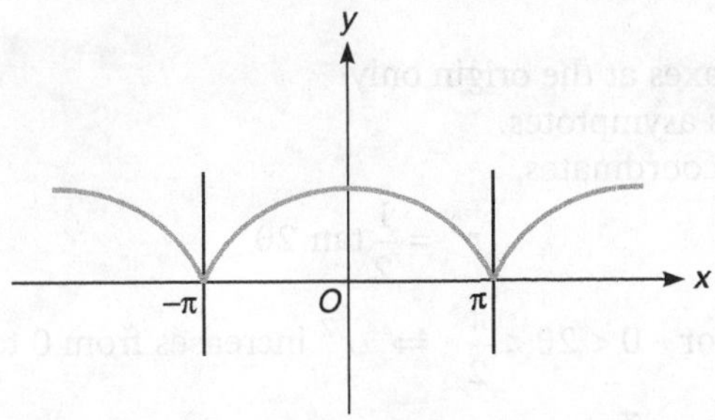

Fig. 3.25

EXAMPLE 9 *Sketch the curve :* $x^5 + y^5 = 5a^2xy^2$.

SOLUTION Here;

1. The curve is symmetrical in opposite quadrants.
2. The curve passes through origin and $x = 0$, $y = 0$ are tangents. Thus, origin is node.
3. It meets coordinate axis at origin.
4. $x + y = 0$ is an asymptote.
5. On transfering to polar coordinates, we get.

$$r^2 = \frac{5a^2 \cos\theta \sin\theta}{\cos^5\theta + \sin^5\theta}$$

when, $\theta = 0$, $r = 0$ when, $\theta = \frac{\pi}{2}$, $r = 0$

As θ increases from $\frac{\pi}{2}$ to $\frac{3\pi}{4}$, r^2 is negative and hence, r is imaginary.

$\therefore$ no portion of the curve lies in this region.

At $\theta = \frac{3\pi}{4}$, $r = \infty$ as θ increases from $\frac{3\pi}{4}$ to π $\Rightarrow$ r decreases from ∞ to 0.

$\therefore$ Curve $x^5 + y^5 = 5a^2xy^2$

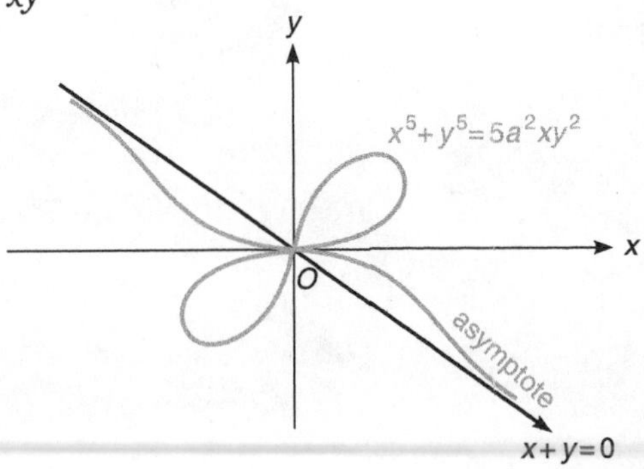

Fig. 3.26

EXAMPLE 10 *Sketch the curve* $y^4 - x^4 + xy = 0.$

SOLUTION Here, $y^4 - x^4 + xy = 0$

1. No line of symmetry.
2. It passes through origin two tangents at (0, 0) as $x = 0$ and $y = 0$,

   $\therefore$ **origin is node.**
3. It cuts the coordinate axes at the origin only.
4. $y = x$, $y = -x$ are its asymptotes.
5. Converting into polar coordinates,

$$r^2 = \frac{1}{2}\tan 2\theta$$

6. When, $0 < \theta < \frac{\pi}{4}$ or $0 < 2\theta < \frac{\pi}{2}$ $\Rightarrow$ r^2 increases from 0 to ∞.

   When, $\frac{\pi}{4} < \theta < \frac{\pi}{2}$ or $\frac{\pi}{2} < 2\theta < \pi$ $\Rightarrow$ r^2 is negative,

   $\therefore$ no curve when $\frac{\pi}{4} < \theta < \frac{\pi}{2}$.

   Hence, the curve

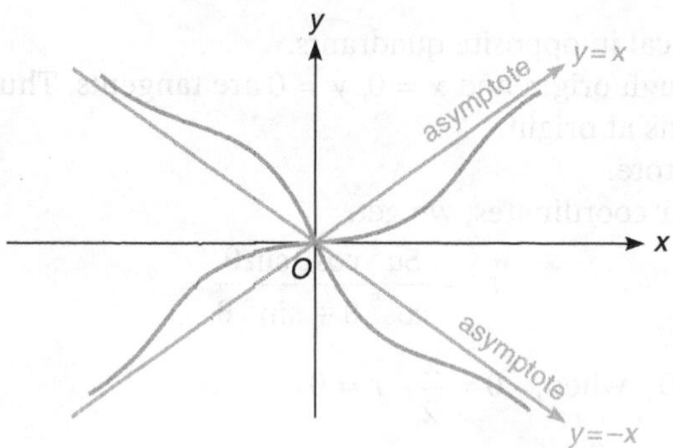

Fig. 3.27

EXERCISE

Plot the Curves :

1. $y = 1 + x^2 - \frac{1}{2}x^4.$
2. $y = (x+1)(x-2)^2$
3. $y = \frac{2}{5}x - \frac{1}{2}x^3 + \frac{1}{10}x^5.$
4. $y = (1-x^2)^{-1}$
5. $y = \frac{x^4}{(1+x)^3}$
6. $y = \frac{(1+x)^4}{(1-x)^4}$
7. $y = \frac{x^2(x-1)}{(x+1)^2}$
8. $y = \frac{x}{(1-x^2)^2}$
9. $y = 2x - 1 + \frac{1}{(x+1)}$
10. $y = \frac{x^2+1}{x^2-1}$
11. $y = \frac{a^2x}{a^2+x^2}$
12. $y^2 = x^2\left(\frac{a+x}{b-x}\right)$
13. $y = \frac{8a^3}{x^2+4a^2}$
14. $y = \frac{\cos x}{\cos 2x}$
15. $y = \text{arc cos}\left(\frac{1-x^2}{1+x^2}\right) = \cos^{-1}\left(\frac{1-x^2}{1+x^2}\right)$
16. $y = \text{arc sin}(\sin x) = \sin^{-1}(\sin x)$
17. $y = \sin(\text{arc sin}\, x)$
18. $y = \text{arc tan}(\tan x)$
19. $y = \text{arc tan}\left(\frac{1}{x}\right)$
20. $y = (x+2)\,e^{1/x}$
21. $y = \frac{1}{2}(\sqrt{x^2+x+1} - \sqrt{x^2-x+1})$
22. $y = \sqrt{x^2+1} - \sqrt{x^2-1}$
23. $y = (x+2)^{2/3} - (x-2)^{2/3}$
24. $y = (x+1)^{2/3} + (x-1)^{2/3}$
25. $y^2 = 8x^2 - x^4$
26. $y^2 = (x-1)(x-2)(x-3)$
27. $y^2 = \frac{x-1}{x+1}$
28. $y^2 = \frac{x^2(1-x)}{(1+x)^2}$
29. $y^2 = x^4(x+1)$
30. $x^2(y-2)^2 + 2xy - y^2 = 0$
31. $x = \frac{1}{4}(t+1)^2,\; y = \frac{1}{4}(t-1)^2$
32. $x = \frac{t^2}{1-t^2},\; y = \frac{1}{1+t^2}$
33. $x = \frac{t^2}{t-1},\; y = \frac{t}{t^2-1}$
34. $x = -5t^2 + 2t^5,\; y = -3t^2 + 2t^3$
35. $x = \frac{t^2+1}{4(1-t)},\; y = \frac{t}{t+1}$
36. $x = \frac{(t+2)^2}{(t+1)},\; y = \frac{(t-2)^2}{t-1}$
37. $x = \frac{t-t^2}{1+t^2},\; y = \frac{t^2-t^3}{1+t^2}$
38. $x^3 + y^3 = 3axy$, where $a > 0$.
39. $(x-a)^2(x^2+y^2) = b^2x^2$, where $a, b > 0$.
40. $x^{2/3} + y^{2/3} = a^{2/3}$, where $a > 0$
41. $x^6 + 2x^3y = y^3$
42. $4y^2 = 4x^2y + x^5$
43. $x^4 + 2y^3 = 4x^2y$
44. $x^3 - 2x^2y - y^2 = 0$
45. $x^2y^2 + y = 1$
46. $x^3 + y^3 = 3x^2$
47. $y^5 + x^4 = xy^2$
48. $x^4 - y^4 + xy = 0$
49. $x^5 + y^5 = xy^2$
50. $x = a\sin 2\theta(1+\cos 2\theta),$
    $y = a\cos 2\theta(1-\cos 2\theta)$

Hints and Solutions

INTRODUCTION OF GRAPHS

1. (i)

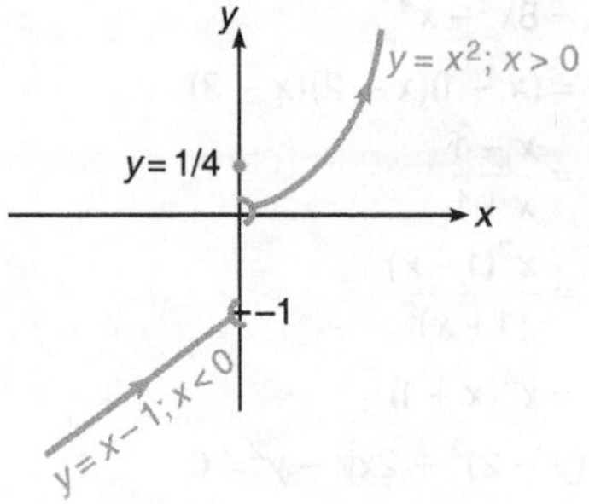

(ii)

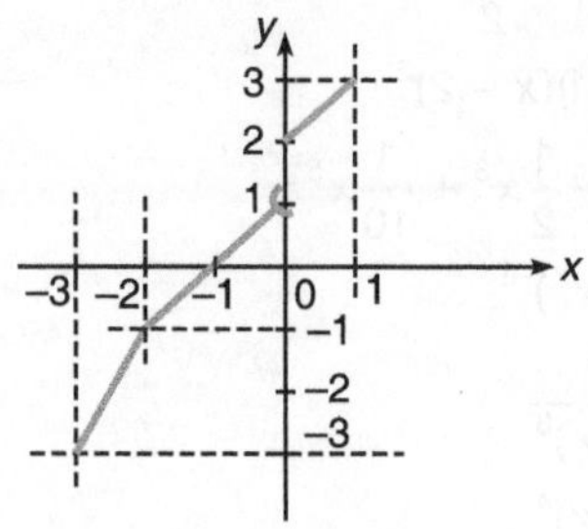

2. (i)

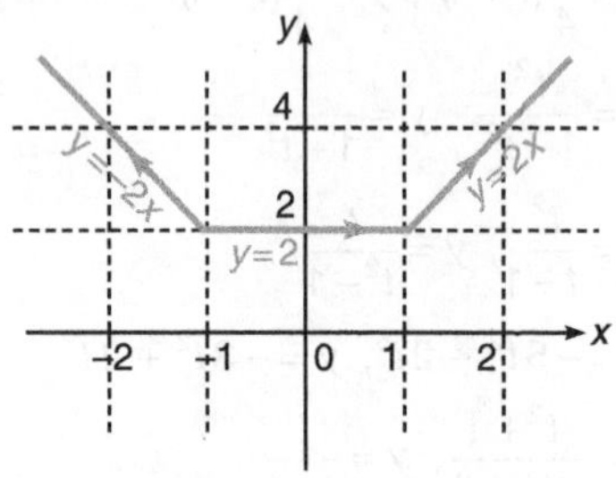

Here; $f(x)=\begin{cases}-2x; & x<-1\\ 2; & -1\le x\le 1\\ 2x; & x>1\end{cases}$

(ii)

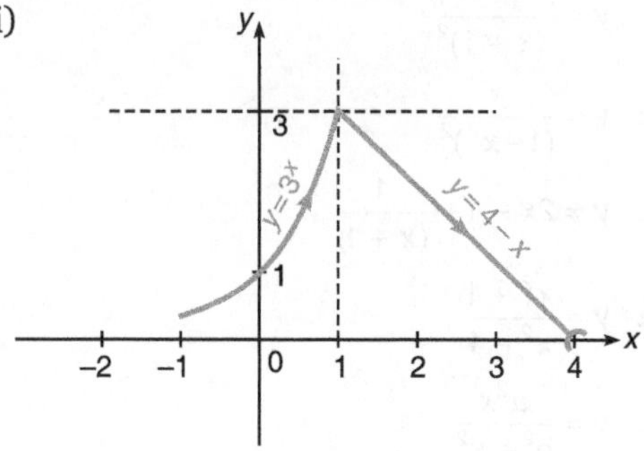

(iii)

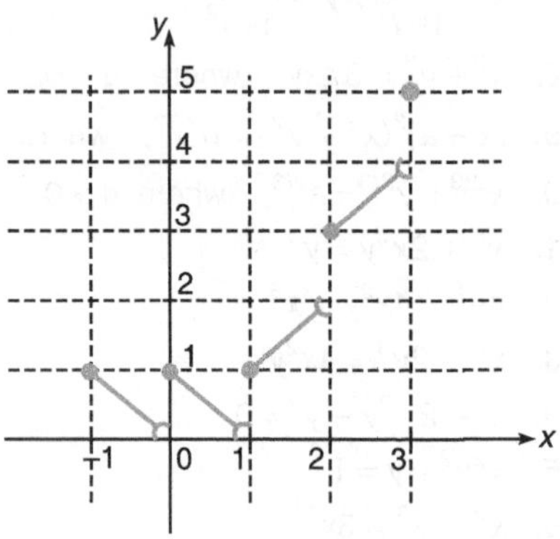

Here, $f(x)=\begin{cases}-x; & -1\le x<0\\ -x+1; & 0\le x<1\\ x; & 1\le x<2\\ x+1; & 2\le x<3\\ 5; & x=3\end{cases}$

(iv)

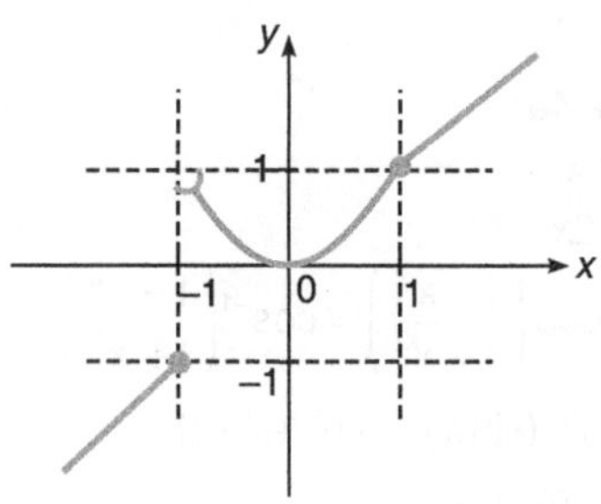

Here, $f(x)=\begin{cases}x^4; & -1<x<1\\ x; & x\le -1 \text{ or } x\ge 1\end{cases}$

3. We know, $y = x^2 + x + 1$ is shown as;

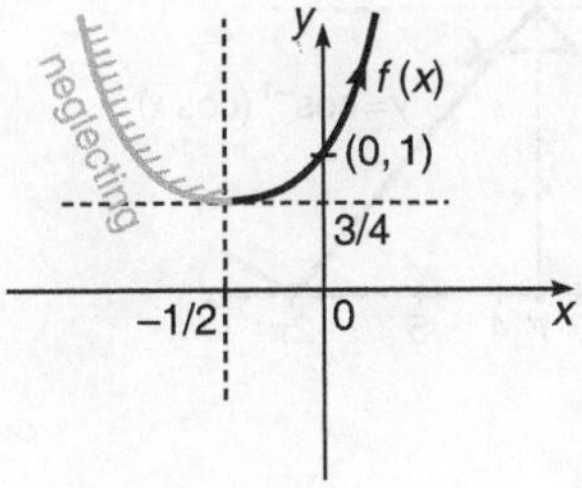

Graph for $f(x)$

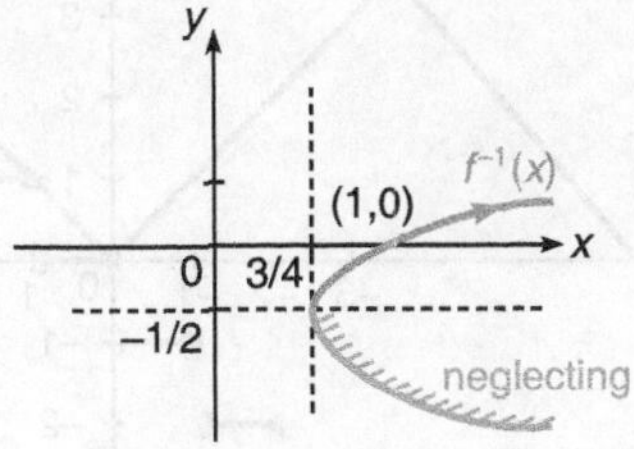

Graph for $f^{-1}(x)$

From above figures it is clear that the graph would exists only when; $x \geq -\frac{1}{2}$.

i.e., inverse for $f(x) = x^2 + x + 1$ would exists only when $\boldsymbol{x \geq -\frac{1}{2}}$.

4.

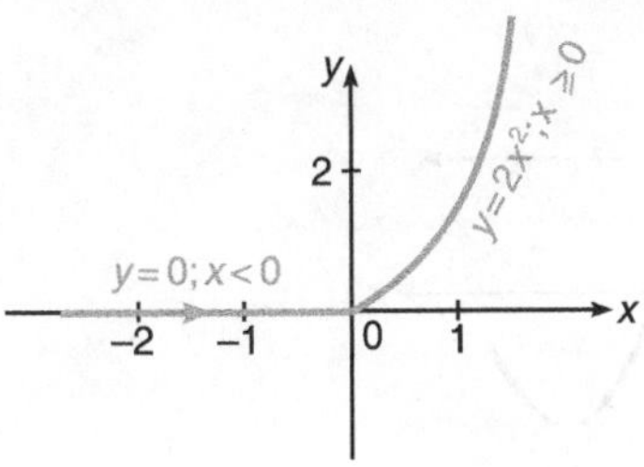

Here, $f(x) = \begin{cases} 2x^2; & x \geq 0 \\ 0; & x \leq 0 \end{cases}$

5.

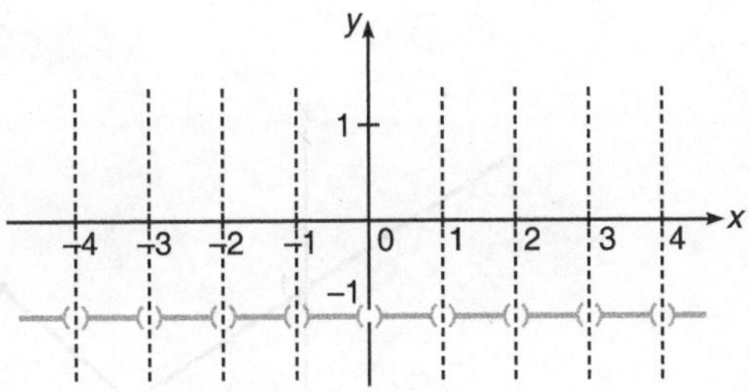

Here, $f(x) = [[x] - x] = [x] + [-x]$

$$= \begin{cases} x - x; & x \in \text{integer} \\ [x] + (-1 - [x]); & x \notin \text{integer} \end{cases}$$

$$= \begin{cases} 0; & x \in \text{integer} \\ -1; & x \notin \text{integer} \end{cases}$$

6.

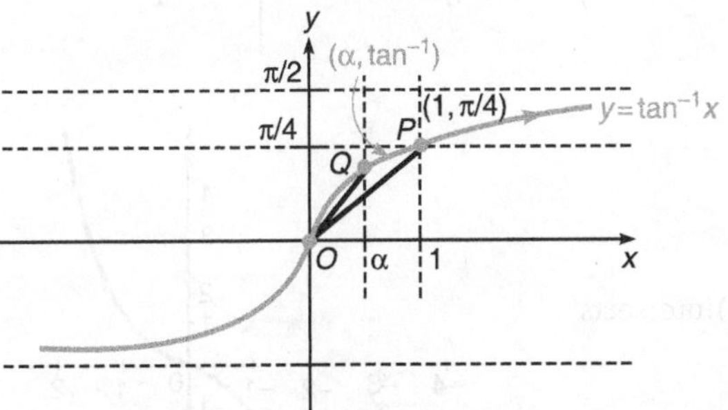

From figure;

Slope of OP < Slope of OQ.

$$\Rightarrow \quad \frac{\frac{\pi}{4} - 0}{1 - 0} < \frac{\tan^{-1}\alpha - 0}{\alpha - 0}$$

$$\Rightarrow \quad \frac{\tan^{-1}\alpha}{\alpha} > \frac{\pi}{4}.$$

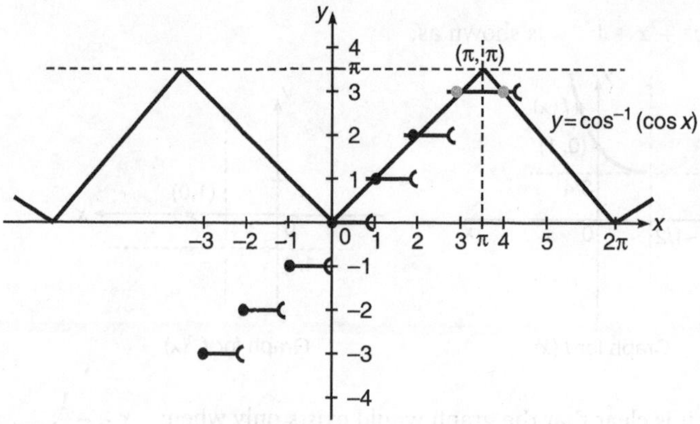

Clearly, the above curves intersect at 5 points;

$\therefore$ number of solutions = 5.

8.

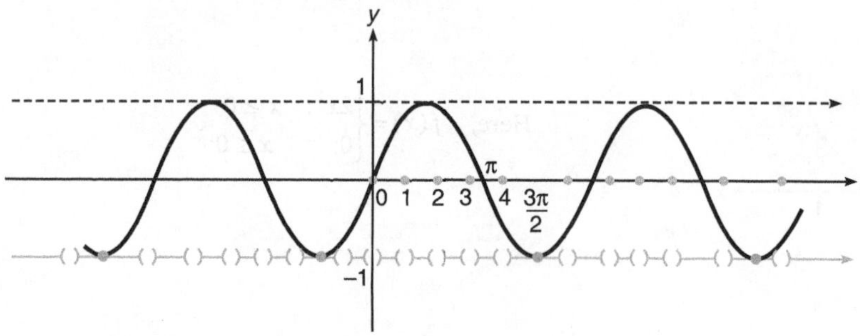

From above figure, number of solutions is infinite.

9. Clearly; $\left|\dfrac{x^2}{x-1}\right| \le 1$

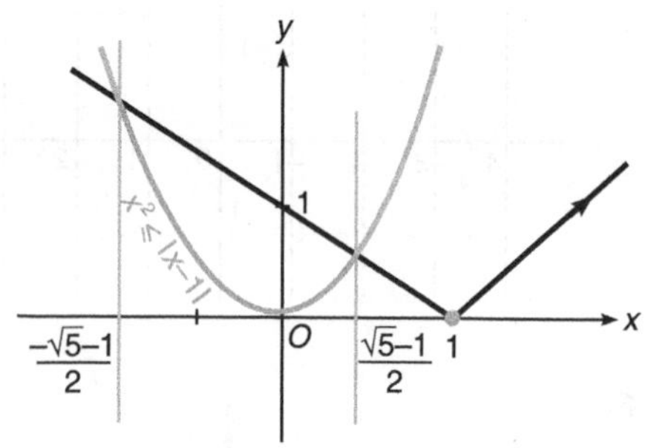

or $\quad x^2 \le |x-1|$

$\Rightarrow \quad x \in \left[\dfrac{-1-\sqrt{5}}{2}, \dfrac{-1+\sqrt{5}}{2}\right]$.

10. Here, $x^3 = 3 + [x]$

$\therefore$ to sketch $f(x) = x^3$

and $\quad g(x) = 3 + [x]$

Clearly; from figure the two curves $f(x)$ and $g(x)$ intersects when $g(x) = 4$.

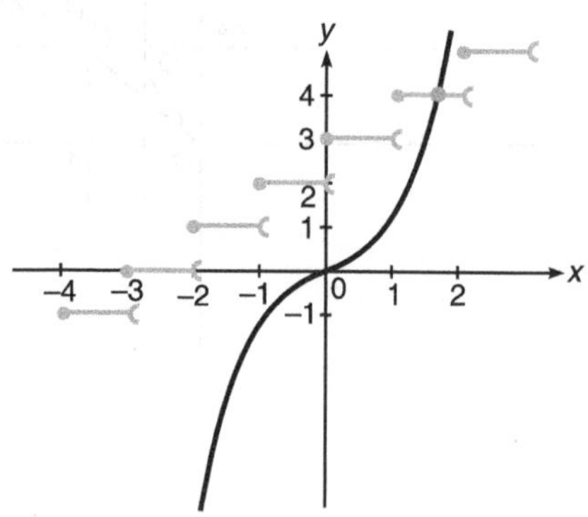

$\therefore \quad f(x) = 4$

$\Rightarrow \quad x^3 = 4$

or $\quad x = 2^{2/3}$.

1. (i)

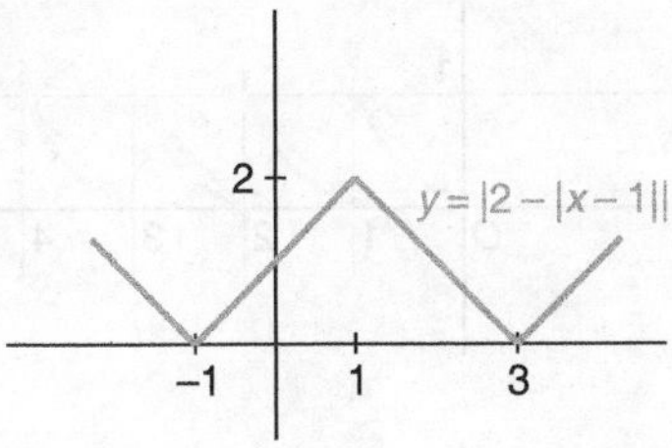

(ii)

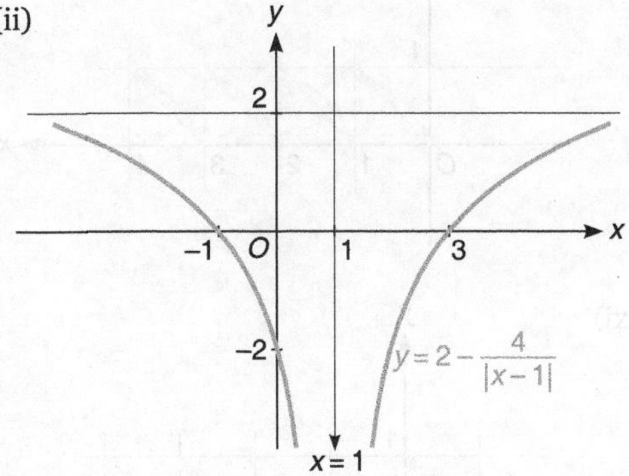

(iii)

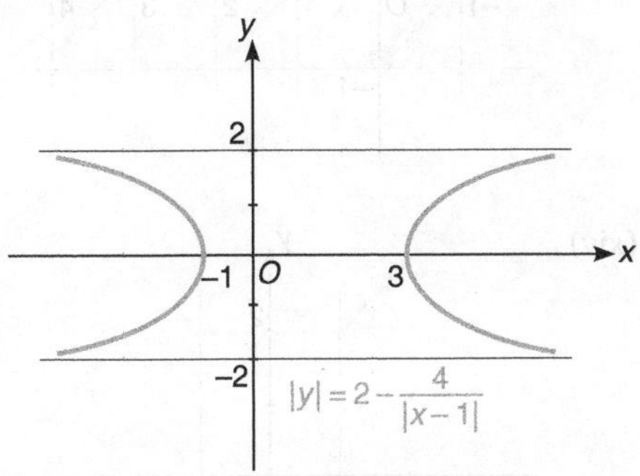

(iv)

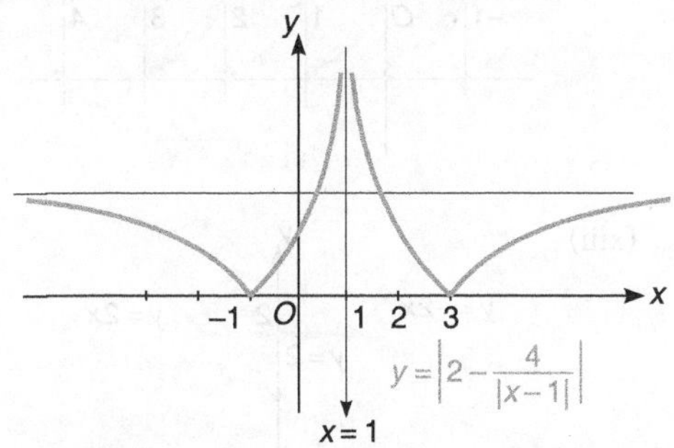

(v)

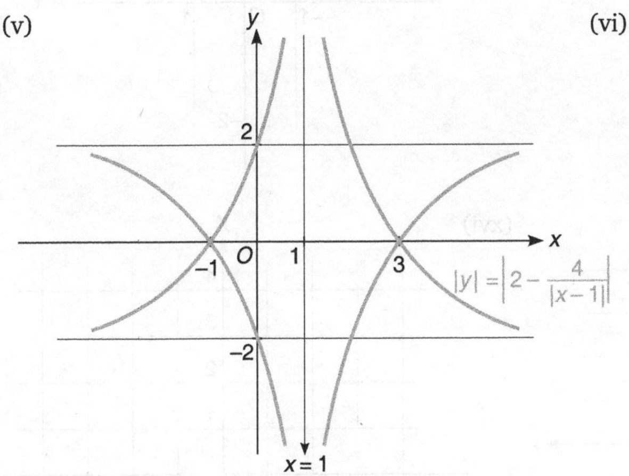

(vi)

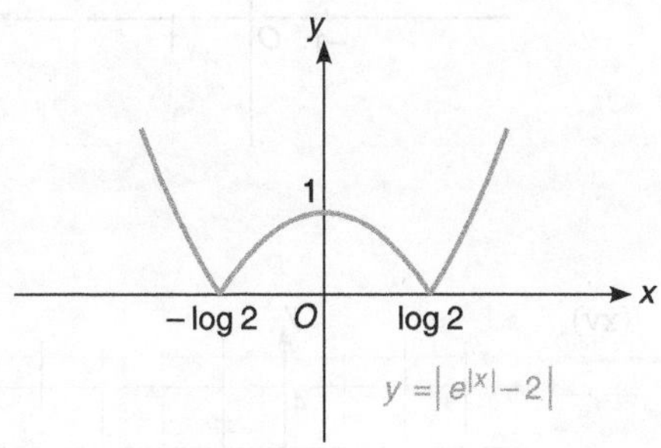

(vii)

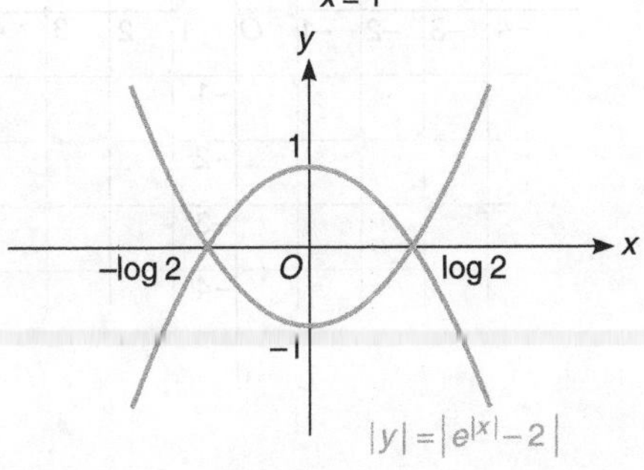

(viii)

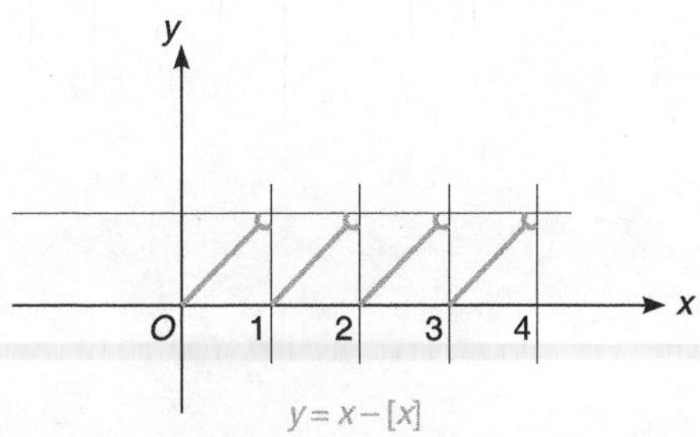

Hints and Solutions

(ix)

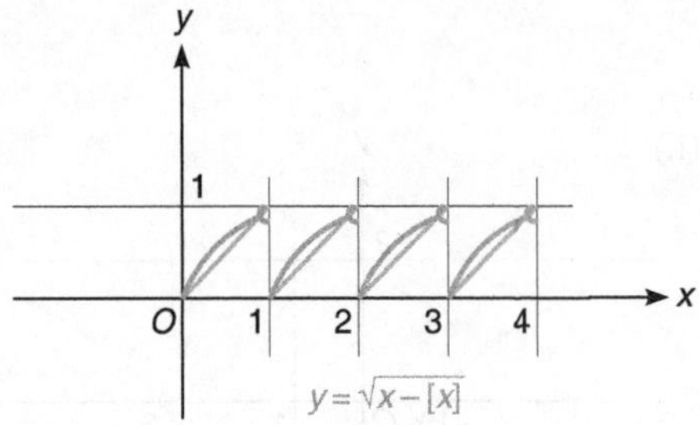

$y = \sqrt{x - [x]}$

(x)

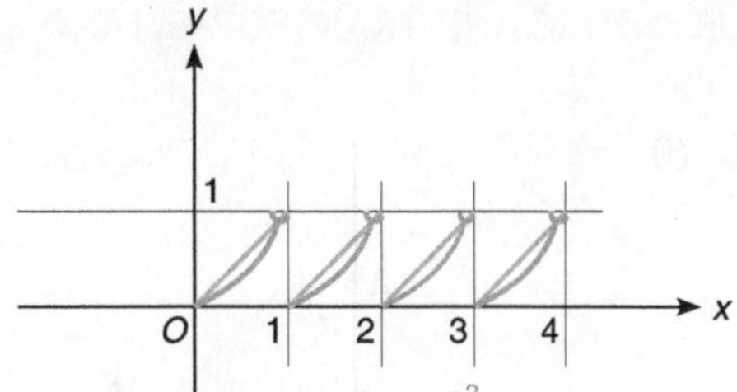

$y = (x - [x])^2$

(xi)

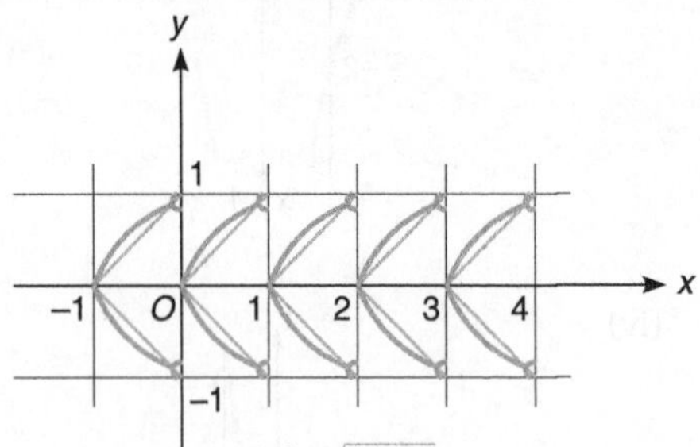

$|y| = \sqrt{x - [x]}$

(xii)

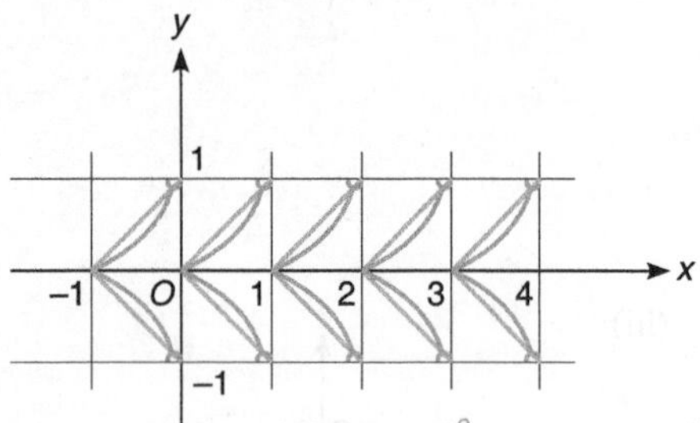

$|y| = (x - [x])^2$

(xiii)

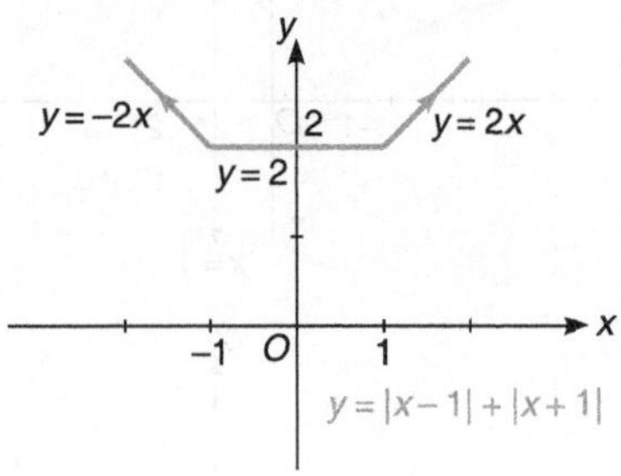

$y = |x - 1| + |x + 1|$

(xiv)

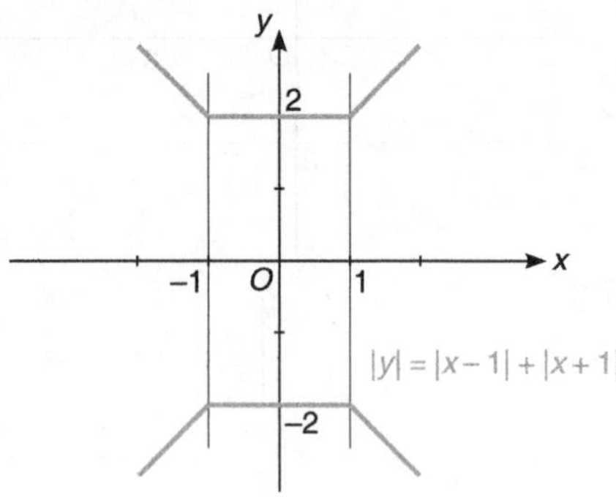

$|y| = |x - 1| + |x + 1|$

(xv)

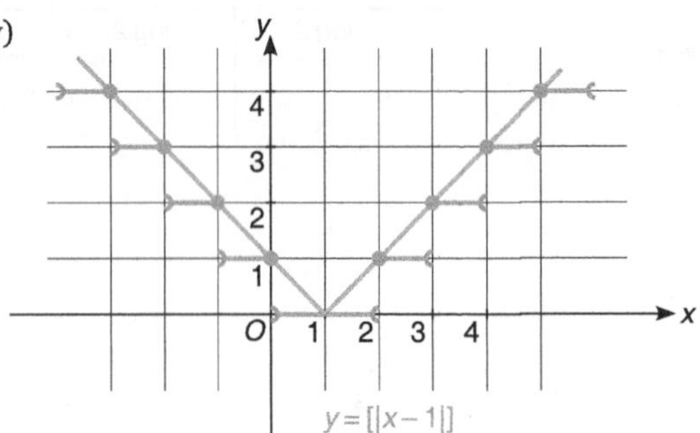

$y = [|x - 1|]$

(xvi)

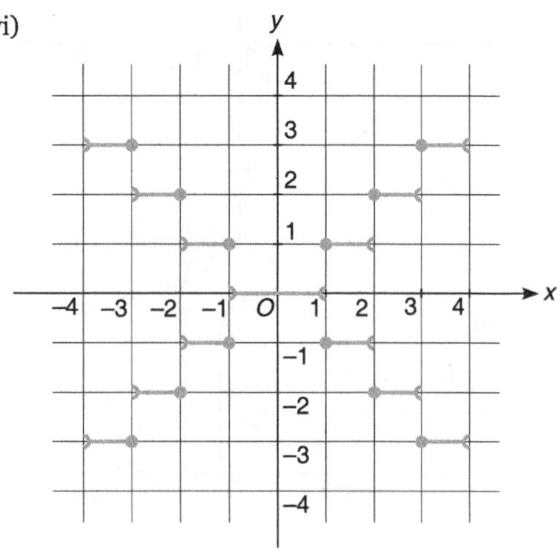

$|y| = [|x - 1|]$

(xvii)

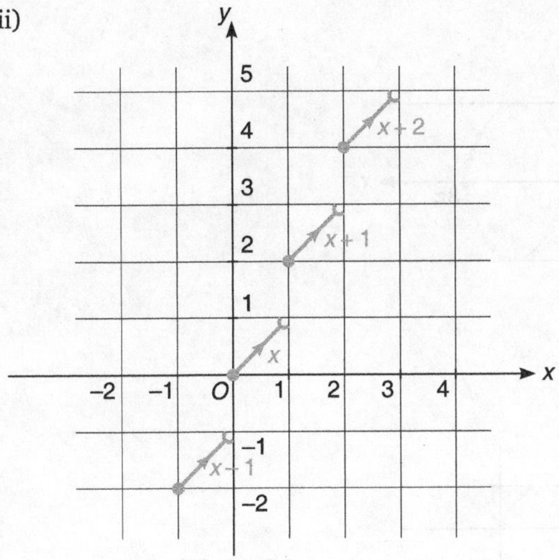

$|y| = x + [x]$

(xviii)

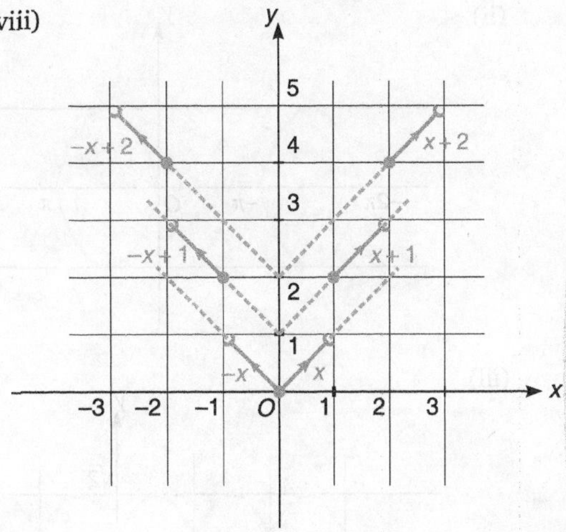

$y = |x| + [|x|]$

(xix)

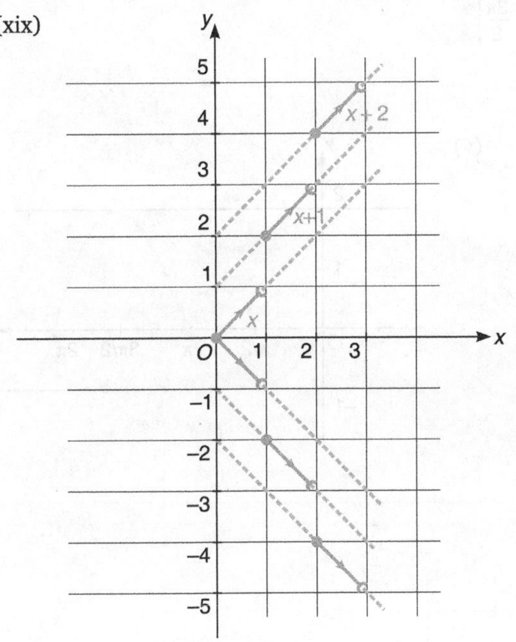

$|y| = x + [x]$

(xx)

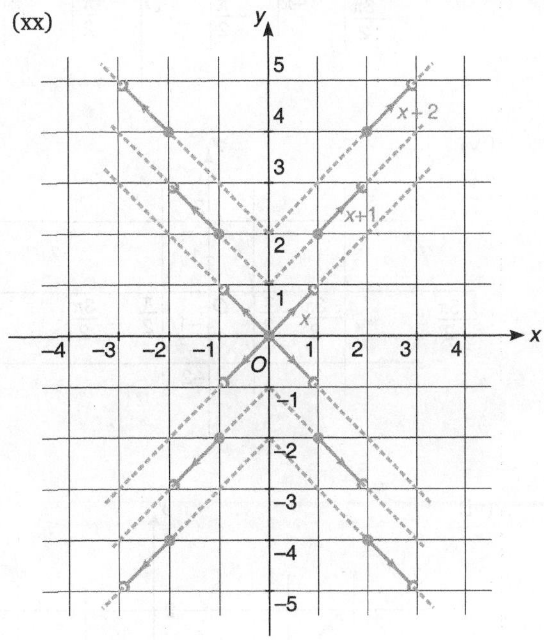

$|y| = |x| + [|x|]$

2. (i)

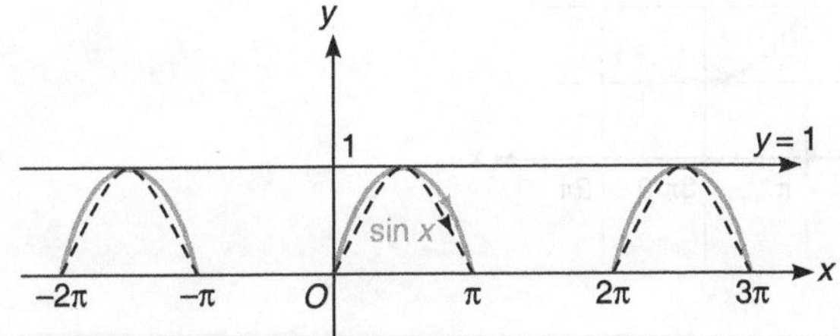

$y = \sqrt{\sin x}$

(ii)

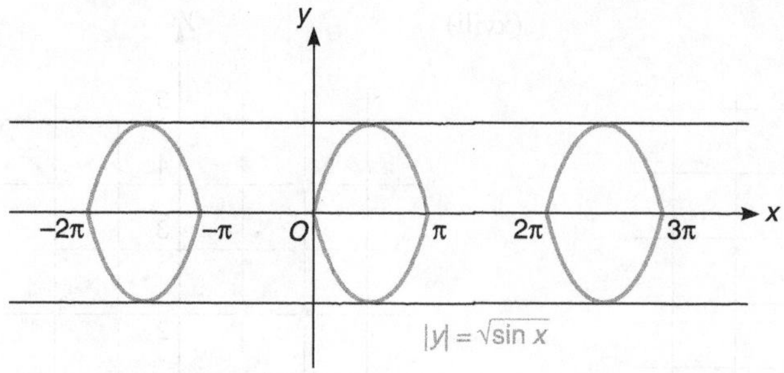

$|y| = \sqrt{\sin x}$

(iii)

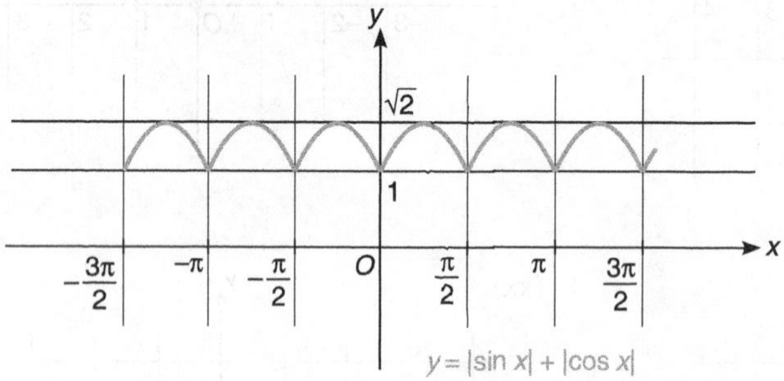

$y = |\sin x| + |\cos x|$

(iv)

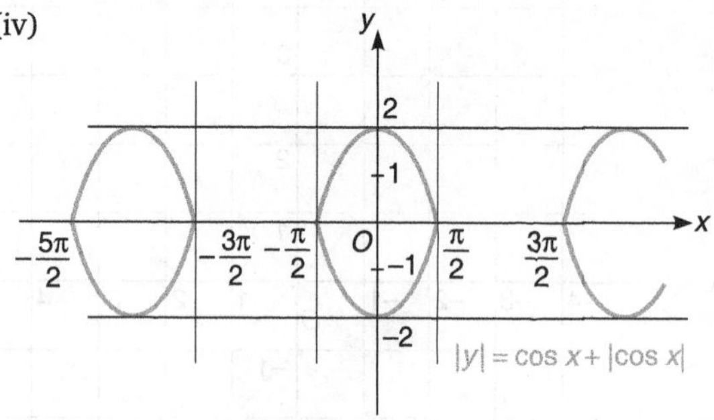

$|y| = \cos x + |\cos x|$

(v)

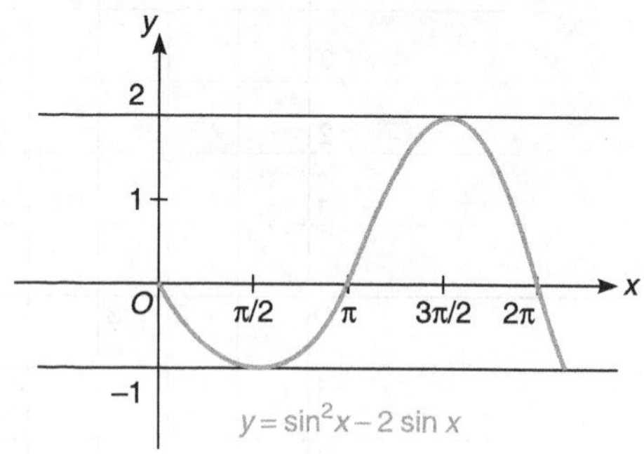

$y = \sin^2 x - 2\sin x$

(vi)

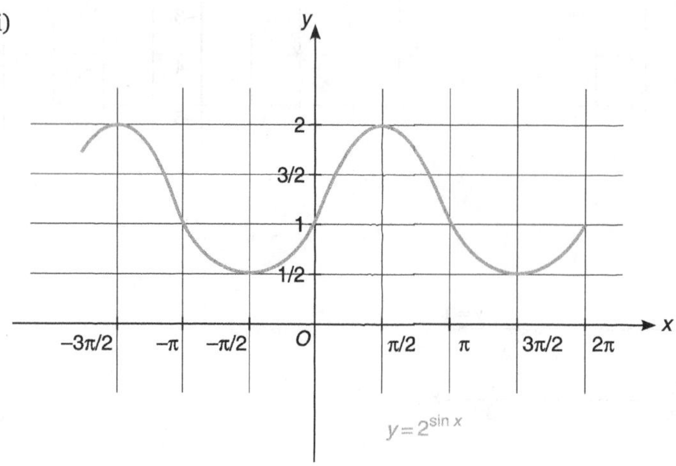

$y = 2^{\sin x}$

(vii)

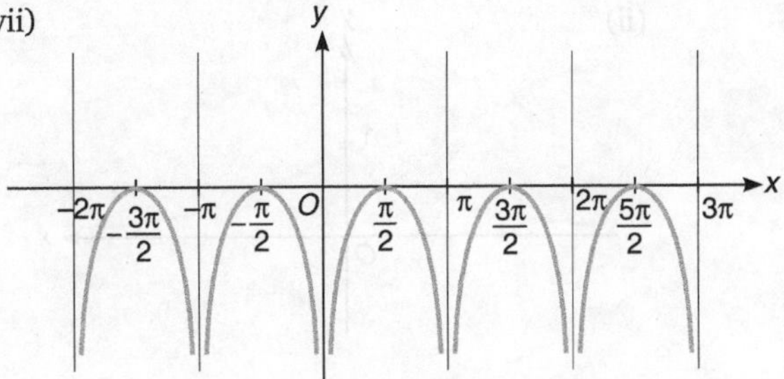

$y = \log_2|\sin x|$

(viii)

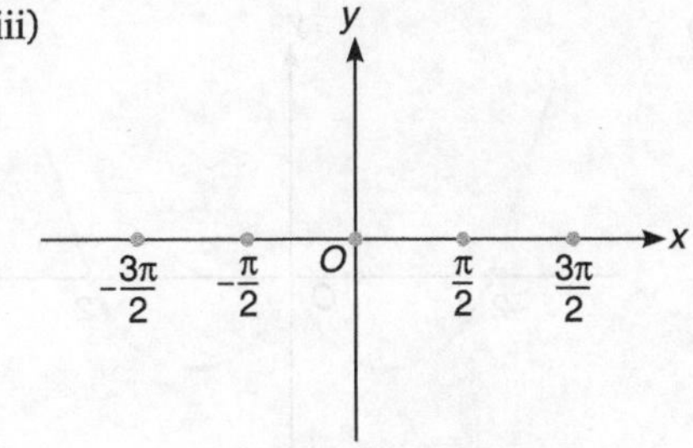

it only represents points

(ix)

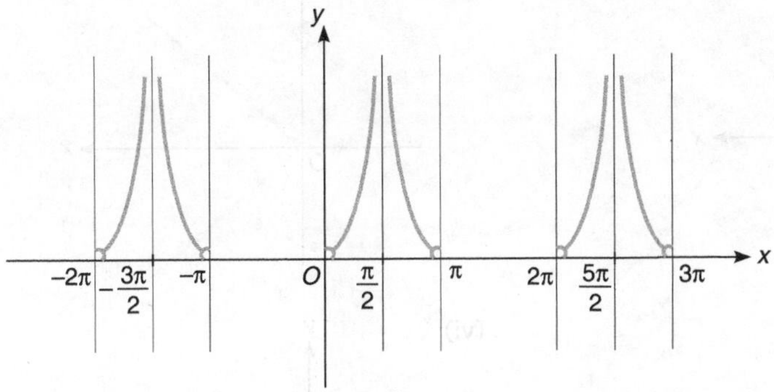

$y = \log_{\sin x} 1/2$

(x)

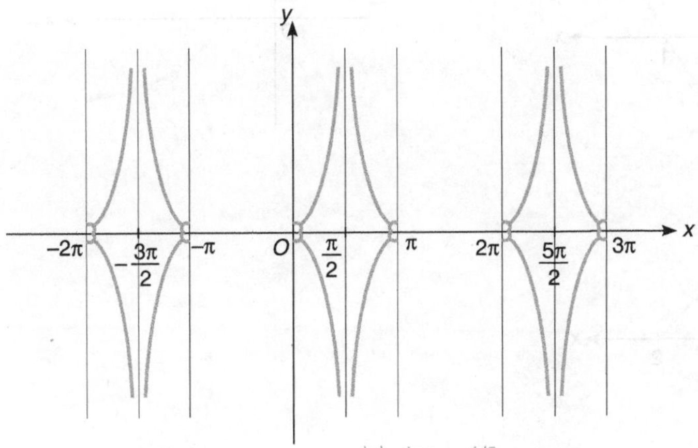

$|y| = \log_{\sin x} 1/2$

3.

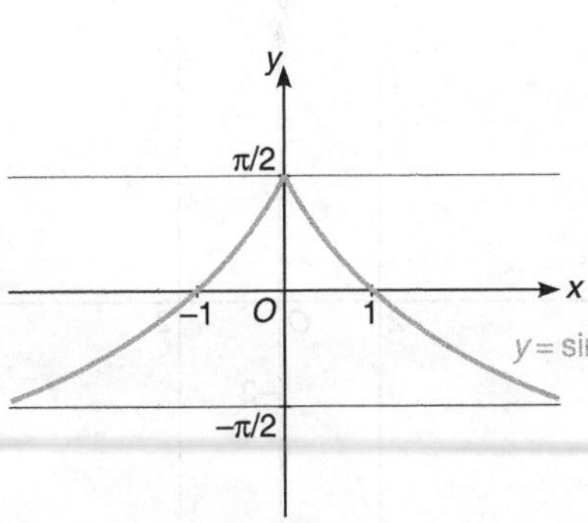

$y = \sin^{-1}\left(\frac{1-x^2}{1+x^2}\right)$

4.

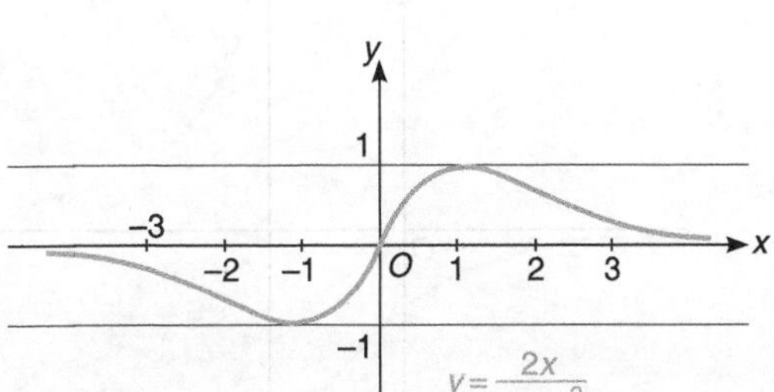

$y = \frac{2x}{1+x^2}$

5. (i)

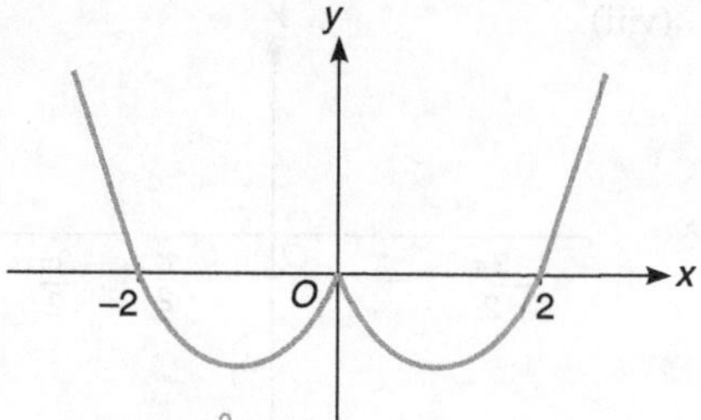

$y = x^2 - 2|x|$

(ii)

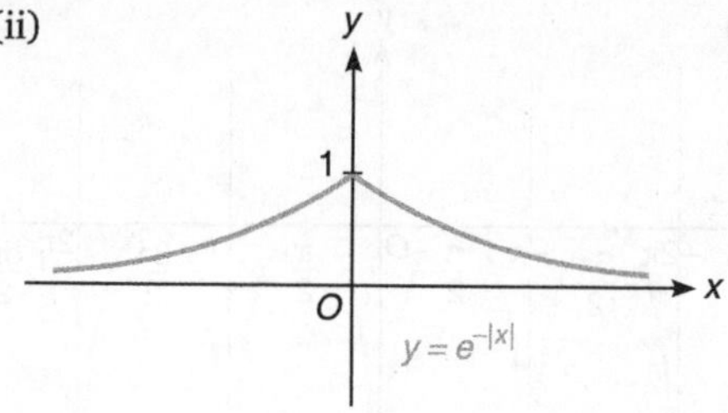

$y = e^{-|x|}$

(iii)

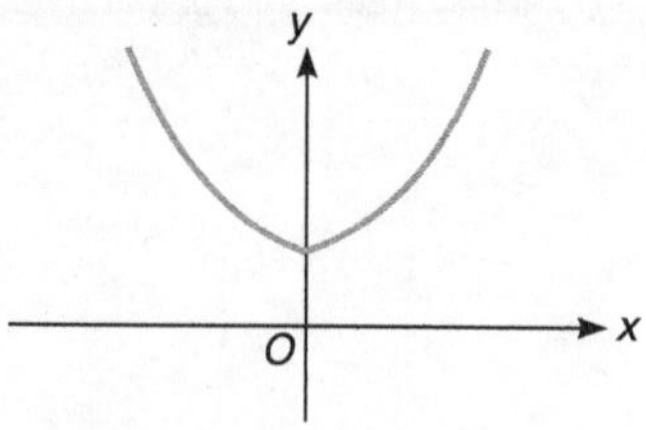

$y = e^{|x|}$

(iv)

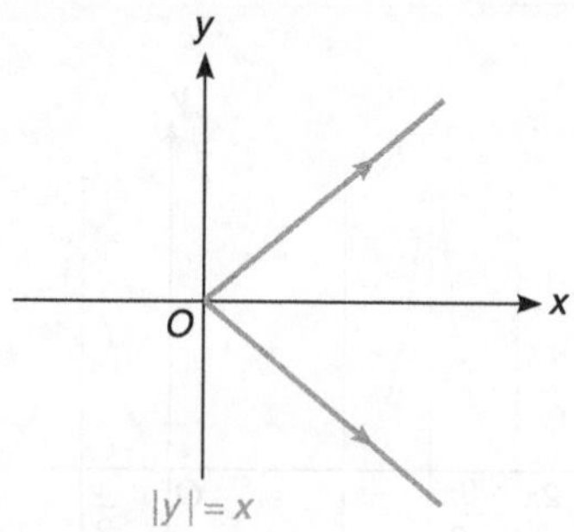

$|y| = x$

(v)

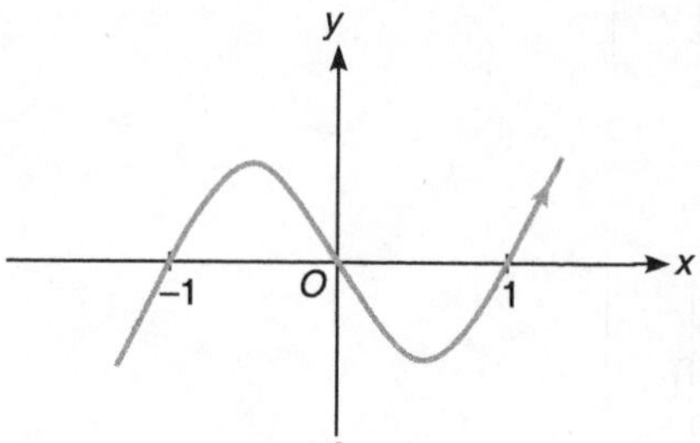

$y = x^3 - x$

(vi)

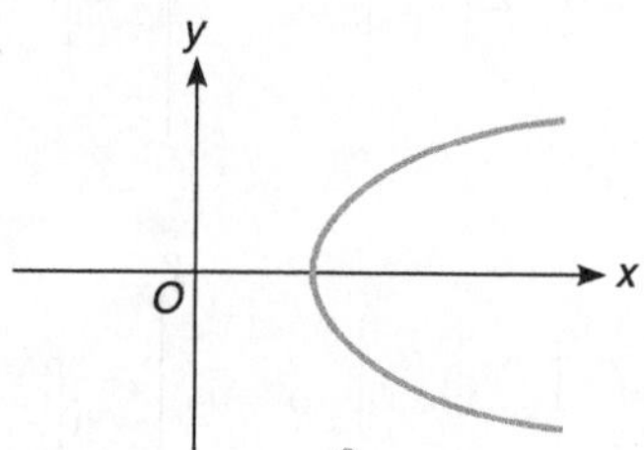

$x = y^2 + 1$

6.

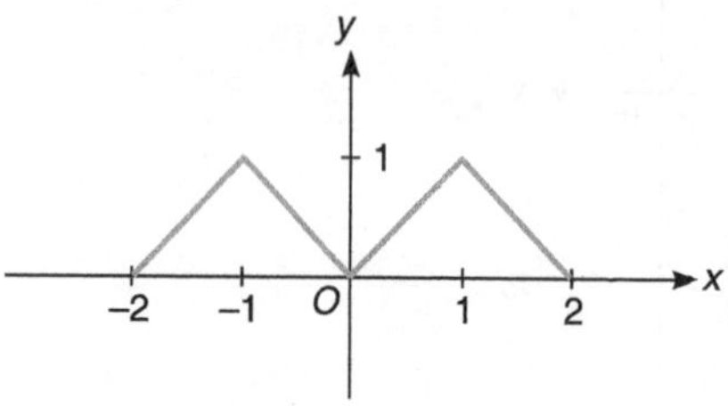

$y = f(x-1) + f(x+1)$

7. (i)

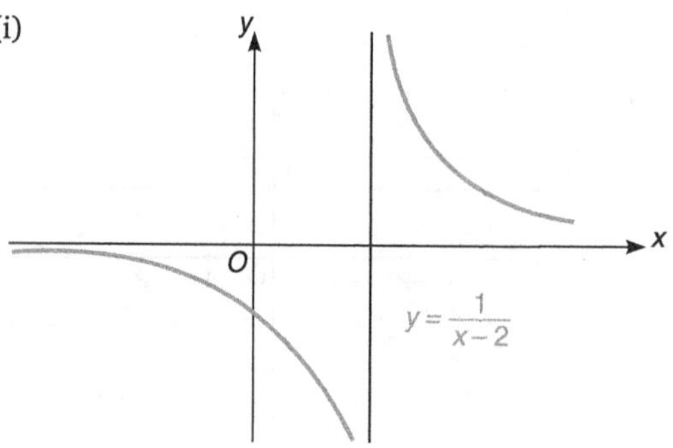

$y = \frac{1}{x-2}$

(ii)

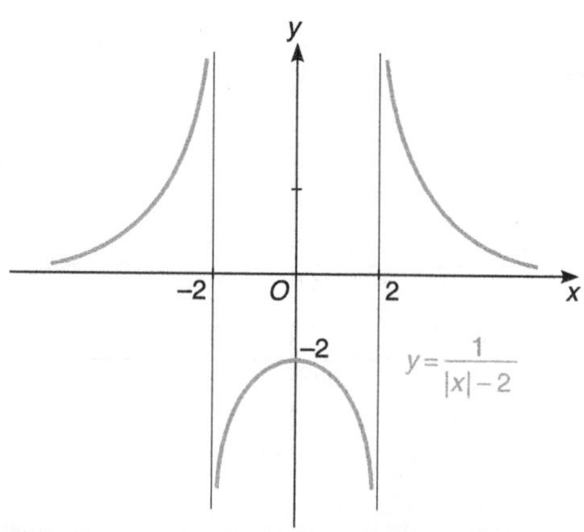

$y = \frac{1}{|x|-2}$

(iii)

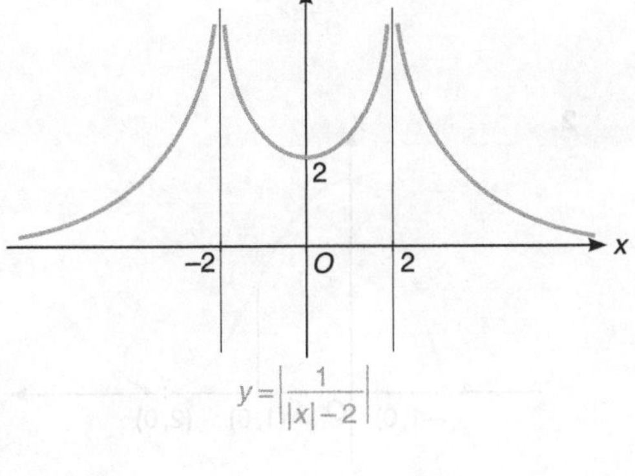

(iv)

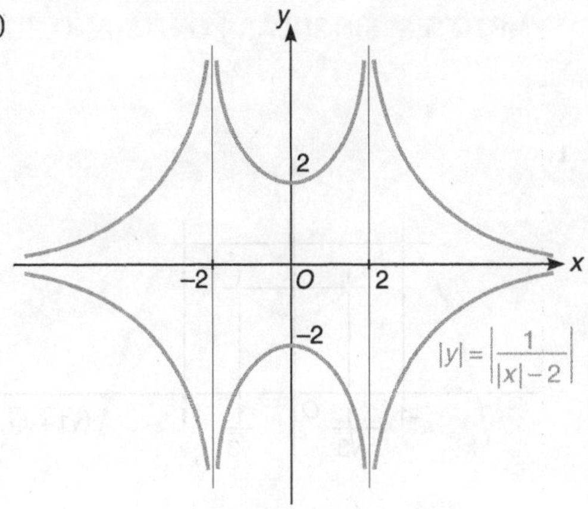

8.

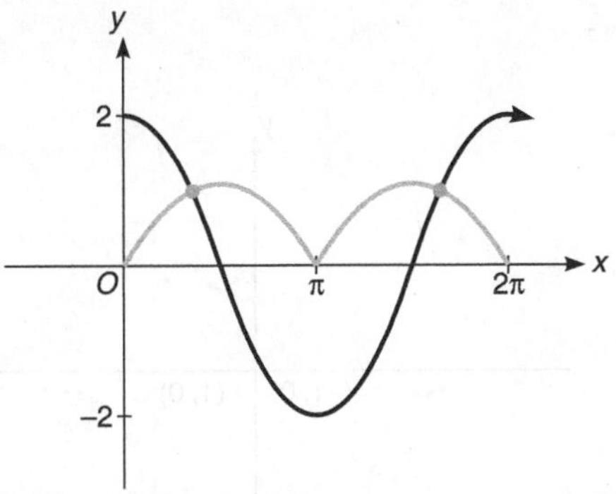

Since, $2\cos x$ and $|\sin x|$, intersects at two points for $x \in [0, 2\pi]$.

$\therefore$ number of solutions are four when $x \in [0, 4\pi]$.

9.

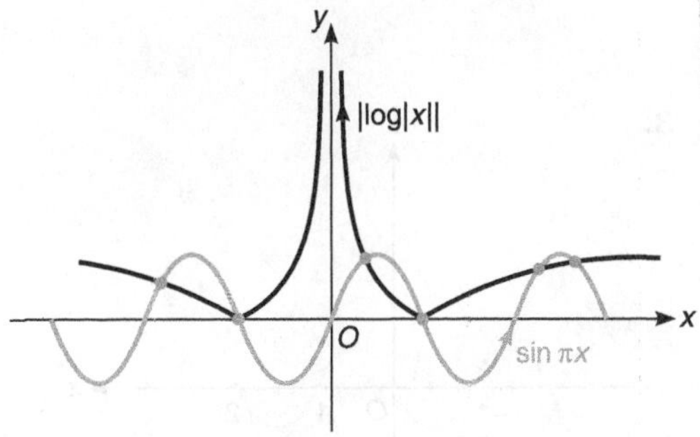

From above figure we have six solutions.

10.

x	0	$\pi/6$	$\pi/2$	$5\pi/6$	x_2	$\left(x_2, \frac{3\pi}{2}\right)$	$\frac{3\pi}{2}$	$\left(\frac{3\pi}{2}, x_4\right)$	x_4		2π
y'		0	– 2	0	9/8		0		9/8		
y''		$\frac{-3\sqrt{3}}{2}$	0	$\frac{3\sqrt{3}}{2}$	0	–	0	+	0	–	
y	1	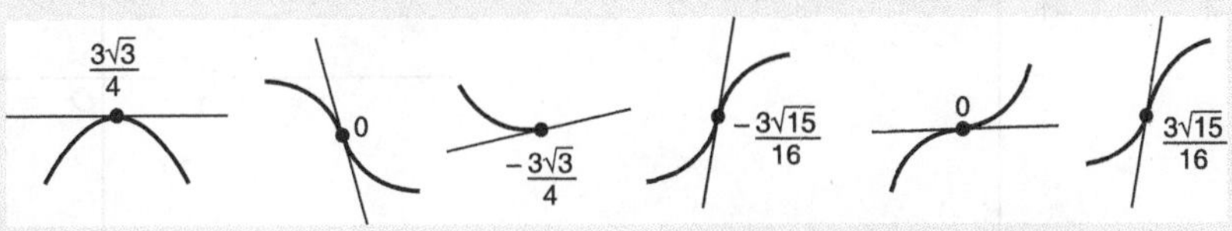									1

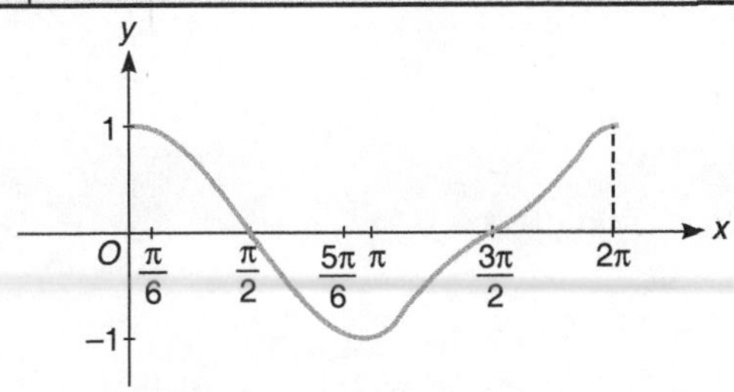

$y = \frac{1}{2}\sin 2x + \cos x,$

periodic with **period 2π**.

1.

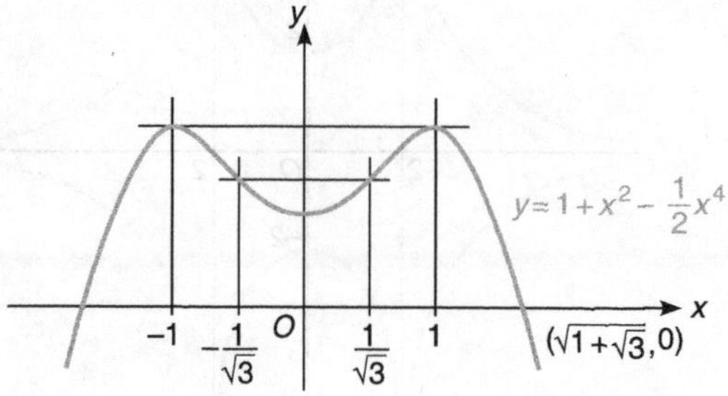
y
x
$y = 1 + x^2 - \frac{1}{2}x^4$
–1
$-\frac{1}{\sqrt{3}}$
O
$\frac{1}{\sqrt{3}}$
1
$(\sqrt{1+\sqrt{3}}, 0)$

2.

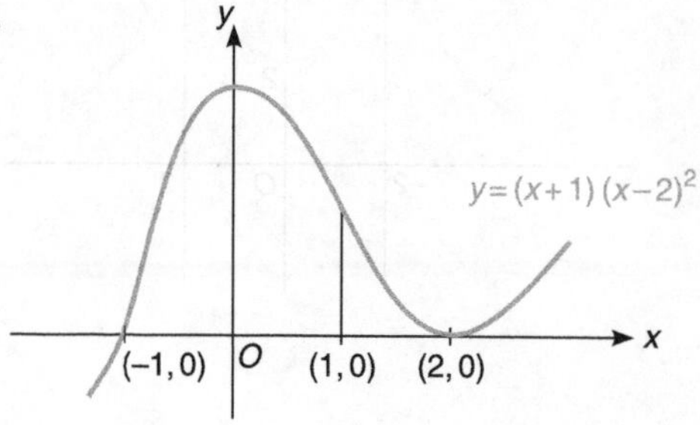
y
x
$y = (x+1)(x-2)^2$
(–1, 0)
O
(1, 0)
(2, 0)

3.

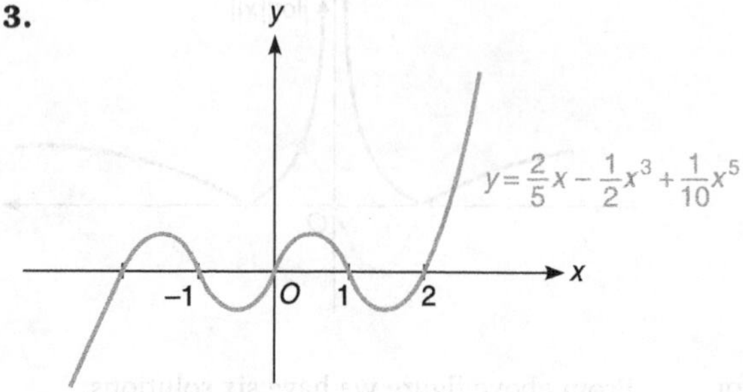
y
x
$y = \frac{2}{5}x - \frac{1}{2}x^3 + \frac{1}{10}x^5$
–1
O
1
2

4.

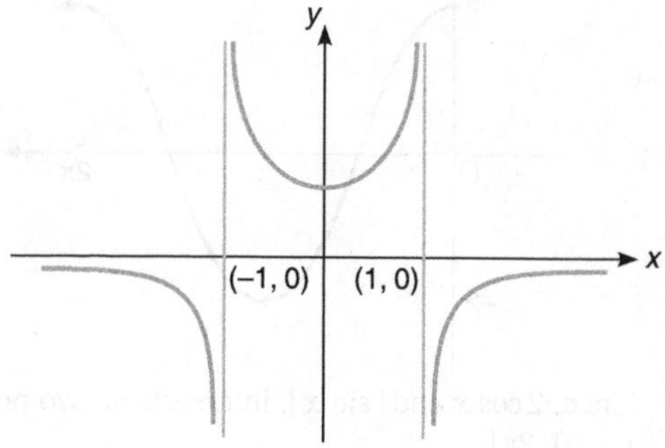
y
x
(–1, 0)
(1, 0)

5.

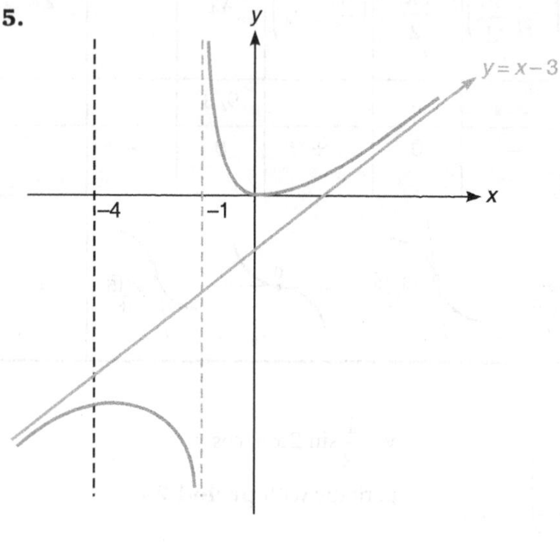
y
x
$y = x - 3$
–4
–1

6.

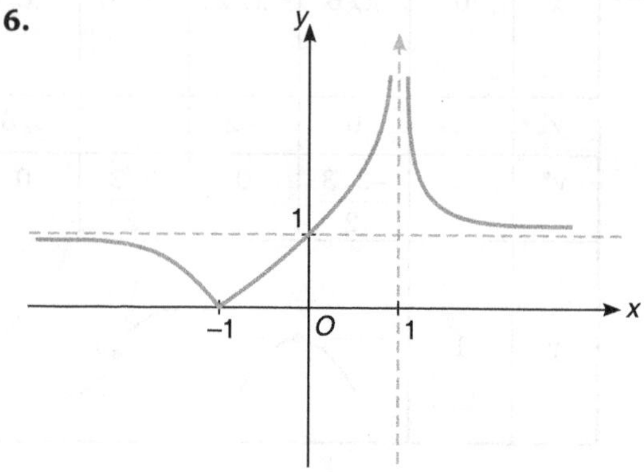
y
x
1
–1
O
1

7.

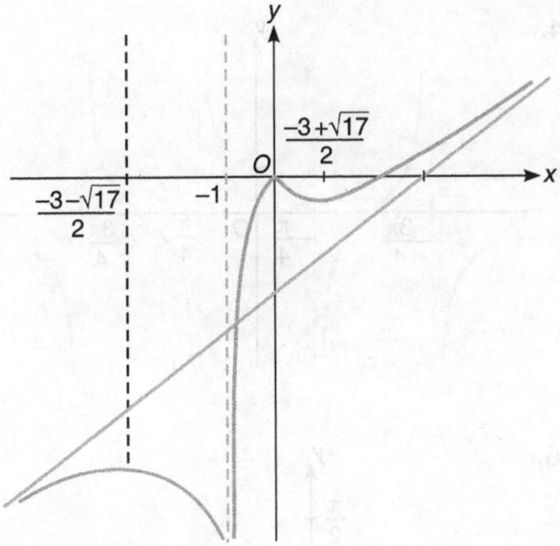
y
x
O
$\frac{-3+\sqrt{17}}{2}$
$\frac{-3-\sqrt{17}}{2}$
–1

8.

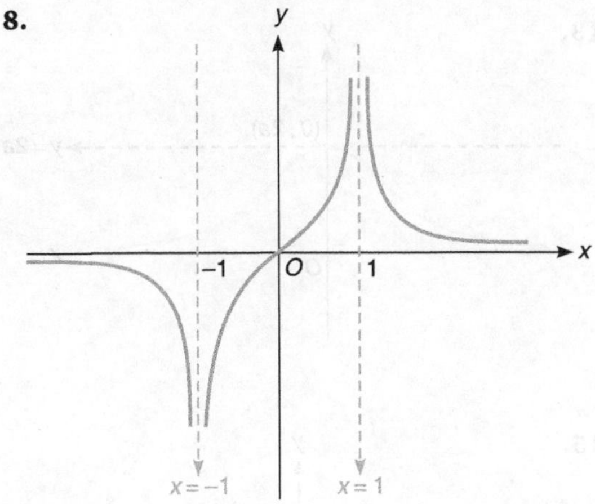
y
x
–1
O
1
x = –1
x = 1

9.

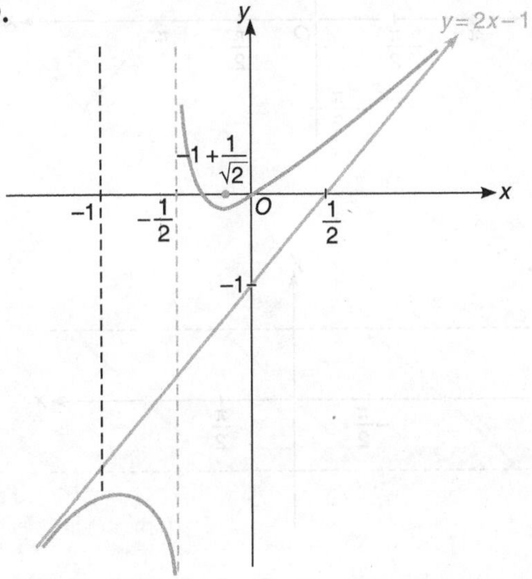
y
y = 2x – 1
$-1+\frac{1}{\sqrt{2}}$
x
–1
$-\frac{1}{2}$
O
$\frac{1}{2}$
–1

10.

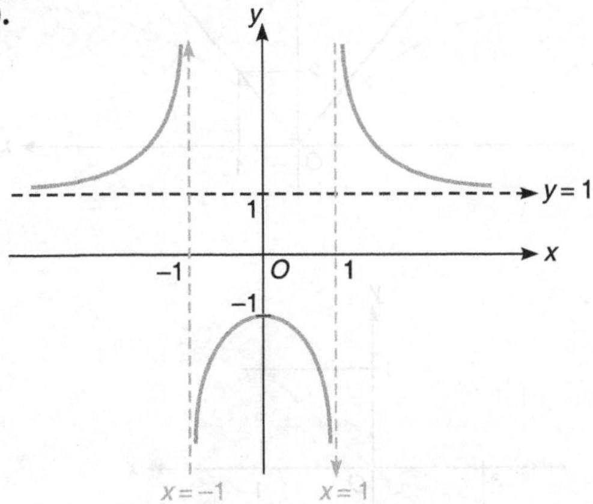
y
y = 1
1
x
–1
O
1
–1
x = –1
x = 1

11.

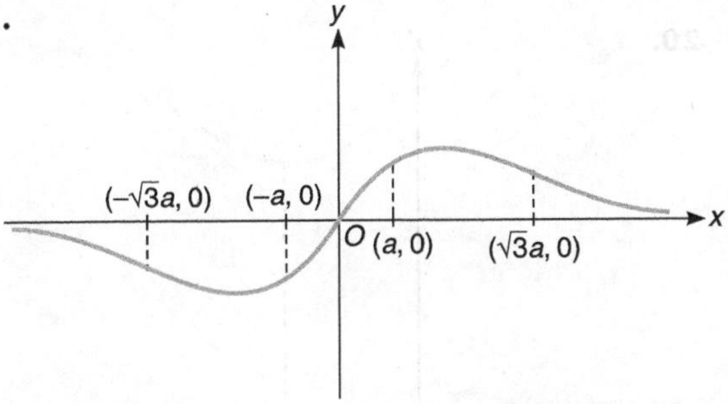
y
$(-\sqrt{3}a, 0)$
$(-a, 0)$
x
O
$(a, 0)$
$(\sqrt{3}a, 0)$

12.

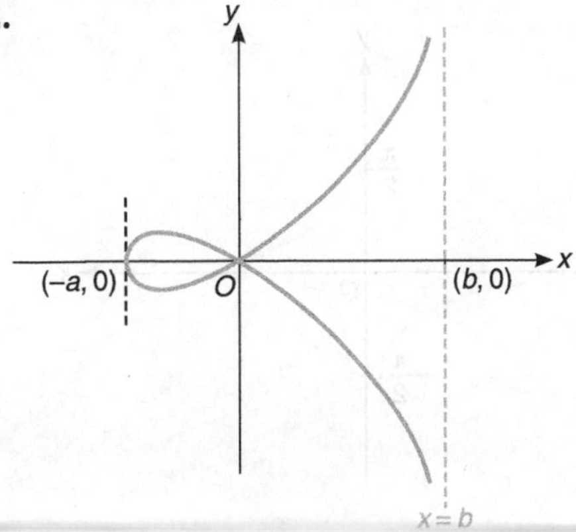
y
x
$(-a, 0)$
O
$(b, 0)$
x = b

13.

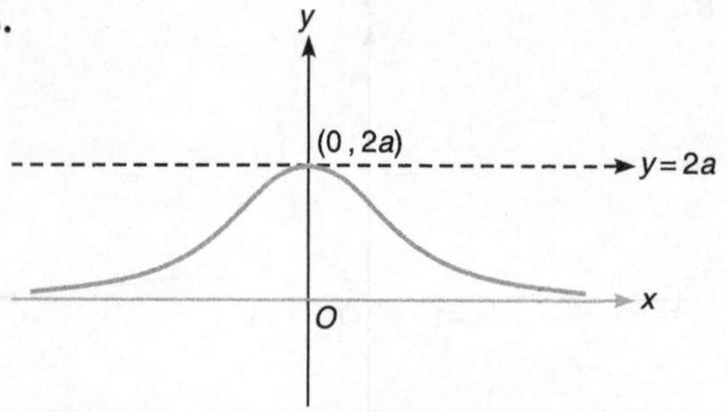

14.

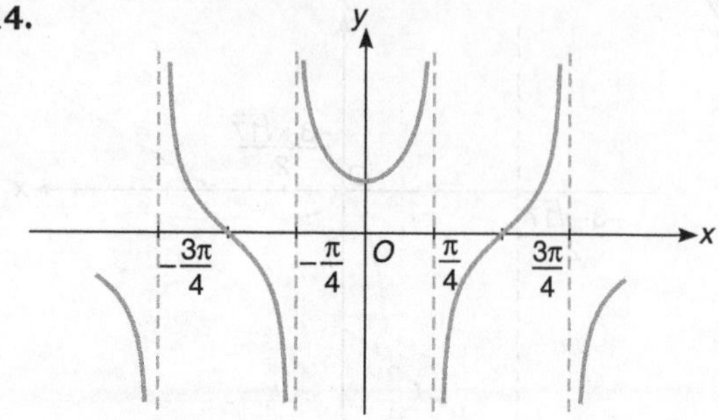

15.

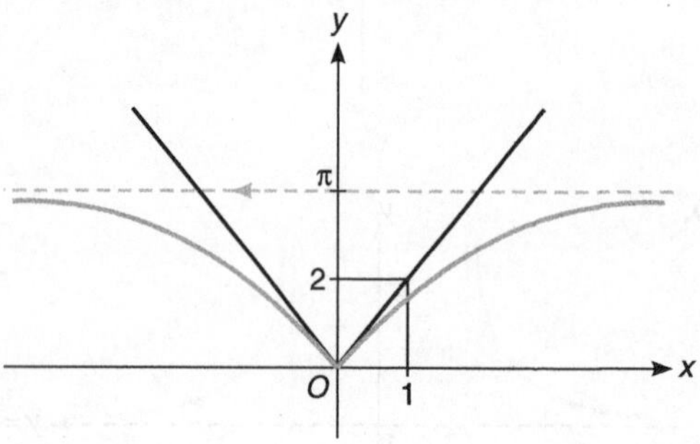

16.

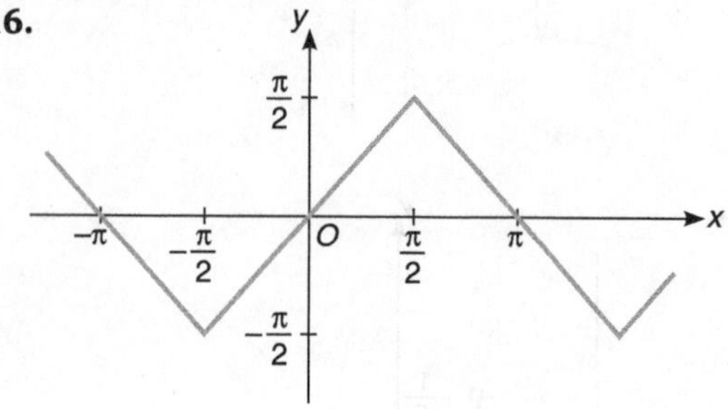

17.

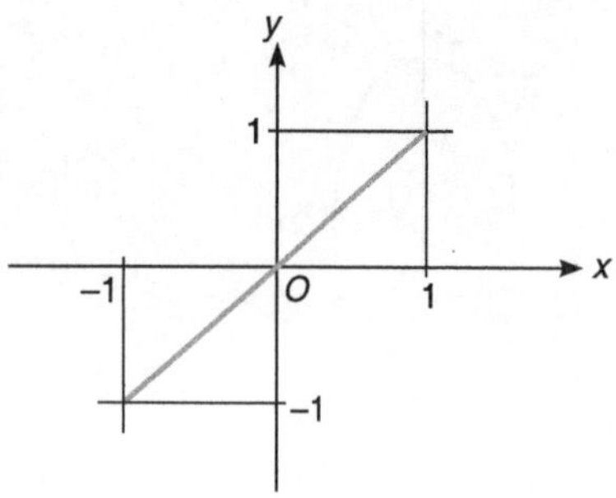

18.

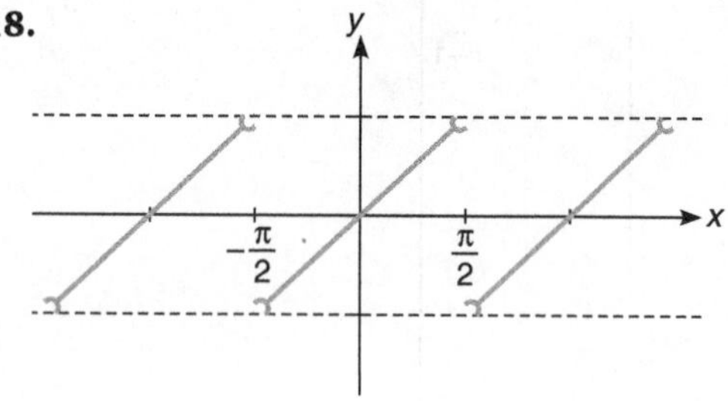

19.

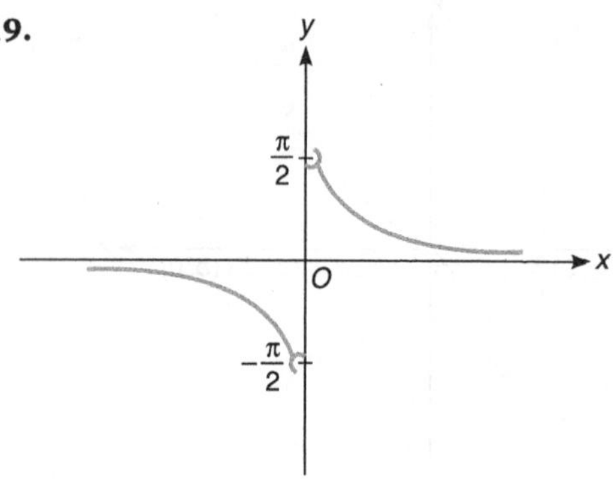

20.

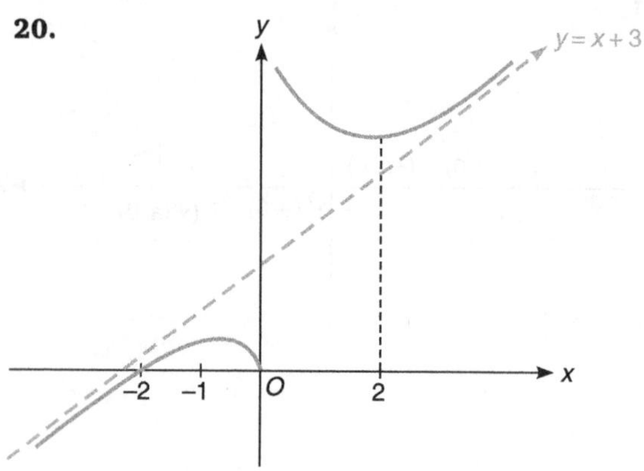

21.

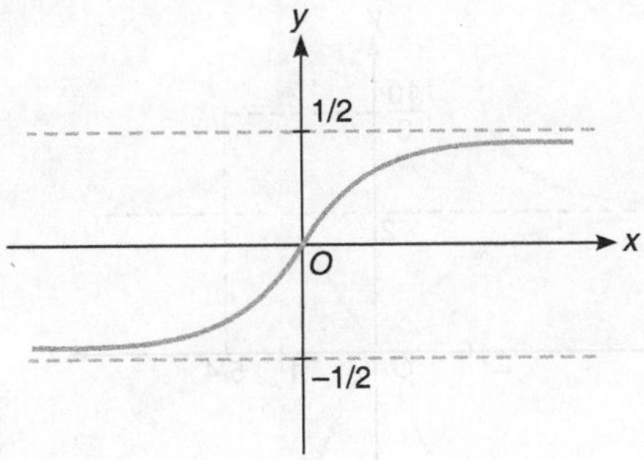

22.

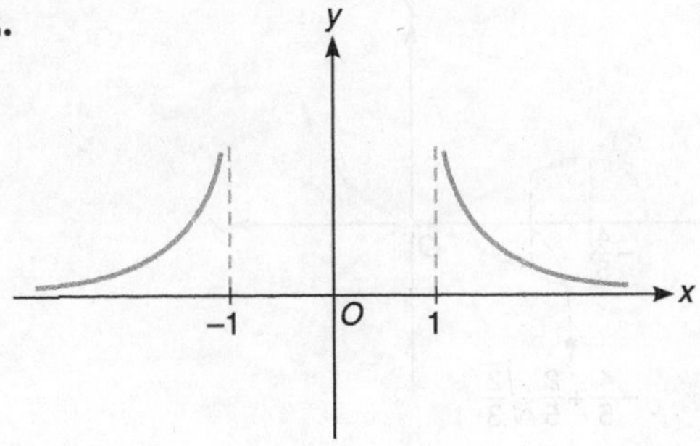

23.

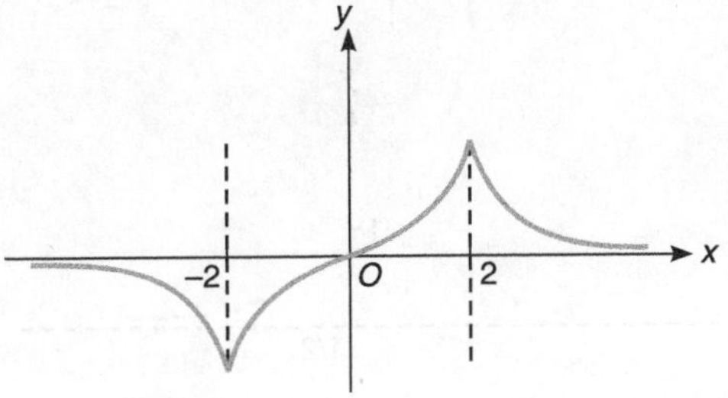

24.

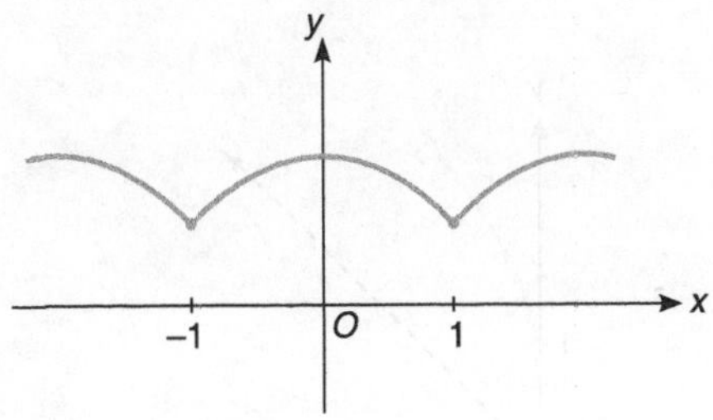

25.

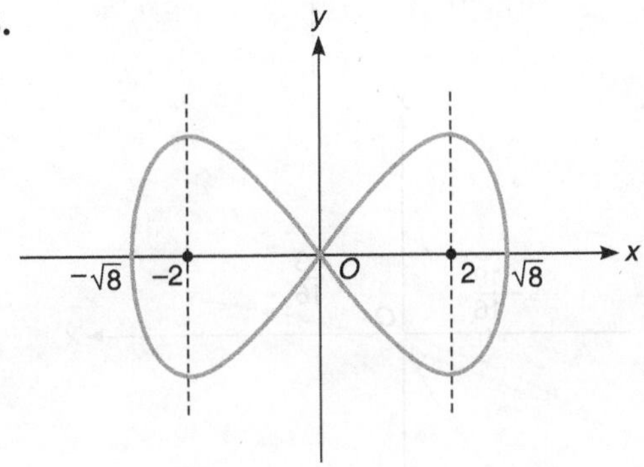

26.

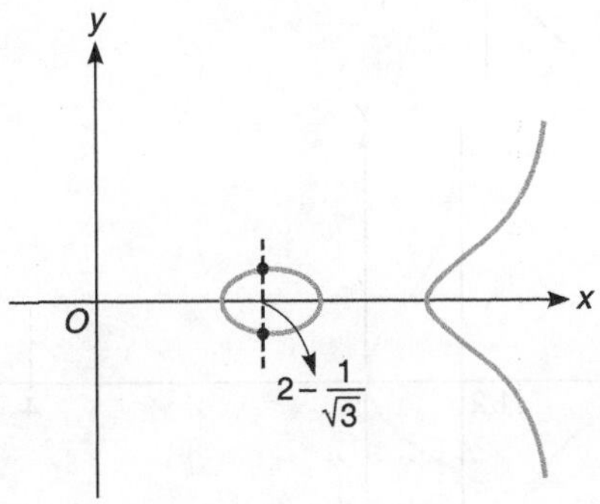

27.

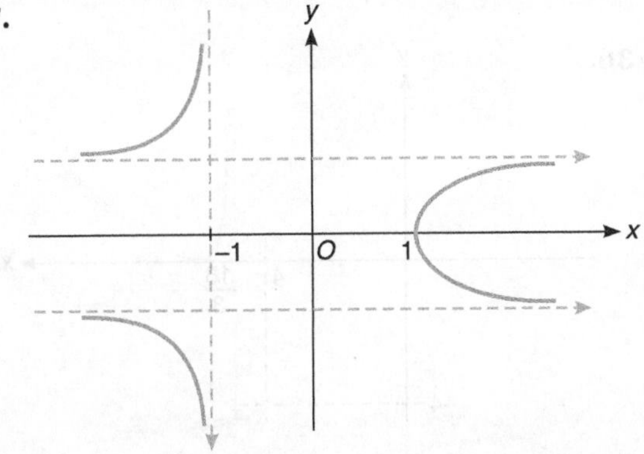

28.

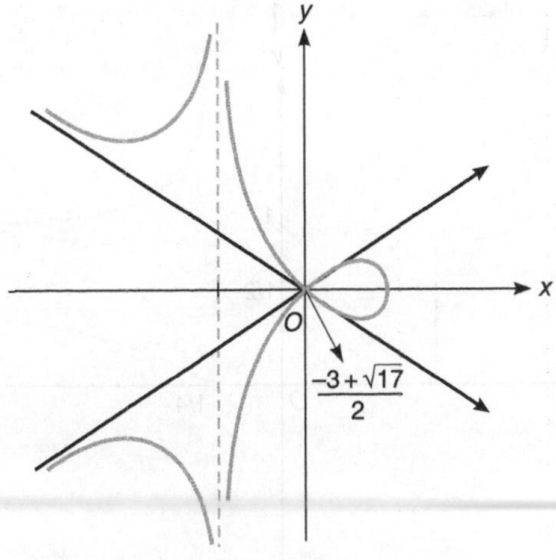

29.

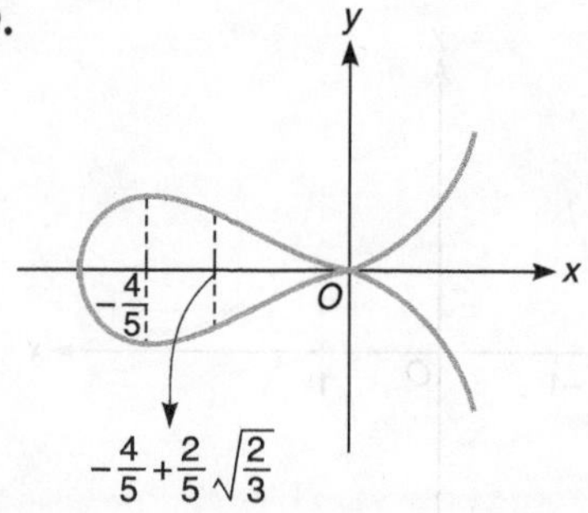

30.

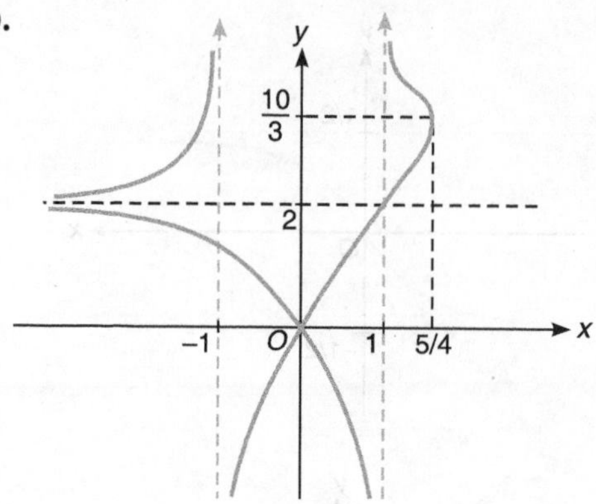

31.

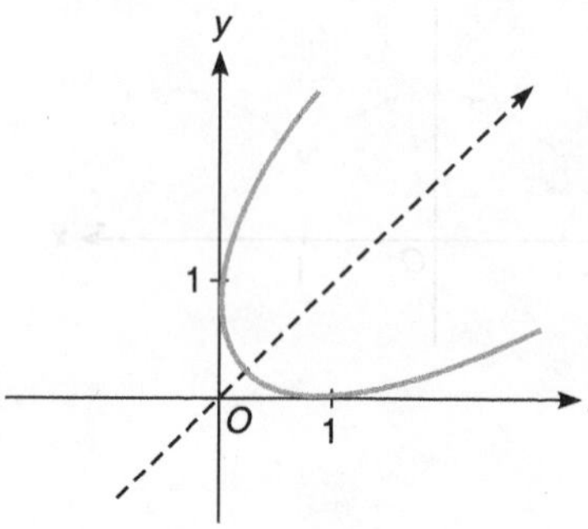

32.

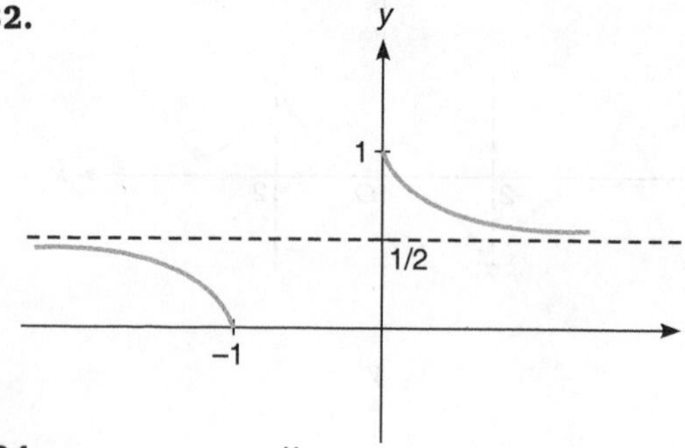

33.

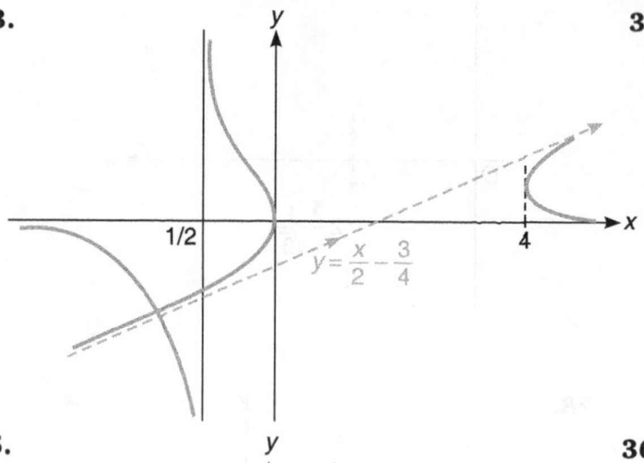

34.

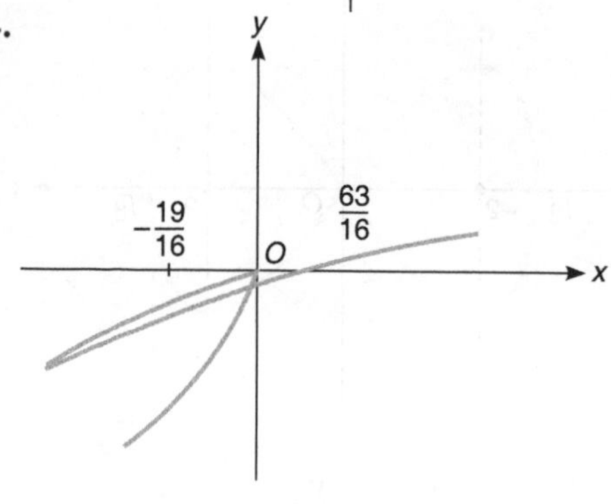

35.

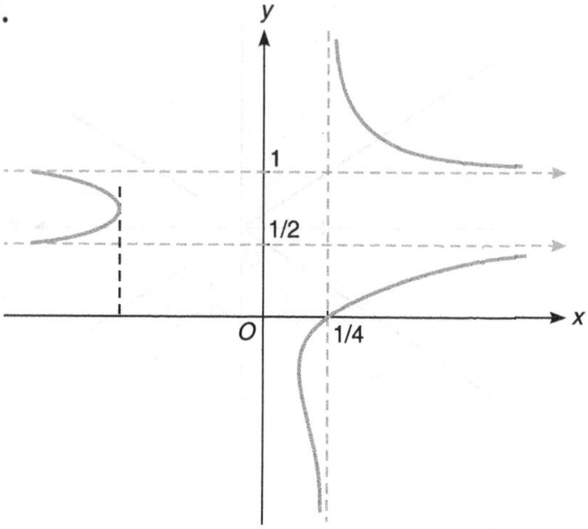

36.

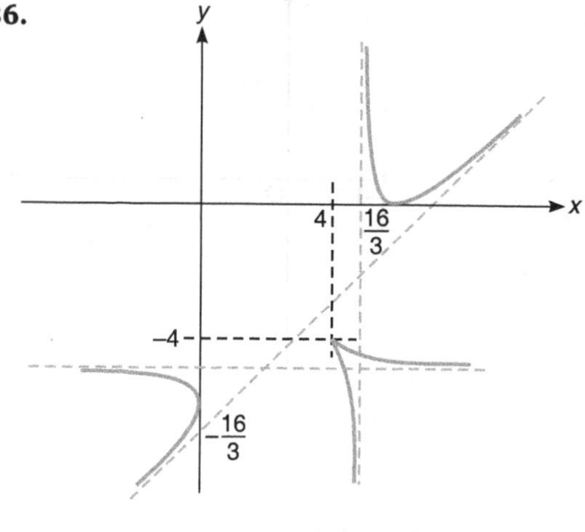

37.

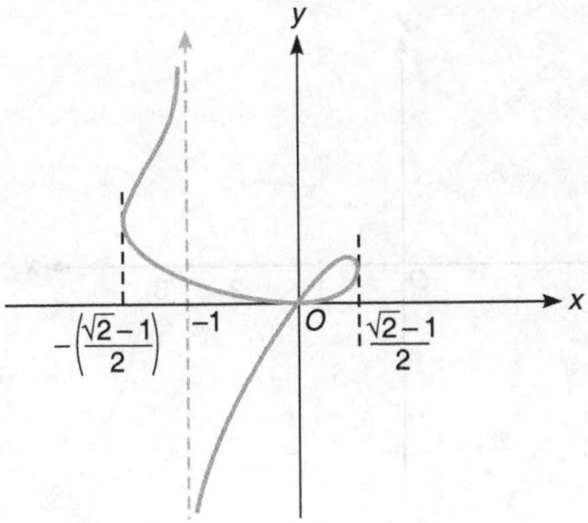

38.

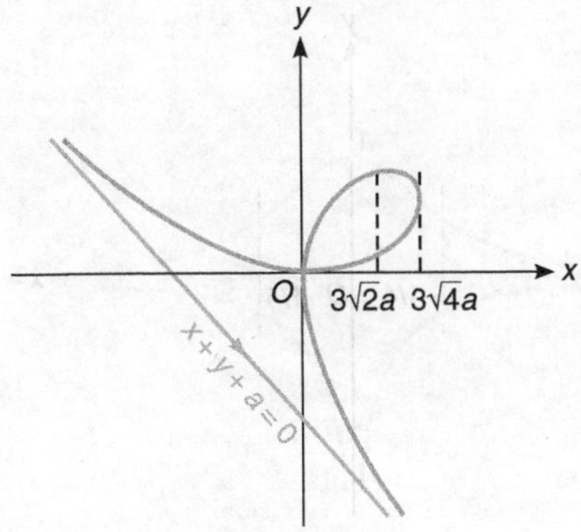

39.

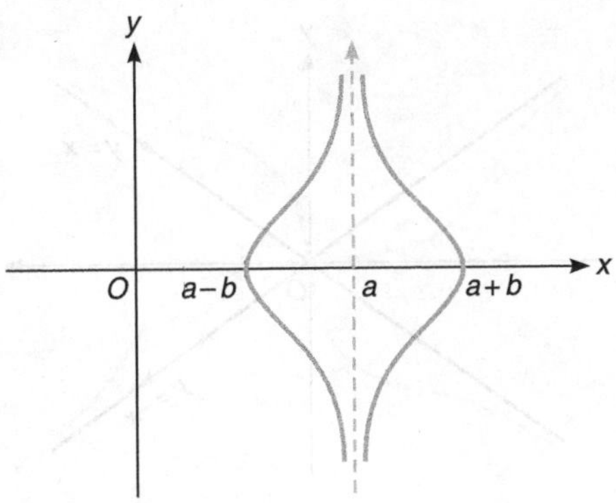

40.

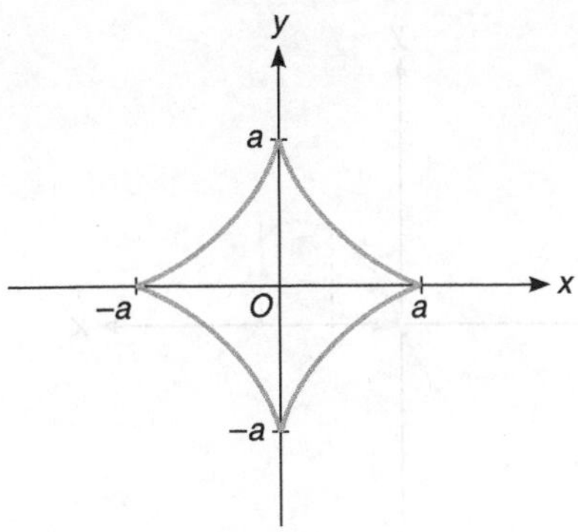

41.

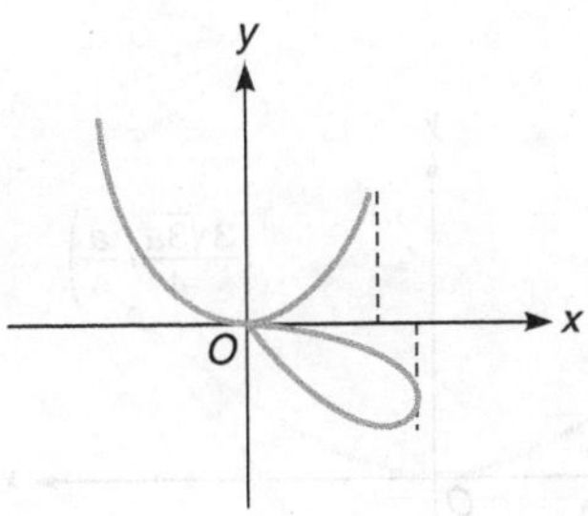

42.

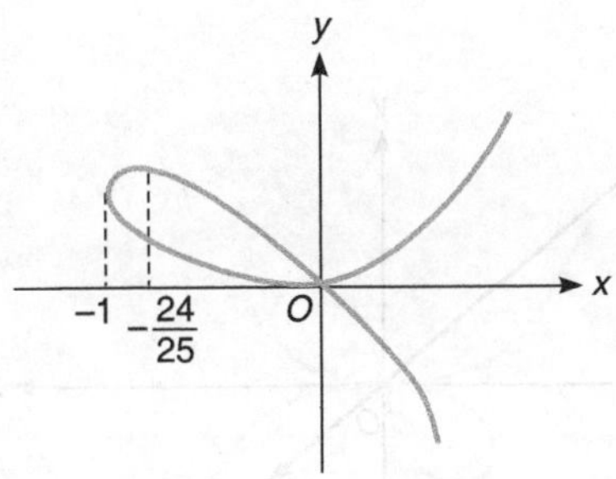

43.

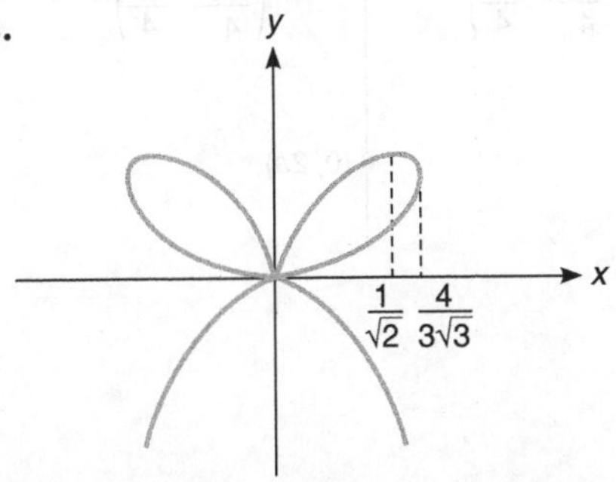

44.

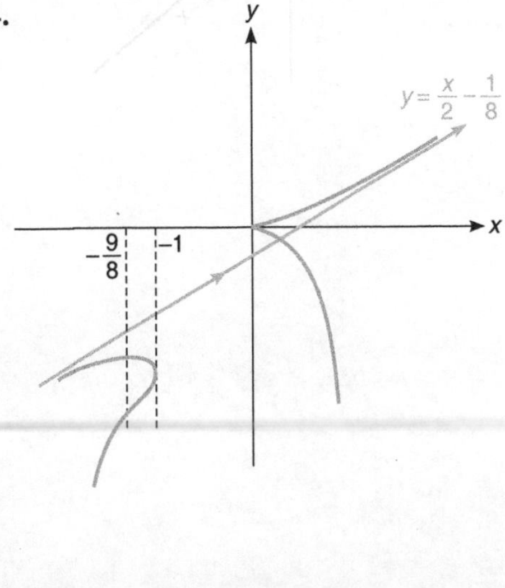

45.

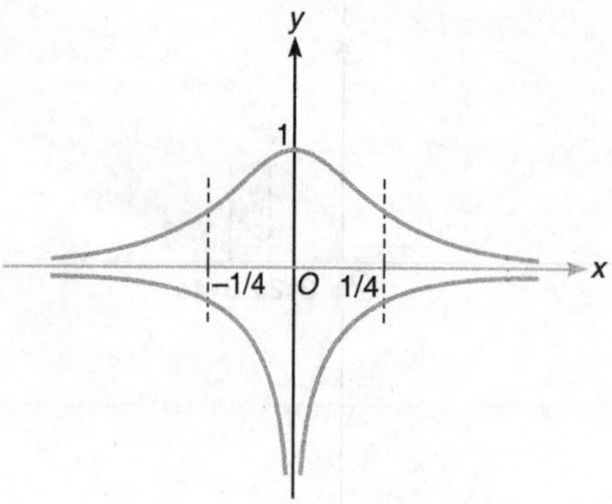

46.

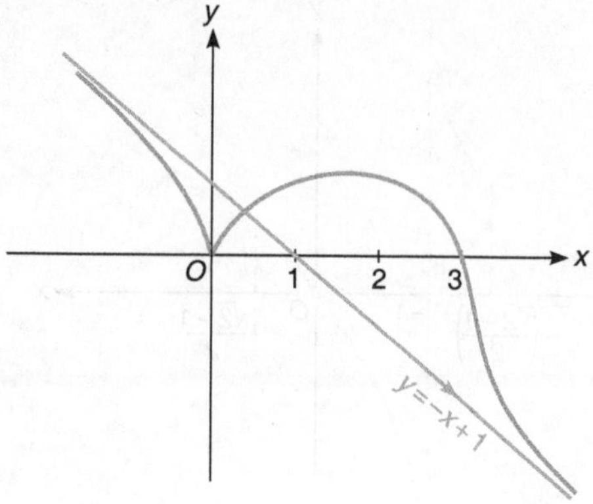

47.

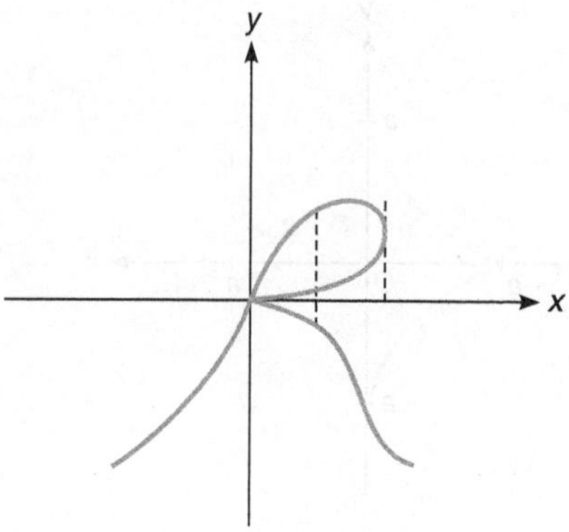

48.

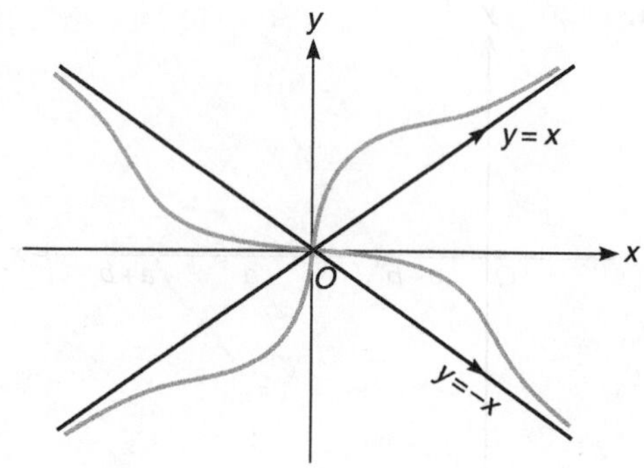

49.

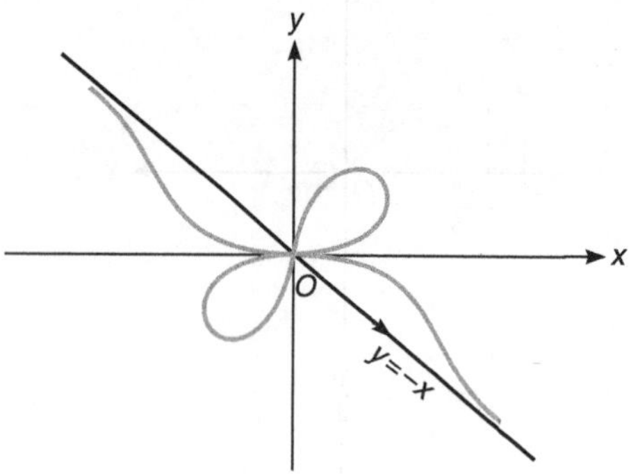

50.

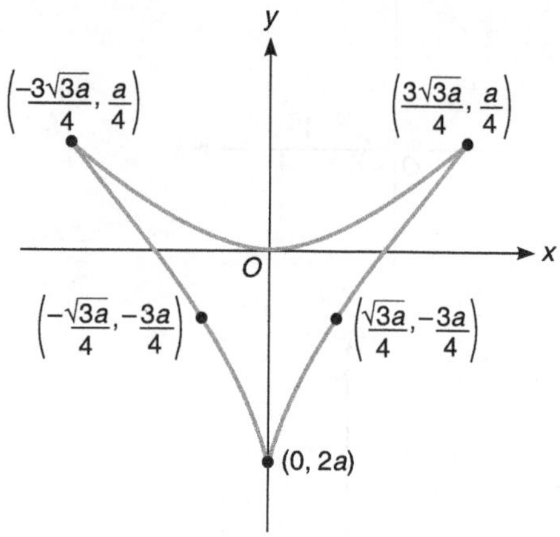